12th International Conference on Topics in Astroparticle and Underground Physics

(TAUP 2011)

Journal of Physics: Conference Series Volume 375

Munich, Germany
5-9 September 2011

ISBN: 978-1-62276-809-7
ISSN: 1742-6588

Printed from e-media with permission by:

Curran Associates, Inc.
57 Morehouse Lane
Red Hook, NY 12571

Printed by Curran Associates, Inc. (2013)

Institute of Physics
Dirac House, Temple Back
Bristol BS1 6BE UK

Phone: 44 1 17 929 7481
Fax: 44 1 17 920 0979

techtracking@iop.org

Additional copies of this publication are available from:

Curran Associates, Inc.
57 Morehouse Lane
Red Hook, NY 12571 USA
Phone: 845-758-0400
Fax: 845-758-2634
Email: curran@proceedings.com
Web: www.proceedings.com

TABLE OF CONTENTS

INDIRECT SEARCHES

THEORY

Journal of Physics: Conference Series **375** (2012) 012001 doi:10.1088/1742-6596/375/1/012001

Status of XENON100

Rafael F. Lang for the XENON100 Collaboration

Department of Physics, Purdue University, 525 Northwestern Avenue, West Lafayette, IN 47907, USA

E-mail: `rafael@purdue.edu`

Abstract. XENON100 is a two-phase time projection chamber with a 62 kg liquid xenon target to search for Dark Matter interactions. Both scintillation and ionization signals are recorded to allow interaction vertex reconstruction in three dimensions. Fiducialization of the target volume results in the lowest background level of any running Dark Matter search experiment. In a 48 kg fiducial target and 100.9 days of live time, no evidence for Dark Matter is found. This leads to the strongest limit on elastic spin-independent WIMP-nucleon interactions for WIMP masses above $\sim 10\,\mathrm{GeV/c^2}$. Also, this data excludes inelastic Dark Matter scattering off sodium or iodine as an explanation of the DAMA modulation.

1. Introduction

Various astrophysical observations at all cosmological length scales suggest that the vast majority of the content of the Universe is non-baryonic [1]. While the presence of large quantities of Dark Matter is well established, its nature is still unknown. Weakly Interacting Massive Particles (WIMPs) are well-motivated particle candidates [2], and a variety of experiments search for signals from WIMPs scattering off a target in a laboratory setting.

2. XENON100

The XENON100 Dark Matter experiment is installed underground at the Laboratory Nazionali del Gran Sasso of INFN, Italy. A 62 kg liquid xenon target is operated as a dual phase (liquid/gas) time projection chamber to search for WIMP interactions. A schematic is shown in figure 1: an interaction in the target generates scintillation light which is recorded as a prompt S1 signal by two arrays of photomultiplier tubes at the top and bottom of the chamber. In addition, each interaction liberates electrons, which are drifted by an electric field to the liquid-gas interface with a speed of about $2\,\mathrm{mm}/\mu\mathrm{s}$. There, a strong electric field extracts the electrons and generates proportional scintillation which is recorded by the same photomultiplier arrays as a delayed S2 signal. The time difference between these two signals gives the depth of the interaction in the time-projection chamber with a resolution of $0.3\,\mathrm{mm}$ (1σ). The hit pattern of the S2 signal on the top array allows to reconstruct the horizontal position of the interaction vertex with a resolution <3 mm (1σ). Taken together, XENON100 is able to precisely localize events in all three coordinates. This enables the fiducialization of the target, yielding a dramatic reduction of external radioactive backgrounds due to the self-shielding capability of liquid xenon. In addition, the ratio S2/S1 allows to discriminate electronic recoils, which are the dominant background, from nuclear recoils, which are expected from Dark Matter interactions. Details of the experimental setup can be found in [3].

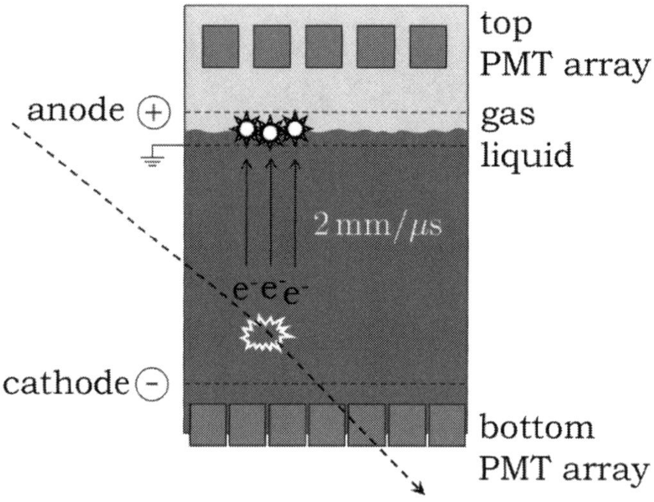

Figure 1. The principle of the two-phase liquid noble gas detector XENON100: the prompt scintillation signal (S1) is recorded by photomultiplier arrays in the liquid and the gas phase. Electrons are liberated in the interaction and drifted to the liquid/gas interface by an electric field. An electric field in the gas phase creates proportional scintillation light (S2) from these electrons, which is also recorded. The depth of the interaction is reconstructed from time difference between the two signals, and the horizontal position is reconstructed from the hit pattern of the S2 signal on the top photomultiplier array.

3. Spring 2010 Data

The Dark Matter search data presented here was acquired between January 13 and June 8, 2010 and represents a total live time of 100.9 days [4]. The WIMP search is restricted to the energy window 4 – 30 photoelectrons, which in nuclear recoil equivalent energy corresponds to 8.4 – 44.6 keV_{nr}. The dominant background from electronic recoils is discriminated from the expected nuclear recoil signal by the parameter $\log_{10}(S2_b/S1)$, where $S2_b$ is the S2 signal summed over the photomultipliers in the bottom array. The electronic recoil background of this data set is affected by a relatively high contamination with ^{nat}Kr of (700 $\pm$ 100) ppt, higher than in both the data acquired before [5] as well as after this particular run (see below). As a consequence, the optimum sensitivity to Dark Matter interactions is achieved for a relatively large fiducial volume with a mass of 48 kg.

The expected electronic recoil background of $< 22 \times 10^{-3}$ events$\cdot(keV_{ee}\cdot kg\cdot day)^{-1}$ before S2/S1 discrimination [6] is lower than that from other direct Dark Matter detection experiments by more than an order of magnitude. The electronic recoil band in $\log_{10}(S2_b/S1)$-energy-space is flattened by subtracting its mean, which was inferred from ^{60}Co Compton calibration data. The resulting low-energy event distribution is shown in figure 2.

Indeed, this distribution is consistent with a sum of three expected background contributions. First, a Gaussian component that is dominated by the ^{85}Kr background. Second, non-Gaussian tails to the electronic recoil band from double-scattered gamma events, where only one interaction happens between the anode and cathode grids of the TPC. The second interaction then adds to the S1 of this event, but not to the S2, thus reducing its S2/S1 ratio. This background has been estimated using ^{60}Co calibration data, normalized to the exposure and taking into account the amount of ^{85}Kr events which can not contribute to such event topologies. Third, nuclear recoils are expected from neutron interactions and are estimated by Monte Carlo simulations using the detailed detector and shield geometry as well as measured radioactivity concentrations [7]. This third population contributes $(0.31^{+0.22}_{-0.11})$ single scatter nuclear recoils in the fiducial volume and the above energy range. The combined background expectation has been verified with the high energy sideband from 30 – 130 photoelectrons.

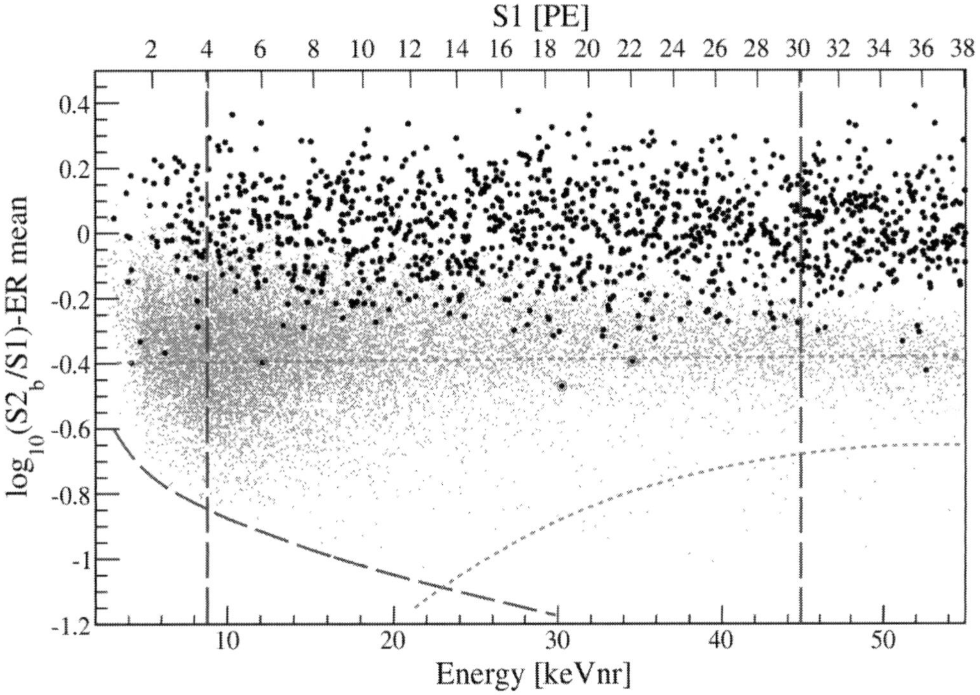

Figure 2. Observed event distribution using the discrimination parameter $\log_{10}(S2_b/S1)$ as a function of nuclear recoil equivalent energy. Gray points indicate the nuclear recoil distribution that has been measured with an ^{241}AmBe neutron source, thus simulating a Dark Matter signal. The energy window as well as the software threshold S2 > 300 photoelectrons are indicated as blue dashed lines and define the WIMP search region. Together with the green dashed lines they define a benchmark region in which (1.8 ± 0.6) events are expected from background, and 3 events observed, as indicated by the red circles. The observed distribution is consistent with the background expectation.

4. Dark Matter Limits

The above background expectation has been used for a Profile Likelihood analysis [8] of this data set [4]. The resulting p-value of the background-only hypothesis is 31%, i.e., there is no significant signal excess. The corresponding limit on the spin-independent WIMP-nucleon elastic scattering cross-section σ is calculated under standard assumptions of the Dark Matter halo [4]. It is shown in figure 3 at 90% confidence level together with the expected sensitivity, that is, the 1σ and 2σ region where the limit should be expected in the absence of any Dark Matter signal, solely due to the expected background. As can be seen, this represents the strongest limit on elastic spin-independent WIMP-nucleon interactions for WIMP masses above $\sim 10\,\mathrm{GeV/c^2}$ and cuts into expectations from constrained minimal supersymmetric models. The same data set has also been used to exclude the possibility that the observed DAMA modulation [9] could be due to inelastic Dark Matter [10] scattering off sodium or iodine [11].

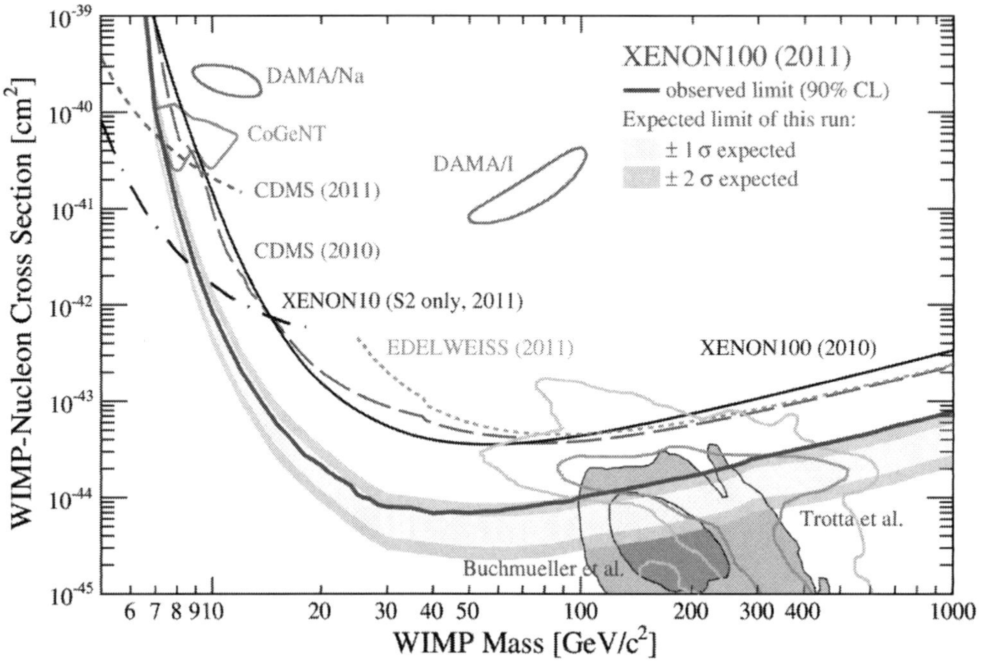

Figure 3. Spin-independent elastic WIMP-nucleon cross-section σ as function of WIMP mass m_χ together with results from various groups [4]. The XENON100 limit at 90% CL is shown as the thick blue line together with the expected sensitivity of this run as yellow and green bands.

5. Ongoing Data Taking

After the data presented here was taken, the xenon has been re-processed through a dedicated cryogenic distillation column. This processing reduced the amount of krypton at least to the lower level already reported in [5]. Electronegative impurities have also been removed from the xenon, and the improved purity results in a larger S2 signal for a given energy. In addition, the hardware S2 trigger threshold was lowered to improve the acceptance of the detector at low nuclear recoil energies. Data taking under these improved conditions is ongoing, and the live time of this upgraded run now exceeds the live time of the data discussed above. Analysis of this data is also ongoing and results are expected in the near future.

6. References

[1] Nakamura K et al. (Particle Data Group) 2010 J. Phys. **G37** 075021
[2] Jungman G, Kamionkowski M and Griest K 1996 Phys. Rept. **267** 195–373 (*Preprint* hep-ph/9506380)
[3] Aprile E et al. (XENON100 Collaboration) 2011 (*Preprint* 1107.2155)
[4] Aprile E et al. (XENON100) 2011 Phys.Rev.Lett. **107** 131302 (*Preprint* 1104.2549)
[5] Aprile E et al. (XENON100) 2010 Phys. Rev. Lett. **105** 131302 (*Preprint* 1005.0380)
[6] Aprile E et al. (XENON100) 2011 Phys.Rev. **D83** 082001 (*Preprint* 1101.3866)
[7] Aprile E et al. 2011 Astropart. Phys. **35** 43–49 (*Preprint* 1103.5831)
[8] Aprile E et al. (XENON100) 2011 Phys. Rev. **D84** 052003 (*Preprint* 1103.0303)
[9] Bernabei R et al. 2010 Eur. Phys. J. **C67** 39–49 (*Preprint* 1002.1028)
[10] Tucker-Smith D and Weiner N 2001 Phys. Rev. **D64** 043502 (*Preprint* hep-ph/0101138)
[11] Aprile E et al. (XENON100) 2011 Phys. Rev. **D84** 061101 (*Preprint* 1104.3121)

12th International Conference on Topics in Astroparticle and Underground Physics (TAUP 2011) IOP Publishing
Journal of Physics: Conference Series **375** (2012) 012002 doi:10.1088/1742-6596/375/1/012002

DAMA/LIBRA results and perspectives

R Bernabei[1,2], P Belli[2], F Cappella[3,4], V Caracciolo[5], R Cerulli[5], C J Dai[6], A d'Angelo[3,4], A Di Marco[1,2], H L He[6], A Incicchitti[4], X H Ma[6], F Montecchia[2,7], X D Sheng[6], R G Wang[6] and Z P Ye[6,8]

[1] Dip. di Fisica, Università di Roma "Tor Vergata", I-00133 Rome, Italy
[2] INFN, sez. Roma "Tor Vergata", I-00133 Rome, Italy
[3] Dip. di Fisica, Università di Roma "La Sapienza", I-00185 Rome, Italy
[4] INFN, sez. Roma, I-00185 Rome, Italy
[5] Laboratori Nazionali del Gran Sasso, I.N.F.N., Assergi, Italy
[6] IHEP, Chinese Academy, P.O. Box 918/3, Beijing 100039, China
[7] Laboratorio Sperimentale Policentrico di Ingegneria Medica, Università degli Studi di Roma "Tor Vergata"
[8] University of Jing Gangshan, Jiangxi, China

E-mail: antonella.incicchitti@roma1.infn.it

Abstract.
The DAMA project operates as an observatory for rare processes, located deep underground at the Gran Sasso National Laboratory of the I.N.F.N.. The main apparatus, DAMA/LIBRA, is investigating the presence of Dark Matter particles in the galactic halo by exploiting the model independent DM annual modulation signature. The present DAMA/LIBRA and the former DAMA/NaI (exposed masses: $\simeq$ 250 kg and $\simeq$ 100 kg of highly radiopure NaI(Tl), respectively) experiments have released so far a total exposure of 1.17 ton $\times$ yr, collected over 13 annual cycles. They have obtained a positive model independent result for the presence of DM particles in the galactic halo at 8.9 σ C.L.. The main results are here briefly summarised and the future perspectives are considered.

1. Introduction

The DAMA project is based on the development and use of low background scintillators. The main experimental set-ups are: i) DAMA/NaI ($\simeq$ 100 kg of highly radiopure NaI(Tl)) that took data for 7 annual cycles and completed its data taking on July 2002 [1, 2, 3, 4, 5, 6, 7, 8, 9, 10, 11, 12]; ii) DAMA/LXe, $\simeq$ 6.5 kg liquid Kr-free Xenon enriched either in ^{129}Xe or in ^{136}Xe [13]; iii) DAMA/R&D, a facility dedicated to tests on prototypes and to perform experiments developing and using various kinds of low background crystal scintillators in order to investigate various rare processes [14]; iv) DAMA/Ge, where sample measurements are carried out and where dedicated measurements on rare events are performed [15]; v) the second generation DAMA/LIBRA set-up, $\simeq$ 250 kg highly radiopure NaI(Tl)) [16, 17, 18, 19] mainly devoted to the investigation of the presence of Dark Matter (DM) particles in the galactic halo. Profiting of the low background features of these set-ups, many rare processes have been studied.

The main apparatus, DAMA/LIBRA, is investigating the presence of Dark Matter particles in the galactic halo by exploiting the model independent DM annual modulation signature originally suggested in the mid 80's[20].

In fact, as a consequence of its annual revolution around the Sun, which is moving in the Galaxy traveling with respect to the Local Standard of Rest towards the star Vega near the constellation of Hercules, the Earth should be crossed by a larger flux of Dark Matter particles around ~2 June (when the Earth orbital velocity is summed to the one of the solar system with respect to the Galaxy) and by a smaller one around ~2 December (when the two velocities are subtracted). Thus, this signature has a specific origin with peculiarities not correlated with seasons. This DM annual modulation signature is very distinctive since the effect induced by DM particles must simultaneously satisfy all the following requirements: (1) the rate must contain a component modulated according to a cosine function; (2) with one year period; (3) with a phase that peaks roughly around ~ 2nd June; (4) this modulation must be present only in a well-defined low energy range, where DM particles can induce signals; (5) it must be present only in those events where just a single detector, among all the available ones in the used set-up, actually "fires" (*single-hit* events), since the probability that DM particles experience multiple interactions is negligible; (6) the modulation amplitude in the region of maximal sensitivity has to be less about 7% in case of usually adopted halo distributions, but it may be significantly larger in case of some particular scenarios such as e.g. those in refs. [21, 22]. Only systematic effects or side reactions able to simultaneously fulfil all the six requirements given above and to account for the whole observed modulation amplitude might mimic this DM signature; no one has been found or suggested. At present status of technology it is the only model independent signature available in direct Dark Matter investigation that can be effectively exploited.

The DAMA/LIBRA data released so far correspond to six annual cycles for an exposure of 0.87 ton×yr [17, 18]. Considering these data together with those previously collected by DAMA/NaI over 7 annual cycles (0.29 ton×yr), the total exposure collected over 13 annual cycles is 1.17 ton×yr; this is orders of magnitude larger than the exposures typically collected in the field.

2. DAMA/LIBRA results

The DAMA/NaI set up and its performances are described in ref.[1, 3, 4, 5], while for the DAMA/LIBRA set-up and its performances see refs. [16, 18]. In particular, the software energy threshold of the experiments is 2 keV. Several analyses on the model-independent DM annual modulation signature have been performed [17, 18]. Figure 1 shows the time behaviour of the experimental residual rates of the *single-hit* events collected by DAMA/NaI and by DAMA/LIBRA in the (2–6) keV energy interval [17, 18].

The superimposed curve is the cosinusoidal function: $A \cos \omega(t-t_0)$ with a period $T = \frac{2\pi}{\omega} = 1$ yr, with a phase $t_0 = 152.5$ day (June 2nd), and modulation amplitude, A, obtained by best fit over the 13 annual cycles. The hypothesis of absence of modulation in the data can be discarded [17, 18] and, when the period and the phase are released in the fit, values well compatible with those expected for a DM particle induced effect are obtained [18]. For example, in the cumulative (2–6) keV energy interval one gets: $A = (0.0116 \pm 0.0013)$ cpd/kg/keV, $T = (0.999 \pm 0.002)$ yr and $t_0 = (146 \pm 7)$ day. So, the analysis of the *single-hit* residual rate favours the presence of a modulated cosine-like behaviour with proper features at 8.9 σ C.L.[18].

The same data of figure 1 have also been investigated by a Fourier analysis, obtaining a clear peak corresponding to a period of 1 year [18]; this analysis in other energy regions shows only aliasing peaks instead. Moreover, in order to verify absence of annual modulation in other energy regions and, thus, to also verify the absence of any significant background modulation, the energy distribution in energy regions not of interest for DM detection has also been investigated. This allowed the exclusion of a background modulation in the whole energy spectrum at a level much lower than the effect found in the lowest energy region for the *single-hit* events [18]. A further relevant investigation has been done by applying the same hardware and software procedures, used to acquire and to analyse the *single-hit* residual rate, to the *multiple-hits* events in which

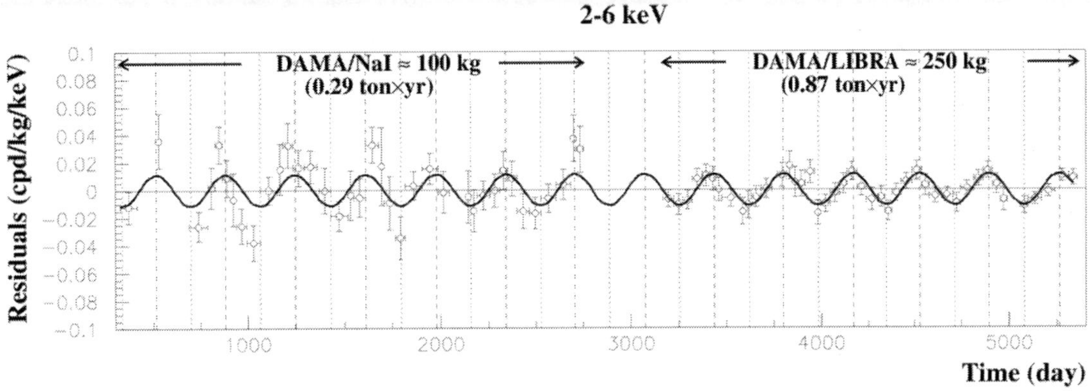

Figure 1. Experimental model-independent residual rate of the *single-hit* scintillation events, measured by DAMA/NaI over seven and by DAMA/LIBRA over six annual cycles in the (2 – 6) keV energy interval as a function of the time [4, 5, 17, 18]. The zero of the time scale is January 1st of the first year of data taking. The experimental points present the errors as vertical bars and the associated time bin width as horizontal bars. The superimposed curve is $A\cos\omega(t-t_0)$ with period $T = \frac{2\pi}{\omega} = 1$ yr, phase $t_0 = 152.5$ day (June 2nd) and modulation amplitude, A, equal to the central value obtained by best fit over the whole data. The dashed vertical lines correspond to the maximum expected for the DM signal (June 2nd), while the dotted vertical lines correspond to the minimum. See Refs. [17, 18] and text.

more than one detector "fires". In fact, since the probability that a DM particle interacts in more than one detector is negligible, a DM signal can be present just in the *single-hit* residual rate. Thus, this allows the study of the background behaviour in the same energy interval of the observed positive effect. The result of the analysis is reported in figure 2 where it is shown the residual rate of the *single-hit* events measured over the six DAMA/LIBRA annual cycles, as collected in a single annual cycle, together with the residual rates of the *multiple-hits* events, in the same (2–6) keV energy interval. A clear modulation is present in the *single-hit* events, while the fitted modulation amplitudes for the *multiple-hits* residual rate are well compatible with zero [18]. Similar results were previously obtained also for the DAMA/NaI case [5]. Thus, again evidence of annual modulation with proper features, as required by the DM annual modulation signature, is present in the *single-hit* residuals (events class to which the DM particle induced events belong), while it is absent in the *multiple-hits* residual rate (event class to which only background events belong). Since the same identical hardware and the same identical software procedures have been used to analyse the two classes of events, the obtained result offers an additional strong support for the presence of a DM particle component in the galactic halo further excluding any side effect either from hardware or from software procedures or from background.

The annual modulation present at low energy has also been analyzed by depicting the differential modulation amplitudes, S_m, as a function of the energy; the S_m is the modulation amplitude of the modulated part of the signal obtained by maximum likelihood method over the data, considering $T = 1$ yr and $t_0 = 152.5$ day.

The S_m values are reported as function of the energy in figure 3. It can be inferred that a positive signal is present in the (2–6) keV energy interval, while S_m values compatible with zero are present just above; in particular, the S_m values in the (6–20) keV energy interval have random fluctuations around zero with χ^2 equal to 27.5 for 28 degrees of freedom. It has been also verified that the measured modulation amplitudes are statistically well distributed in all the crystals, in all the annual cycles and energy bins; these and other discussions can be found

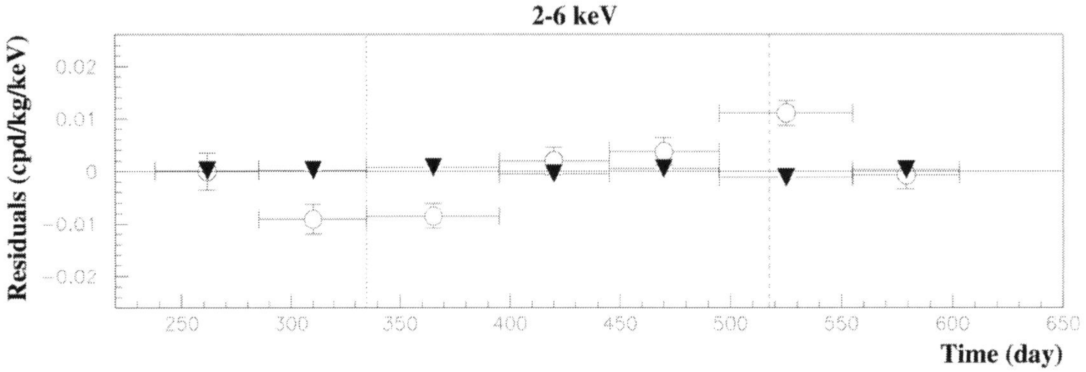

Figure 2. Experimental residual rates over the six DAMA/LIBRA annual cycles for *single-hit* events (open circles) (class of events to which DM events belong) and for *multiple-hit* events (filled triangles) (class of events to which DM events do not belong). They have been obtained by considering for each class of events the data as collected in a single annual cycle and by using in both cases the same identical hardware and the same identical software procedures. The initial time of the figure is taken on August 7th. The experimental points present the errors as vertical bars and the associated time bin width as horizontal bars[17, 18]. See text.

in ref. [18]. Other interesting results have been obtained by analysing the data introducing a

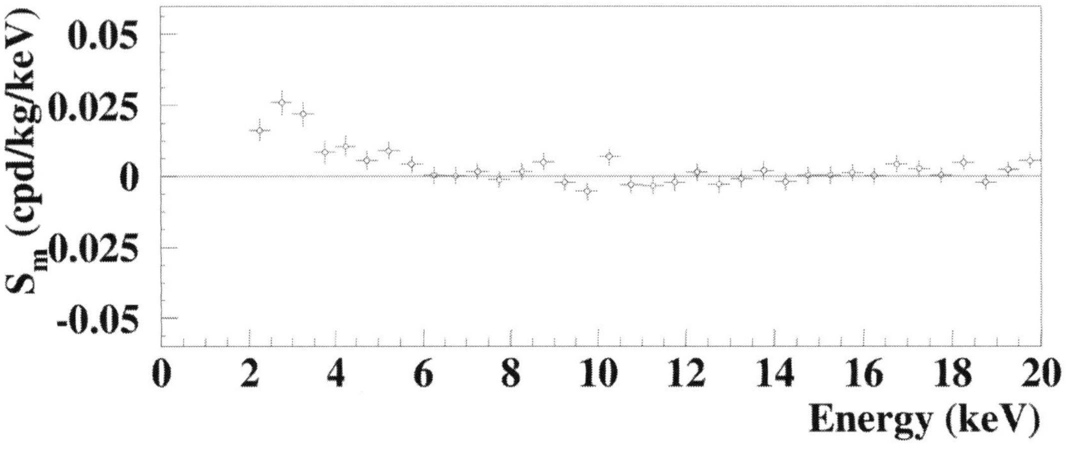

Figure 3. Energy distribution of the modulation amplitudes S_m for the total cumulative exposure 1.17 ton×yr. The energy bin is 0.5 keV. A clear modulation is present in the lowest energy region, while S_m values compatible with zero are present just above. In fact, the S_m values in the (6–20) keV energy interval have random fluctuations around zero with χ^2 equal to 27.5 for 28 degrees of freedom. See Refs. [17, 18].

sine-like modulation in the maximum likelihood procedure. In this case alternatively the signal can be written as: $S_{0,k} + S_{m,k} \cos\omega(t-t_0) + Z_{m,k} \sin\omega(t-t_0) = S_{0,k} + Y_{m,k} \cos\omega(t-t^*)$, where $S_{0,k}$, $S_{m,k}$ and $Z_{m,k}$ are the constant part, the cosine-like and sine-like modulation amplitudes of the signal in k-th energy interval. Obviously, for signals induced by DM particles one would expect: i) $Z_{m,k} \sim 0$ (because of the orthogonality between the cosine and the sine functions); ii) $S_{m,k} \simeq Y_{m,k}$; iii) $t^* \simeq t_0 = 152.5$ day. In fact, these conditions hold for most of the dark halo models; however, it is worth noting that slight differences in the phase could be expected in case of possible contributions from non-thermalized DM components, such as e.g. the SagDEG stream [7] and the caustics [23]. The 2σ contours in the plane (S_m, Z_m) for the (2–6) keV and (6–14) keV energy intervals and those in the plane (Y_m, t^*) are reported in figure 4 [18]. The best fit values for the (2–6) keV energy interval are (1σ errors): $S_m = (0.0111 \pm 0.0013)$ cpd/kg/keV; $Z_m = -(0.0004 \pm 0.0014)$ cpd/kg/keV; $Y_m = (0.0111 \pm 0.0013)$ cpd/kg/keV; $t^* = (150.5 \pm 7.0)$ day; while for the (6–14) keV energy interval are: $S_m = -(0.0001 \pm 0.0008)$ cpd/kg/keV; $Z_m = (0.0002 \pm 0.0005)$ cpd/kg/keV; $Y_m = -(0.0001 \pm 0.0008)$ cpd/kg/keV and t^* obviously not determined. These results confirm those achieved by other kinds of analyses. In particular, a modulation amplitude is present in the *single-hit* events of the lower energy intervals and the period and the phase agree with those expected for DM induced signals [18].

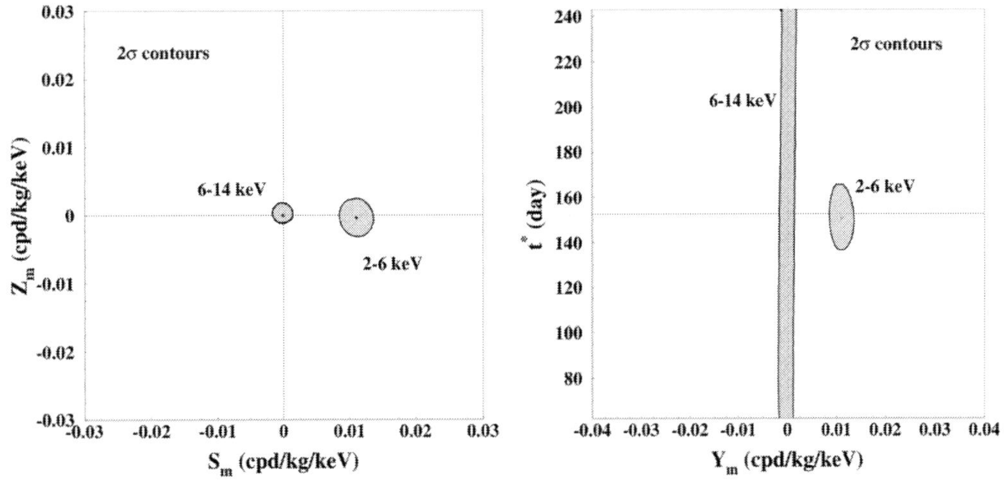

Figure 4. 2σ contours in the plane (S_m, Z_m) (*left*) and in the plane (Y_m, t^*) (*right*) for the (2–6) keV and (6–14) keV energy intervals. The contours have been obtained by the maximum likelihood method, considering the cumulative exposure of 1.17 ton × yr. See text.

Both the data of DAMA/LIBRA and of DAMA/NaI fulfil all the requirements of the DM annual modulation signature.

Sometimes naive statements were put forwards as the fact that in nature several phenomena may show some kind of periodicity. It is worth noting that the point is whether they might mimic the annual modulation signature in DAMA/LIBRA (and former DAMA/NaI), i.e. whether they might be not only quantitatively able to account for the observed modulation amplitude but also able to contemporaneously satisfy all the requirements of the DM annual modulation signature; the same is also for side reactions.

Careful investigations on absence of any significant systematics or side reaction have been quantitatively carried out (see e.g. ref. [4, 5, 16, 17, 18, 24], and refs therein). No systematics or side reactions able to mimic the signature (that is, able to account for the measured modulation amplitude and simultaneously satisfy all the requirements of the signature) has been found or suggested by anyone over more than a decade.

3. About interpretation

In literature it is present a rich and plenteous variety of theoretical patterns that are stimulating and intriguing to interpret in model dependent ways the results coming from different experiments on dark matter investigation.

Anyhow, it is worth noting that DAMA has firstly a model independent result due to the exploitation of a DM signature with specific peculiarities.

About the interpretation of this model independent result in terms of a possible DM candidate and in order to consider a comparative overview with measurements by other experiments, it is important to refresh that there is neither an unique reference theoretical model of interpretation, nor a single set of assumptions for parameters in astrophysical, nuclear and particle Physics. In addition often comparisons are not performed in a fully consistent way. So every model dependent analysis chooses a model framework by fixing many parameters and assumptions, but many uncertainties are still present. Uncertainties are present on the models about the nature of the candidate particle, on the interaction coupling, on the form factors for each target material, on spin factors, on the scaling laws, on the nuclear Physics framework, on the dark halo model and related parameters, on possible galactic streams and so on. Uncertainties can also exist in some cases on some experimental aspects as: marginal exposures, poorly known detector response, extrapolated energy threshold, extrapolated energy scale, extrapolated energy resolution, definition of a fiducial volume, possible non-uniformity of some detector response, not regular and/or suitable calibration procedures, on quenching factors, and on stability of the operating conditions, etc.. In addition, in some cases, there are significant uncertainties on subtraction/rejection procedures when applied and on the stability of all the selected acceptance windows and related quantities. All these aspects can affect results and comparisons at various extent, precluding the comparisons from an universal value.

The obtained DAMA model independent evidence is compatible with a wide set of scenarios regarding the nature of the DM candidate and related astrophysical, nuclear and particle Physics. For examples some given scenarios and parameters are discussed e.g. in Refs. [2, 4, 5, 6, 7, 8, 9, 10, 11] and in Appendix A of Ref. [17]. Further large literature is available on the topics [25]; other possibilities are open and we just recall the recent paper [26] on the case of a light-mass Dark Matter candidate particle interacting with the detector nuclei by coherent elastic process; comparison with recent possible positive hint [27] is also given.

It is worth noting that no other experiment exists, whose result can be directly compared in a model-independent way with those by DAMA/NaI and DAMA/LIBRA. Moreover, concerning those activities claiming model dependent exclusion under some largely arbitrary assumptions (see for example discussions in [4, 5, 17, 28, 29]) and generally using marginal exposures, it is worth noting that often important critical points exist in some of their experimental aspects (energy threshold, energy scale, multiple selection procedures, non uniformity of the detectors response, absence of suitable periodical calibrations in the same running conditions and in the claimed low energy region, stabilities, etc.); in addition existing experimental and theoretical uncertainties are not considered.

Similar considerations hold for the indirect detection searches, in fact also in this case no direct model-independent comparison can be performed between the results obtained in direct and indirect activities, since it does not exist a biunivocal correspondence between the observables in the two kinds of experiments. Moreover, these searches are restricted to some DM candidates and scenarios and their results are strongly model dependent. Anyhow, measurements published up to now are not in conflict with the effect observed by DAMA experiments.

4. Upgrades and perspectives

A first upgrade of the DAMA/LIBRA set-up was performed in September 2008.

A further and more important upgrade has been performed in the end of 2010 (see figure 5). In fact, all the PMTs have been replaced with new ones with higher quantum efficiency; this will allow the lowering of the software energy threshold and, hence, the improvement of the performance and of the sensitivity also for deeper corollary information on the nature of the DM candidate particle(s) and on the various related astrophysical, nuclear and particle Physics scenarios.

Since January 2011 the DAMA/LIBRA experiment is again in data taking in the new configuration. Further improvements are foreseen with new preamplifiers and trigger modules realised to further implement the low energy studies. In the future DAMA/LIBRA will also continue its study on several other rare processes [19] as also the former DAMA/NaI apparatus did [12].

Figure 5. Replacement of the PMTs in DAMA/LIBRA during the upgrade of 2010. All procedures involving detectors, photomultipliers (PMTs), etc. have been carried out in HP (High Purity) Nitrogen atmosphere.

5. Conclusion

The positive evidence for the presence of DM particles in the galactic halo is now supported at 8.9σ C.L. (on a cumulative exposure of 1.17 ton$\times$yr i.e. 13 annual cycles of DAMA/NaI and DAMA/LIBRA). An update of corollary analyses in some of the many possible scenarios for DM candidates, interactions, halo models, nuclear/atomic properties, etc. is in progress as well as analyses/data taking to investigate other rare processes. The last upgrade in fall 2010 was successfully concluded, while further improvements are planned. The strictly quality control allows DAMA/LIBRA to be still the highest radiopure set-up in the field with the largest exposed sensitive mass, the full control of running conditions, the largest duty-cycle and an exposure orders of magnitude larger than any other activity in the field.

References

[1] Bernabei R et al 1999 *Il Nuovo Cim.* A **112** 545
[2] Bernabei R et al 1996 *Phys. Lett.* B **389** 757; 1998 *Phys. Lett.* B **424** 195; 1999 *Phys. Lett.* B **450** 448; 2000 *Phys. Rev.* D **61** 023512; 2000 *Phys. Lett.* B **480** 23; 2001 *Phys. Lett.* B **509** 197; 2002 *Eur. Phys. J.* C **23** 61; 2002 *Phys. Rev.* D **66** 043503
[3] Bernabei R et al 2000 *Eur. Phys. J.* C **18** 283
[4] Bernabei R et al 2003 *La Rivista del Nuovo Cimento* **26** n.1, 1
[5] Bernabei R et al 2004 *Int. J. Mod. Phys.* D **13** 2127
[6] Bernabei R et al 2006 *Int. J. Mod. Phys.* A **21** 1445
[7] Bernabei R et al 2006 *Eur. Phys. J.* C **47** 263
[8] Bernabei R et al 2007 *Int. J. Mod. Phys.* A **22** 3155
[9] Bernabei R et al 2008 *Eur. Phys. J.* C **53** 205
[10] Bernabei R et al 2008 *Phys. Rev.* D **77** 023506

[11] Bernabei R et al 2008 *Mod. Phys. Lett.* A **23** 2125

[12] See in the publication list in: http://people.roma2.infn.it/dama

[13] Belli P et al 1996 *Astropart. Phys.* **5** 217; Belli P et al 1996 *Nuovo Cim.* C **19** 537; Belli P et al 1996 *Phys. Lett.* B **B387** 222; 1996 *Phys. Lett.* B **389** 783 (err.); Bernabei R et al 1998 *Phys. Lett.* B **436** 379; Belli P et al 1999 *Phys. Lett.* B **465** 315; Belli P et al 2000 *Phys. Rev.* D **61** 117301; Bernabei R et al 2000 *New J. of Phys.* **2** 15.1; Bernabei R et al 2000 *Phys. Lett.* B **493** 12; Bernabei R et al 2002 *Nucl. Instr. & Meth* A **482** 728; Bernabei R et al 2001 *Eur. Phys. J. direct* C **11** 1; Bernabei R et al 2002 *Phys. Lett.* B **527** 182; Bernabei R et al 2002 *Phys. Lett.* B **546** 23; Bernabei R et al 2003 *Beyond the Desert 2003* (Berlin: Springer) p. 365; Bernabei R et al 2006 *Eur. Phys. J.* A **27** s01 35

[14] Bernabei R et al 1997 *Astropart. Phys.* **7** 73; Bernabei R et al 1997 *Nuovo Cim.* A **110** 189; Belli P et al 1999 *Astropart. Phys.* **10** 115; Belli P et al 1999 *Nucl. Phys.* B **563** 97; Bernabei R et al 2002 *Nucl. Phys.* A **705** 29; Belli P et al 2003 *Nucl. Instr. & Meth* A **498** 352; Cerulli R et al 2004 *Nucl. Instr. & Meth* A **525** 535; Bernabei R et al 2005 *Nucl. Instr. & Meth* A **555** 270; Bernabei R et al 2006 *Ukr. J. Phys.* **51** 1037; Belli P et al 2007 *Nucl. Phys.* A **789** 15; Belli P et al 2007 *Phys. Rev.* C **76** 064603; Belli P et al 2008 *Phys. Lett.* B**658** 193; Belli P et al 2008 *Eur. Phys. J.* A **36** 167; Belli P et al 2009 *Nucl. Phys.* A**826** 256 Belli P et al 2010 *Nucl. Instr. & Meth* A **615** 301; Belli P et al 2011 *Nucl. Instr. & Meth* A **626-627** 31; Belli P et al 2011 *J. Phys. G: Nucl. Part. Phys.* **38** 015103

[15] Belli P et al 2007 *Nucl. Instr. & Meth.* A **572** 734; Belli P et al 2008 *Nucl. Phys.* A **806** 388; Belli P et al 2009 *Nucl. Phys.* A **824** 101; Belli P et al 2009 *Proceed. of the Int. Conf. NPAE 2008* (ed. INR-Kiev, Kiev) p. 473; Belli P et al 2009 *Eur. Phys. J.* A **42** 171; Belli P et al 2010 *Nucl. Phys.* A **846** 143; Belli P et al 2011 *Nucl. Phys.* A **859** 126; Belli P et al 2011 *Phys. Rev.* C **83** 034603; Belli P et al 2011 *Eur. Phys. J.* A **47** 91

[16] Bernabei R et al 2008 *Nucl. Instr. & Meth.* A **592** 297

[17] Bernabei R et al 2008 *Eur. Phys. J.* C **56** 333

[18] Bernabei R et al 2010 *Eur. Phys. J.* C **67** 39.

[19] R. Bernabei et al., *Eur. Phys. J.* **C62**, 327–332 (2009).

[20] Drukier K A et al 1986 *Phys. Rev.* D **33** 3495; Freese K et al 1988 *Phys. Rev.* D **37** 3388

[21] Smith D and Weiner N 2001 *Phys. Rev.* D **64** 043502; Tucker-Smith D and Weiner N 2005 *Phys. Rev.* D **72** 063509

[22] Freese K et al 2005 *Phys. Rev.* D **71** 043516; 2004 *Phys. Rev. Lett.* **92** 111301

[23] Ling F S, Sikivie P and Wick S 2004 *Phys. Rev.* D **70** 123503

[24] Bernabei R et al 2010 *AIP Conf. Proceed.* **1223** 50 (arXiv:0912.0660); *J. Phys.: Conf. Ser.* **203** 012040 (arXiv:0912.4200); http://taup2009.lngs.infn.it/slides/jul3/nozzoli.pdf, talk given by F. Nozzoli; 2011 *Can. J. Phys.* **89** 11; 2011 *SIF Atti Conf.* **103** (arXiv:1007.0595); pre-print ROM2F/2011/12, TIPP2011 Conf., Chicago, USA (2011).

[25] Belli P et al 2011 *Phys. Rev.* D **84** 055014; Bottino A et al 2010 *Phys. Rev.* D **81** 107302; Fornengo N et al 2011 *Phys. Rev.* D **83** 15001; Fitzpatrick A L et al 2010 arXiv:1003.0014; Hooper D et al 2010 arXiv:1007.1005v2; Cerdeno D G and Seto 0 2009 *JCAP* **0908** 032; Cerdeno D G, Munoz C and Seto 0 2009 *Phys. Rev.* D **79** 023510; Cerdeno D G et al 2007 *JCAP* **0706** 008; Gunion J F et al 2009 arXiv:1009.2555; Belikov A V et al 2010 arXiv:1009.0549; Arina C and Fornengo N 2007 *JHEP* **11** 029; Blanger G et al arXiv:1105.4878; Chang S et al 2009 Phys. Rev. D **79** 043513; Chang S et al 2010 arXiv:1007.2688; Foot R 2010 arXiv:1001.0096, arXiv:1106.2688, *Phys. Rev.* D **82** 095001; Mambrini Y 2011 *JCAP* **1107** 009, *JCAP* 2010 **1009** 022; Bai Y and Fox P J 2009 *JHEP* **0911** 052; Alwall J et al 2010 arXiv:1002.3366; Khlopov M. Yu et al arXiv:1003.1144; Andreas S et al 2010 *Phys. Rev.* D **82** 043522; Boucenna M S, Profumo S 2011 arXiv:1106.3368 Graham P W et al 2010 Phys. Rev. D **82** 063512; Batell B, Pospelov M and Ritz A 2009 Phys. Rev. D **79** 115019; Del Nobile E et al 2011 Phys. Rev. D **84** 027301; Kopp J et al 2010 *JCAP* **1002** 014; Barger V et al 2010 Phys. Rev. D **82** 035019; Chang S et al 2010 *JCAP* **1008** 018; Feng J L et al 2011 *Phys. Lett.* B **703** 124 Frandsen M T et al 2011 Phys. Rev. D **84** 041301; Kim Y G and Shin S 2009 *JHEP* **0905** 036; Shin S 2010 arXiv:1011.6377; Buckley M R 2011 arXiv:1106.3583 Fornengo N et al 2011 arXiv:1108.4661 Gondolo P et al 2011 arXiv:1106.0885 Kuflik E et al 2010 arXiv:1003.0682; Arina C et al 2011 arXiv:1105.5121; Buckley M R et al 1011.1499

[26] Belli P et al 2011 *Phys. Rev.* D **84** 055014

[27] Aalseth C E et al 2010 arXiv:1002.4703; 2011 arXiv:1106.0650

[28] Bernabei R et al 2009 "Liquid Noble gases for Dark Matter searches: a synoptic survey", Exorma Ed., Roma, ISBN 978-88-95688-12-1, pp. 1–53 (arXiv:0806.0011v2).

[29] Collar J I and McKinsey D N 2010 arXiv:1005.0838; 2010 arXiv:1005.3723; Collar J I 2010 arXiv:1006.2031; 2010 arXiv:1010.5187; 2011 arXiv:1103.3481; 2011 arXiv:1106.0653; 2011 arXiv:1106.3559

12th International Conference on Topics in Astroparticle and Underground Physics (TAUP 2011) IOP Publishing
Journal of Physics: Conference Series **375** (2012) 012003 doi:10.1088/1742-6596/375/1/012003

Latest results from the CRESST-II Dark Matter Search

F Petricca[1], G Angloher[1], M Bauer[3], I Bavykina[1], A Bento[1,5], C Bucci[4], C Ciemniak[2], G Deuter[3], F v Feilitzsch[2], D Hauff[1], P Huff[1], C Isaila[2], J Jochum[3], M Kiefer[1], M Kimmerle[3], J C Lanfranchi[2], S Pfister[2], W Potzel[2], F Pröbst[1], F Reindl[1], S Roth[2], K Rottler[3], C Sailer[3], K Schäffner[1], J Schmaler[1], S Scholl[3], W Seidel[1], M v Sivers[2], L Stodolsky[1], C Strandhagen[3], R Strauss[2], A Tanzke[1], I Usherov[3], S Wawoczny[2], M Willers[2] and A Zöller[2]

[1] Max-Planck-Institut für Physik, Föhringer Ring 6, D-80805 München, Germany
[2] Physik-Department E15, Technische Universität München, D-85747 Garching, Germany
[3] Eberhard-Karls-Universität Tübingen, D-72076 Tübingen, Germany
[4] INFN, Laboratori Nazionali del Gran Sasso, I-67010 Assergi, Italy
[5] CI, Physics Department, University of Coimbra, P-3004 516 Coimbra, Portugal

E-mail: petricca@mpp.mpg.de

Abstract. The CRESST-II experiment is searching for Dark Matter particles in the form of WIMPs via their elastic scattering off nuclei in a target material. The CRESST target consists of scintillating $CaWO_4$ crystals which are operated as cryogenic calorimeters at millikelvin temperatures and read out by transition edge sensors. Each interaction in $CaWO_4$ produces a phonon signal in the target crystal and also a light signal that is measured by a secondary cryogenic calorimeter, allowing a very efficient discrimination between electron recoils from radioactive e/γ background and nuclear recoils. Moreover, to some extent, the different types of recoiling nuclei (O, Ca, W) can be distinguished.
In the paper we present the latest results of the experiment, obtained from a net exposure of 730kg days acquired with 8 detectors between July 2009 and March 2011. The data has shown a considerable number of events in our signal region. Since this large number of events is not consistent with the known sources of background in our experiment, we discuss the compatibility of this excess of events with a possible WIMP-signal.

1. Introduction

The CRESST-II (Cryogenic Rare Event Search with Superconducting Thermometers) experiment, located at the Laboratori Nazionali del Gran Sasso in Italy, aims for direct detection of WIMP Dark Matter via their elastic scattering off nuclei in a target material. The main challenges of this kind of measurement are to achieve a sufficient background suppression to be sensitive to the extremely low expected rates and to detect the small recoil energies anticipated in WIMP-nucleus scatterings (see Figure 1).
To fulfil these requirements, CRESST uses scintillating $CaWO_4$ crystals operated as cryogenic detectors in a low-background environment. The scintillation produced at the target material allows for a discrimination of the type of interacting particle on an event-by-event basis. This

enables us to distinguish between potential rare WIMP interactions producing nuclear recoils and events induced by the dominant radioactive backgrounds which are mainly constituted of e/γ-events.
In the following we report on the latest data collected with the CRESST setup between 2009 and 2011, corresponding to a total net exposure of 730 kg d.

2. CRESST detectors

2.1. Detector modules

The CRESST target for WIMP interactions consists of scintillating $CaWO_4$ crystals cut in a cylindrical shape with 40 mm in diameter and height. The present experimental setup is capable to accommodate up to 33 of these crystals for a maximum target mass of about 10 kg.
For coherently scattering WIMPs, the total scattering rate in $CaWO_4$ is dominated by the heavy tungsten nuclei ($\sigma \propto A^2$, with A being the mass number of the target nucleus). However, given the limitation imposed by the finite energy threshold of the detectors, scatterings off other constituents may become relevant considering the kinematics involved. To illustrate this point Figure 2 shows, for different WIMP masses, the contributions of the three elements in $CaWO_4$ in the energy interval between 12 and 40 keV which is a typical energy range of interest for the CRESST detectors. Tungsten completely dominates only for WIMPs with masses above

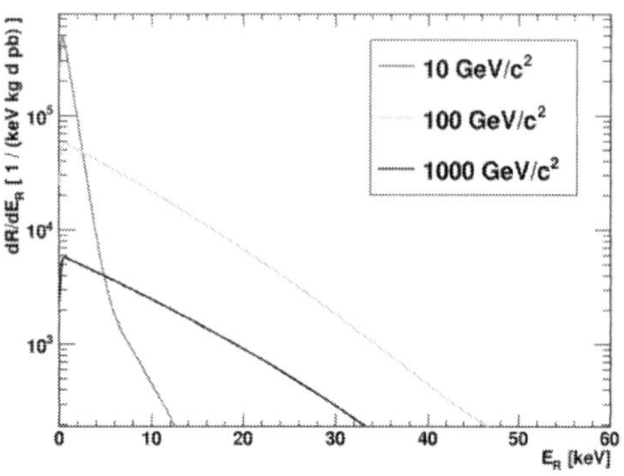

Figure 1. Differential event rates for various WIMP masses in $CaWO_4$ detectors versus the nuclear recoil energy E_R. The rates are given per keV, kg and day for a WIMP-nucleon cross-section of 1 pb.

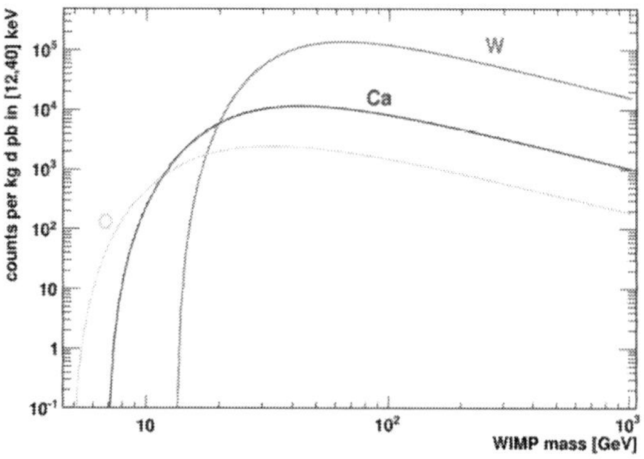

Figure 2. Contributions of the three types of nuclei present in a $CaWO_4$ target to the total rate of WIMP interactions, as a function of the WIMP mass and for a cross-section of 1 pb, assuming coherent WIMP scatterings and a detector with a sensitive energy range between 12 and 40 keV.

30 GeV. For low WIMP masses, up to about 12 GeV, however the scatterings off tungsten are completely below threshold, and oxygen and calcium recoils give the only possible Dark Matter signal. For this reason, in order to be sensitive to the largest possible range of WIMP masses, when looking for possible WIMP interactions, we consider recoils of all three types of nuclei in $CaWO_4$.
The target crystals are operated as cryogenic calorimeters, read out with tungsten transition edge sensors (TES)[1], at temperatures of about 10 mK. The $CaWO_4$ crystals with their TESs are indicated as phonon detectors. The extreme sensitivity of these detectors makes them suitable to measure the low-energy nuclear recoils expected in WIMP-nucleus scatterings.
In order to measure the small fraction of the energy deposited in the target crystal which is converted into scintillation light, each crystal is paired with a separate cryogenic light detector.

Each light detector consists of a light absorber (sapphire wafer of 40 mm diameter and 0.4 mm thickness, with an epitaxially grown silicon layer on one side for photon absorption or pure silicon wafer with the same geometry) equipped, similar to the target crystal, with a tungsten TES. A crystal and the corresponding light detector form a detector module. A schematic view of a

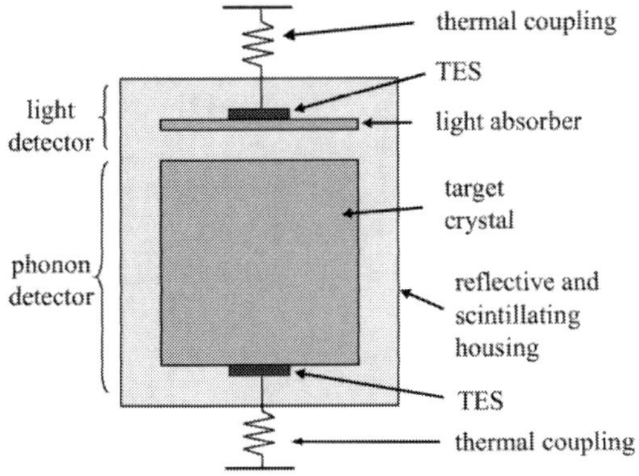

Figure 3. Schematic drawing of a CRESST detector module, consisting of the target crystal and an independent light detector. Both are read out by transition edge sensors (TES) and are enclosed in a common reflective and scintillating housing.

module is shown in Figure 3. Both detectors of a module are held by thin, silver-coated bronze clamps and are enclosed in a common scintillating and highly reflective housing to maximize the collection of the scintillation light [2].

2.2. Event discrimination

Following a particle interaction, a detector module yields two coincident signals (one from the phonon and one from the light detector). While the phonon channel provides a sensitive measurement of the total energy deposition in the target (approximately independent of the type of interacting particle), the light signal is very sensitive to the different types of interactions and can therefore be used for discrimination. We define the light yield (LY) of an event as the ratio of energy measured with the light detector divided by the energy measured with the phonon detector. The energy scale of the light channel is defined such that 122 keV γ's from a ^{57}Co calibration source have a light yield of unity. α's and nuclear recoils are observed to have a lower light yield [2]. As a consequence, interactions from particles of different nature lead to characteristic bands in the light yield-energy plane, allowing for a very efficient discrimination between the potential signals, expected to show up as nuclear recoils, and the dominant radioactive background which is mainly constituted of e/γ events. Furthermore, since the light yield of nuclear recoils varies widely among different nuclei, to the extent that the different recoil bands can be separated in the light yield-energy plane it is possible to identify the type of recoiling nucleus. This distinctive feature of CRESST allows to study possible WIMP interactions with different target nuclei in parallel in the same setup.

3. The latest CRESST run

The latest run of CRESST took place between June 2009 and April 2011. It included a neutron test and γ-calibrations with ^{57}Co and ^{232}Th sources. Out of the 18 detector modules installed in the cryostat only 8 were used for the Dark Matter analysis. One was excluded due to its unusually poor energy resolution and an additional one was excluded being equipped with a test $ZnWO_4$ crystal. The remaining ones mainly had difficulties in cooling the light detectors and could therefore not provide the full information needed for Dark Matter analysis, but still were used to tag coincident events with signals in more than one detector module.
The validity of events which are considered for analysis is ensured with few basic quality cuts applied to the raw data [2]. The data set discussed here was collected by eight detector

modules, between July 2009 and March 2011, corresponding to a total net exposure after cuts of 730 kg days.

3.1. Acceptance region

As any of the nuclei in $CaWO_4$ can be the relevant target for WIMP scattering depending on the mass of a possible WIMP, the acceptance region is chosen such that it extends between the upper boundary of the oxygen band and the lower boundary of the tungsten band and that implicitly includes the calcium band. Since, as a result of the incoming WIMP velocities and nuclear form factors, no significant WIMP signal is expected at energies above 40 keV (see Figure 1), the accepted recoil energies are limited by this value. The lower energy bound, in each detector module, is chosen such that the expected e/γ-leakage into the acceptance region is one event in the whole data set of that module. Due to the different resolutions (width of the bands) and levels of e/γ-background in the crystals, this value is different for each module.
A typical acceptance region is shown in orange in Figure 4 which presents in the light yield-energy plane the data set collected with one detector module (Ch51). The calculated bands for α's, oxygen recoils, and tungsten recoils shown in the figure are determined as described in [2]. With this definition of the acceptance region we find 67 accepted events in the sum over the eight detector modules used for Dark Matter analysis.

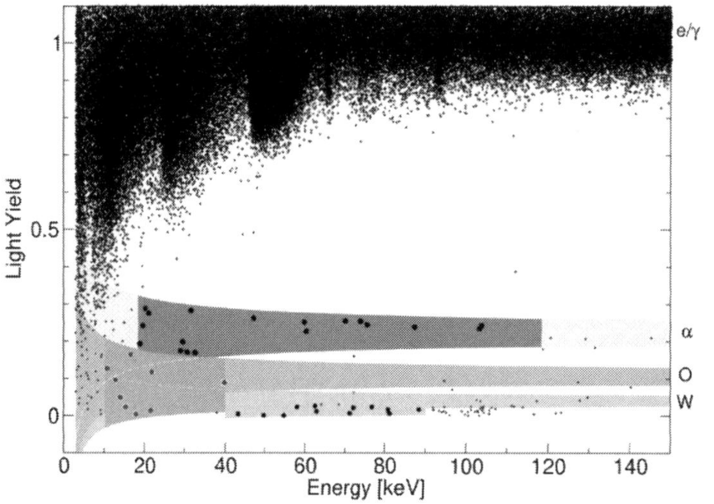

Figure 4. The data of one detector module (Ch51), shown in the light yield vs. recoil energy plane. The large population around a light yield of 1 is due to e/γ-background events. The shaded areas indicate the bands, where alpha (yellow), oxygen (turquoise), and tungsten (grey) recoil events are expected. Additionally highlighted is the acceptance region of this detector (orange) and the reference regions in the α-band (magenta) and in the ^{206}Pb-recoils band (green).

4. Background in the acceptance region

In Figure 4, next to the dominant e/γ-background, which is observed around a light yield of 1, we identify other background sources which could contribute events into the acceptance region. The scope of the analysis is to estimate the contribution of each of the identified backgrounds and to investigate a possible excess above the expectation. This is done using a maximum likelihood analysis which is thoroughly discussed in [2]. Here only its final results will be presented together with a qualitative discussion of the backgrounds in order to clarify the basic arguments and assumptions used in the analysis.

4.1. α-Background

In the data (see Figure 4 as an example) we observe low-energy α's with energies down to few keV. We understand these events to be due to α particles which have lost most of their energy before reaching the target crystals (for details on the housing of the detector modules see [2]). The presence of these events in our data is a consequence of the contamination with α emitters of the non-scintillating clamps holding the crystals. These events are present in all detectors, though the rates vary among modules.
Since the acceptance region has some overlap with the α-band, low energy α-events may be misidentified, leading to a certain number of background events in the acceptance region. In order to estimate this background contribution we select a reference region of the alpha band (highlighted in magenta in Figure 4) which is free of overlap with any other band. For this reference region we use an energy range of 100 keV whose low-energy bound is chosen in each detector module such that the expected e/γ-leakage is 0.1 event in the whole reference region of that module [2]. For each module we estimate in the reference region the energy spectrum dN_α/dE of the low-energy α-events. Using this spectrum we calculate for each individual module the ratio of α-events expected in the reference region and of α-events expected in the acceptance region. This ratio is then used to estimate the expected α-events in the whole acceptance region from the number of events observed in the reference region.

4.2. Lead Recoil Background

In Figure 4 we also identify an event population in and below the tungsten band around 100 keV. This population is caused by the lead nuclei from the α-decays of ^{210}Po present on the surface of the holding clamps and is present, though with different intensities, in all detector modules. The event distribution of the ^{206}Pb recoils peaks at the full recoil energy of 103 keV but it exhibits a decreasing low-energy tail which, in some detectors, reaches the energy range of interest for Dark Matter. Since this lead-recoil band and the acceptance region overlap considerably, a leakage of some ^{206}Pb events into the acceptance region cannot be excluded.
For an estimate of this background, we follow a similar strategy as for the α background. We define a reference region for each detector module which contains predominantly ^{206}Pb recoils. As a reference region, we choose the low-energy tail of the distribution ($\leq$90 keV) at energies above the acceptance region ($\geq$40 keV), where a possible WIMP signal cannot contribute. An example is highlighted in green in Figure 4. In this region we model the spectral energy density dN_{Pb}/dE of the ^{206}Pb events and then extrapolate the fit function into the energy range of the acceptance region to estimate the number of ^{206}Pb events expected there [2].

4.3. Neutron Background

Neutron scatterings, which mainly induce oxygen recoils in the energy range of interest, could -if neutrons are present- also contribute to events observed in the acceptance region. Differently from WIMPs, neutrons can cause coincident events in more than one detector module and this characteristic feature was exploited to estimate the contribution of neutrons to the background. Two classes of neutron production mechanisms are identified as possibly relevant sources of background:

(i) neutrons emitted by radioactive processes inside the neutron shielding (for a detailed description of the CRESST-II setup see [1, 3]);

(ii) neutrons produced by muons in the lead or copper shielding where the muon is missed by the veto.

For each of the two production mechanisms we used a corresponding calibration to infer the characteristic ratio between singles and coincident scatterings and to estimate from the observed coincidences the expected number of single neutron events [2].

Table 1. Results of the maximum likelihood fit. The expected contributions from the considered backgrounds and from a possible WIMP signal are listed for the two solutions together with the corresponding WIMP masses and interaction cross sections. The small statistical error given for the e/γ-background reflects the large number of observed events in the e/γ-band. The other listed errors correspond to a 1σ confidence interval.

	M1	M2
e/γ-events	8.00 ± 0.05	8.00 ± 0.05
α-events	$11.5^{+2.6}_{-2.3}$	$11.2^{+2.5}_{-2.3}$
neutron events	$7.5^{+6.3}_{-5.5}$	$9.7^{+6.1}_{-5.1}$
Pb recoils	$14.8^{+5.3}_{-5.2}$	$18.7^{+4.9}_{-4.7}$
signal events	$29.4^{+8.5}_{-7.7}$	$24.2^{+8.1}_{-7.2}$
m_χ [GeV]	25.3	11.6
$\sigma_{\rm WN}$ [pb]	$1.6 \cdot 10^{-6}$	$3.7 \cdot 10^{-5}$

5. Likelihood Analysis Results

The maximum likelihood fit, described in detail in [2], is based on a parameterized model of the background sources discussed in Section 4 and of a possible signal. The final results are summarized in Table 1. The total likelihood function has two maxima in the parameter space. M1 is the global maximum, but M2 is only slightly disfavored with respect to M1. The background contributions are very similar for M1 and M2 and in both cases the largest contribution is assigned to a possible WIMP signal. The possibility of two different solutions for the WIMP mass is a consequence of the different nuclei present in our target material. The given shape of the observed energy spectrum can be explained by two sets of WIMP parameters. In the case of M1, the WIMPs are heavy enough for the recoils of tungsten nuclei to appear in the energy range of the acceptance region (see Figure 2) and the possible signal is composed of 69 % of recoils on tungsten, 25 % on calcium and 7 % on oxygen. In the case of M2 tungsten recoils appear below the low-energy bound of the acceptance region. The observed signal constitutes of 52 % oxygen and 48 % calcium recoils.

Figure 5 illustrates the fit result, showing an energy spectrum of all accepted events together with the expected contributions of backgrounds and a possible WIMP signal.

The leakage of e/γ-events is the only contribution whose energy spectrum resembles that of a possible WIMP signal. A severe underestimation of this contribution could therefore explain the energy spectrum of the observed excess. However, in addition to the energy spectrum, we have to take into account the information coming from the distribution in the light-yield parameter. Figure 6 shows the corresponding light-yield spectrum of the accepted events (grey) and of all events in the energy range of the acceptance region (brown), together with the expectations from all considered background sources, as inferred from the likelihood fit. The leakage of events from the e/γ-band is naturally expected to be close to the band itself. For this reason the light-yield distribution of the e/γ contribution is expected to rise towards high light yields. However, Figure 6 clearly shows that the distribution of the total accepted events does not allow a relevant increase of the e/γ contribution.

The statistical significance at which we can reject the background-only hypothesis, determined with the likelihood ratio test, is 4.7σ for M1 and 4.2σ for M2, meaning that the backgrounds which have been considered are alone insufficient to explain the data. The properties of the

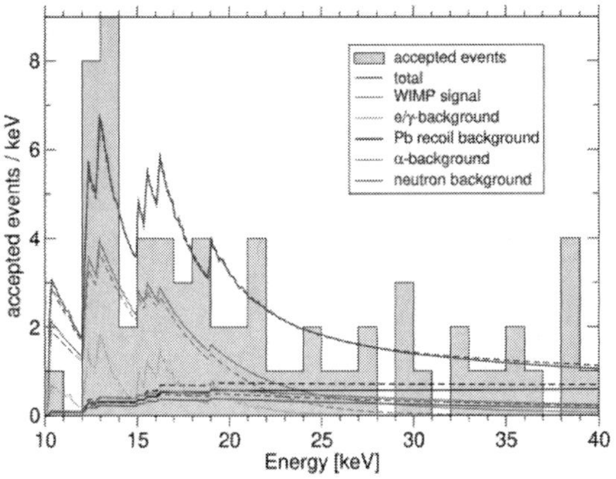

Figure 5. Energy spectrum of the accepted events from all detector modules, together with the expected contributions from the considered backgrounds and a possible WIMP signal, as inferred from the likelihood fit. The solid and dashed lines correspond to the fit results for M1 and M2, respectively.

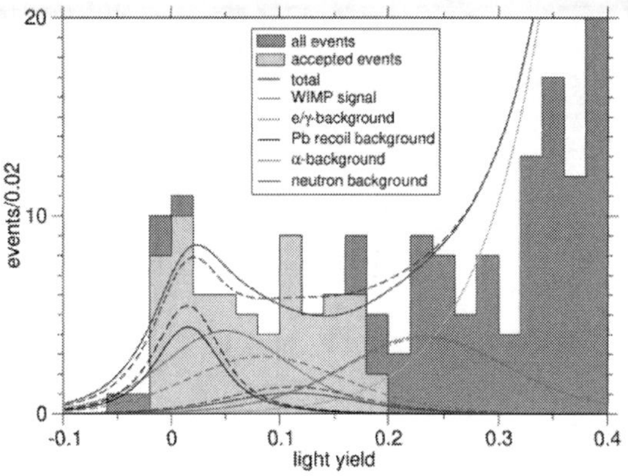

Figure 6. Light-yield distribution of all events in the energy range of the acceptance region (brown) and of the accepted events (grey). The expected contributions of background and of a possible WIMP signal from the likelihood fit are calculated for the fixed yield range of the plot. Outside the (energy dependent) yield boundaries of the acceptance region they are extrapolations of the fit. The solid and dashed lines correspond to the parameter values in M1 and M2, respectively.

additional source of events indicated by the result could be satisfied by Dark Matter particles, in the form of coherently scattering WIMPs. The background contributions are nonetheless still relatively high, therefore a reduction of the overall background rate is necessary to reduce the uncertainties in the modeling of it and to clarify the observation.

Despite the fact that we have an unfavorable situation in terms of background, it is still interesting to study the WIMP parameter space compatible with our observations. Figure 7 shows the location of the two likelihood maxima in the $(m_\chi, \sigma_{\mathrm{WN}})$-plane, together with the 1σ and 2σ confidence regions. The latest result is consistent with the data of an earlier CRESST run [1] where only tungsten nuclei were considered as a possible target for WIMP scatterings and the 1σ region of M1 is slightly disfavored. The same data of [1] have been reanalyzed considering all three nuclear recoil bands like it is done in the analysis of the present data [2] to increase the sensitivity for light WIMPs. The resulting exclusion limit, which is above the 2σ regions of [2], is shown in Figure 7. A similar analysis was presented in [9] where, however, the acceptance region was restricted to the tungsten band as in [1] accepting, as a consequence, only a fraction of the oxygen and calcium recoils.

The parameters derived from our observation are in tension with the limits published by other experiments [4, 5, 6]. Furthermore, the parameter regions compatible with the observation of DAMA/LIBRA (regions taken from [8]) and CoGeNT [7] are located outside the 2σ CRESST region.

6. Future Developments

At present, a new physics run is in preparation and several improvements aimed at a reduction of the overall background level are being implemented. The most important one addresses

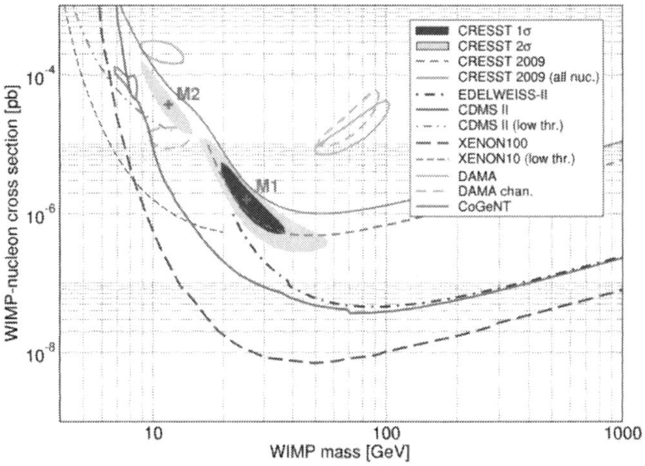

Figure 7. The WIMP parameter space compatible with the latest CRESST results. The CRESST contours have been calculated with respect to the global likelihood maximum M1. In addition the exclusion limits from CDMS-II [4], XENON100 [5], and EDELWEISS-II [6] are shown as well as the 90% confidence regions favored by CoGeNT [7] and DAMA/LIBRA [8]. For comparison we also show the CRESST limit obtained in an earlier run [1] in 2009 and the results of the reanalysis of the same run which takes into account all three nuclei in the target material.

modifications of the clamps holding the target crystals aiming at the reduction of the α and ^{206}Pb-recoil backgrounds. Another modification addresses the neutron background and foresees the installation of an additional layer of polyethylene shielding inside the vacuum can of the cryostat. This additional layer which will complement the present polyethylene shielding which is located outside the lead and copper shielding.

Acknowledgments

This work was supported by funds of the Deutsche Forschungsgemeinschaft DFG (Transregio 27: Neutrinos and Beyond), the Munich Cluster of Excellence (Origin and Structure of the Universe), the Maier-Leibnitz-Laboratorium (Garching), the Science and Technology Facilities Council (STFC) UK, as well as the DFG Internationales Graduiertenkolleg GRK 683. We are grateful to LNGS for their generous support of CRESST, in particular to Marco Guetti for his constant assistance. We would like to thank H. Kraus, S. Henry and A. Brown for their numerous and vital contributions to the experiment. We thank Prof. Volker Soergel for his strong support and his constant interest in all aspects of our work.

References

[1] Angloher G *et al.* 2009 *Astropart. Phys.* **31** 270
[2] Angloher G *et al.* 2011 (*Preprint* `astro-ph/1109.0702`)
[3] Angloher G *et al.* 2005 *Astropart.Phys.* **23** 325
[4] Ahmed Z *et al.* 2010 *Science* **327** 1619
[5] Aprile E *et al.* 2011 **107**(13) 131302
[6] Armengaud E *et al.* 2011 *Phys. Lett. B* **702** 329
[7] Aalseth C E *et al.* 2011 *Phys. Rev. Lett.* **107**(14) 141301
[8] Savage C *et al.* 2009 *JCAP* **0904** 010
[9] Brown A *et al.* 2011 (*Preprint* `astro-ph/1109.2589`)

12th International Conference on Topics in Astroparticle and Underground Physics (TAUP 2011) IOP Publishing
Journal of Physics: Conference Series **375** (2012) 012004 doi:10.1088/1742-6596/375/1/012004

Searching for WIMPs with EDELWEISS

Eric Armengaud, for the EDELWEISS collaboration
IRFU/SPP, CEA Saclay, 91191 Gif-sur-Yvette, France
E-mail: eric.armengaud@cea.fr

Abstract. The EDELWEISS-II experiment, operated in the Frejus laboratory in a low-background environment, uses cryogenic germanium detectors to look for WIMPs. We present the results of a WIMP search carried out recently with ten so-called InterDigit detectors. This technology enables a high level of gamma radioactivity rejection within a controlled fiducial volume. A cross-section of 4.4×10^{-8} pb could be excluded for a WIMP mass of 85 GeV. We also present the status of the EDELWEISS-III project, which will operate 40 newly-designed FID detectors in an upgraded installation to improve significantly the sensitivity to low WIMP scattering cross-sections.

1. Introduction

The nature of what is perceived as dark matter from cosmological and astrophysical observations remains totally unknown. A large experimental effort is currently devoted to test the WIMP hypothesis: dark matter would be constituted of weakly interacting particles with GeV-TeV scale mass, associated to new physics at the electroweak scale which may be probed at colliders like LHC. The best way to test this hypothesis is to achieve the direct detection of WIMPs from our galactic halo, by measuring the nuclear recoil spectrum generated by their interactions in a detector. The interaction rates are less than an event per kg per day, and the recoil energies are at some tens of keV. Dedicated detectors must therefore achieve a low energy threshold combined with high background rejection capabilities.

The EDELWEISS collaboration [1] designs germanium bolometers with a dual heat-and-ionization readout to reach this goal. In the following, we will present the EDELWEISS-II infrastructure and detectors whose novelty rely on the ability to reject surface interaction with an interleaved electrode design. We then review the results of a WIMP search which was recently published in [2], as well as a combination of this search with a similar one by the CDMS collaboration [3]. Finally, we present the status of the currently ongoing EDELWEISS-III program.

2. Experimental setup and detector performances

2.1. The EDELWEISS-II installation at Modane

EDELWEISS-II consists of a dilution cryostat able to host up to 40 kilograms of target material down to 18 mK. Its reversed geometry simplifies the installation of detectors. The whole cryogenic system can be remotely controlled, and operated this way during months without major trouble.

The experiment is located in the Modane Underground Laboratory (LSM), with a 4800 m water-equivalent depth reducing the cosmic muon flux down to $\sim 4\mu/\mathrm{m}^2/\mathrm{day}$. To reduce

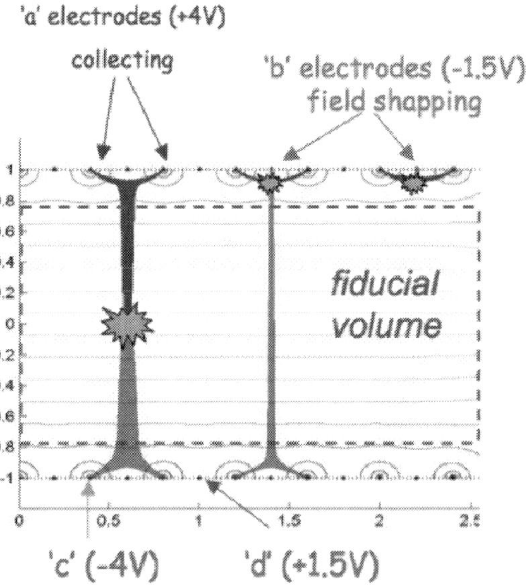

Figure 1. Illustration of the section of an ID detector. The electrode rings are represented as dots on the top and bottom sides, with an example of voltage setting. The thin color lines indicate equipotentials within the crystal. Three examples of interactions at different positions are shown, with simplistic trajectories of electrons (blue) and holes (red) represented.

environmental backgrounds, all materials used in the vicinity of the detectors were tested for their radiopurity with dedicated HPGe detectors. A clean room surrounds the whole setup, and the cryostat environment is under a permanent flow of deradonized air. A 20 cm thick lead shield around the cryostat screens the external gamma background. Fast neutrons generate nuclear recoils that cannot be discriminated against WIMPs when occuring as single-hit events. To protect detectors against neutrons, a 50 cm thick polyethylene shielding attenuates this background by three orders of magnitude. Furthermore a muon veto with a coverage of more than 98% tags neutrons produced by muons within this PE shield, in particular in the lead shield.

2.2. The InterDigit detector principle

EDELWEISS-II detectors are Germanium bolometers of typical mass $\sim$ 400 g, with electrodes deposited on their surfaces. At around 20 mK operating temperature, low energy interactions are detected by neutron transmutation doped (NTD) heat sensors that are glued at the surface of the crystals. A measurement of the ionization yield is provided by the additional electrode signals for each interaction. Since electron recoils yield a factor $\sim$ 3 times more ionization signal than nuclear recoils, an event-by-event discrimination against the gamma background is therefore possible with a high rejection power.

However, this rejection is possible only when the charges generated by an interaction are well collected, which is not always the case for interactions taking place near the electrodes. In particular, beta radioactivity from nearby surfaces, or from the detector surfaces themselves, would generate such a low-energy background. One therefore needs to be able to define a clean inner fiducial volume for each detector. For the InterDigit (ID) detector design, the electrodes used are concentric rings polarized at alternate voltages, as shown on Fig. 1. The resulting geometry of the electric field alters the propagation of charges near the surfaces: surface interactions are tagged by the presence of charge signals on the so-called veto electrodes located at the surface where the interaction takes place. On the other hand, interactions taking place within the fiducial volume generate an equal amount of charge on the two "collecting" electrodes on both sides of the crystal. The resulting discrimination against near-surface interactions is remarkably efficient, in particular because even when an interaction takes place in low electric field regions or just under collecting electrodes, the initial expansion of the charge cloud due to Coulomb interactions is sufficient to generate charges in a veto electrode. In first-generation ID detectors, the interleaved electrodes cover the top and bottom surfaces of the detectors. The cylindrical edge of the crystals are equipped with plain guard electrodes.

The first ID detector was built in 2007, and a few detectors were extensively tested at LSM in 2008. Physics runs were then carried out with 10 detectors in 2009-2010.

2.3. Measurements of detector performances

The ability to reject known sources of background against the fiducial nuclear recoils from WIMPs was tested on ID detectors using different radioactive sources.

The exposure of a 200-g ID detector to a ^{210}Pb source provides a clean measurement of the low-energy beta background rejection factor. Fiducial interactions are selected by requiring both a perfect charge balance between the signals of the top and bottom collecting electrodes, and the absence of charge deposits on the field shapping (veto) and guard electrodes. This dual rejection provides a strong redundancy, and enables the detector operation even when charges cannot be read on one of the field shapping electrodes. A discrimination at a level of $\sim 10^{-5}$ was experimentally proven in [4].

Fig. 2 shows the measured ionization yield Q as a function of recoil energy $E_{\rm rec}$ for fiducial volume events observed during calibrations with a neutron source. The distributions for nuclear recoils (NR) were measured for each detector and were all found to be compatible with the parametrization $Q = 0.16\,E_{\rm rec}^{0.18}$ from less that 10 keV to 200 keV. The region with 90 % acceptance ($1.64\,\sigma$) for NR is well described by a parametrization using the experimental resolutions on the heat and ionization signals. Data from neutron calibrations also enables to test the influence of thresholds on the nuclear recoil efficiency. In particular, Fig. 2 illustrates the full efficiency of the detectors to NR even below 20 keV.

Regular gamma-ray calibrations with external ^{133}Ba sources were carried out during the physics runs, and provided a total statistics of 3.5×10^5 electron recoils in the fiducial volume of the exposed detectors. Fig. 3 displays the measured ionization yield for these events, together with the average 90 % NR region. While the vast majority of electron recoils have an ionization yield far away from the NR region, a small population of "anomalous events", namely 6 events, leaks down in the NR band. We derive a gamma rejection factor of $(3 \pm 1) \times 10^{-5}$ for these detectors in the recoil energy range $20 < E_{\rm rec} < 200$ keV. It is suspected, but not fully proven yet, that these events are in fact due to multiple scatterings, with one interaction in the fiducial region and one in the guard region of the detectors, the latter having a less efficient charge collection.

Finally, the effective detector mass after fiducial selection was measured using low-background data itself: the cosmogenic isotopes ^{68}Ge and ^{65}Zn provide gamma-ray interactions homogeneously distributed within each crystal with an energy of ~ 10 keV. The ratio of event counts before and after the fiducial selection cuts provides a measurement of this cut efficiency in real WIMP search conditions. The average fiducial mass is 160 g. Electrostatic models confirm this value, and demonstrate that it is primarily limited by the presence of guard regions. Neutron calibrations also provided a consistency check for this measurement.

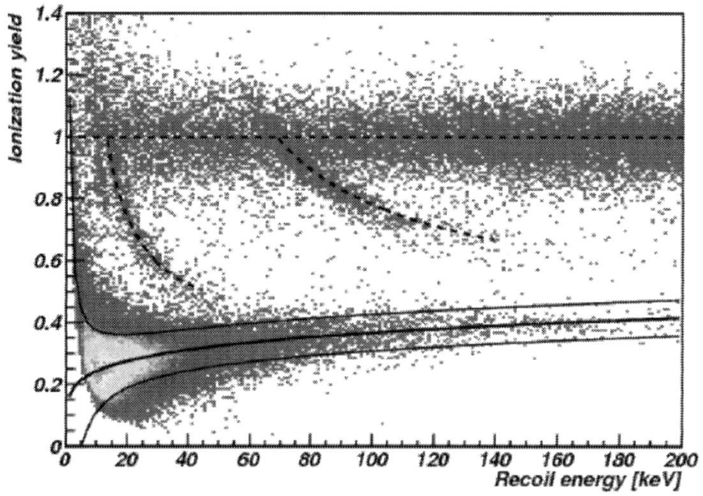

Figure 2. Ionization yield measurement as a function of recoil energy from neutron calibrations using all ID detectors used for WIMP search. The dotted lines represent the expected locations of gamma interactions (yield equal to one) and inelastic neutron scatterings (curved lines). The solid lines are the 90 % elastic nuclear recoil parametrization.

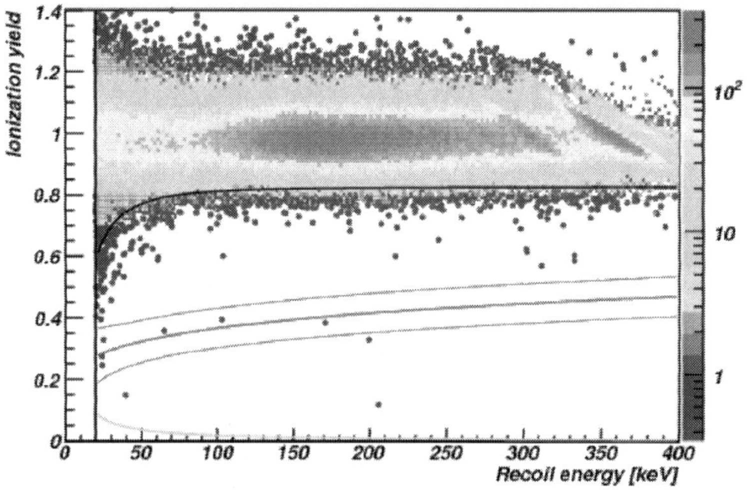

Figure 3. Ionization yield measurement as a function of recoil energy from a large-statistics gamma-ray calibration with all ID detectors used for WIMP search. Red line: average NR band. Green line: average ionization threshold.

3. A search for WIMPs with ten ID detectors

3.1. Data set and backgrounds

A WIMP search was carried out between April 2009 and May 2010 with ten 400-g detectors. While all heat sensors were working correctly, a few veto or guard electrodes were malfunctioning (5 over 60 ionization channels), but the redundancy between channels prevented the loss of detectors for physics. The EDELWEISS-II setup provided a remarkably stable cryogenic environment at 18 mK during all this run. For each detector, an online trigger was applied on the heat sensor timelines. The recorded pulses are then processed offline using optimal filters. The rejection of noisy periods is carried out using the measured FWHM baselines, and requiring them to be below 2 keV for fiducial ionization and 2.5 keV for heat and guard ionization. These cuts imply a 17 % exposure loss. The WIMP search cuts are then simply: fiducial volume selection; coincidence rejection (coincidences between detectors as well as with the muon veto); and finally selection of recoil energies between 20 and 200 keV, and ionization yields within the 90% NR region and outside the 99.99% gamma region. This results in a 384 kg.days net exposure.

The expected backgrounds are gamma and beta radioactivities, and fast neutron interactions.

- The measured gamma rejection factor described above, combined with the observation of 18000 bulk electron recoils during WIMP search, gives an expected background of < 0.9 events (90% CL).
- Similarly, the measured beta rejection factor is combined with the observation of 5000 surface events in WIMP search to give an expected background of < 0.3 events.
- A detailled description of the different sources of neutrons is given in other presentations of this conference [5, 6]. The observed muon-induced event rate combined with the muon veto efficiency gives an upper limit of 0.4 muon-induced neutron events. The measured neutron flux from the rock, and upper limits from radioactivity measurements of the lead, polyethylen and EDELWEISS infrastructure materials are combined to Monte-Carlo simulations to give an upper limit of 0.3 neutron events from these sources. A potentially more important neutron source had been identified as the connectors and cables located inside the cryostat, which could induce up to 1.1 nuclear recoil events.

Summing all these background upper limits gives an estimation of 3.0 background events in the NR region for the WIMP search.

3.2. Constraints on WIMP interactions

The observed distribution shows five events in the WIMP-search region (Fig. 4), one at 172 keV and the other ones between 20 and 23.2 keV. Given the abovementioned backgrounds, there is no clear evidence for WIMPs.

Applying the standard procedure in the field to set an upper limit in the presence of a weak, poorly constrained background [7], we obtain a limit on the spin-independent WIMP-nucleon cross-section of 4.4×10^{-8} pb for a WIMP mass of 85 GeV. The 90 % limit as a function of WIMP mass is presented on Fig. 5. This limit is degraded by the presence of the observed events at low energy.

The same data and analysis can be used to contrain a possible inelastic diffusion cross-section of WIMPs on nucleons. This scenario was invoked to reconcile the DAMA signal with other experiments, since in that case the expected recoil spectrum is globally reduced and suppressed at low recoil energies, and the modulation signal is enhanced. For example, if the WIMP excitation energy is 120 keV, the EDELWEISS data excludes the DAMA region for all WIMP masses above ~ 90 GeV.

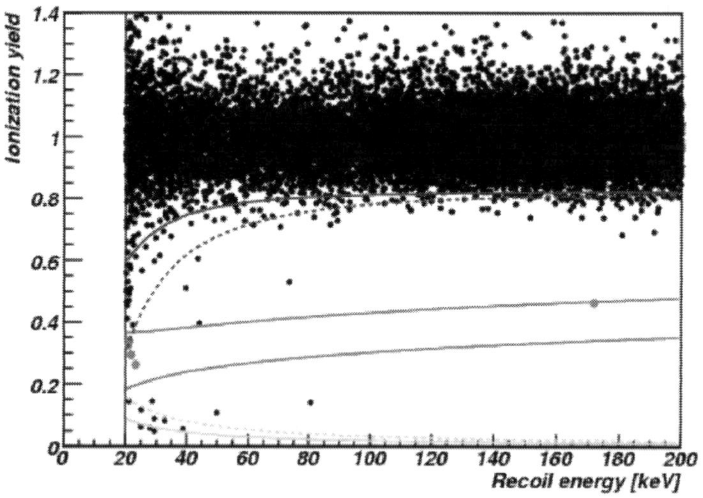

Figure 4. Ionization yield versus recoil energy for the fiducial events recorded during the EDELWEISS-II WIMP search. The five candidates in the WIMP search band (red) are highlighted in red. Blue lines: 99.99% gamma discrimination line in average (continuous) or worst noise conditions (dashed). Green lines: ionization threshold in average (continuous) or worst noise conditions (dashed).

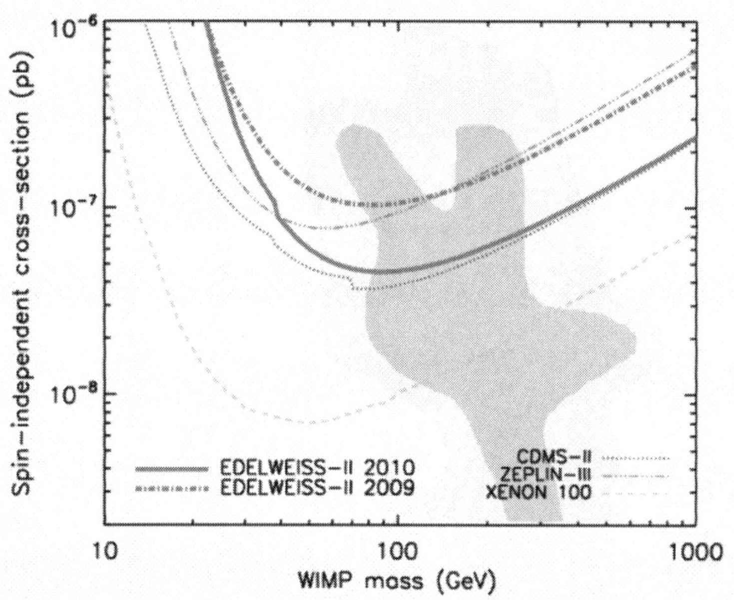

Figure 5. Limit at 90% CL on the WIMP-nucleon cross-section as a function of WIMP mass, obtained by the EDELWEISS-II search. Limits from other experiments, and a pre-LHC prediction in the CMSSM framework are also represented.

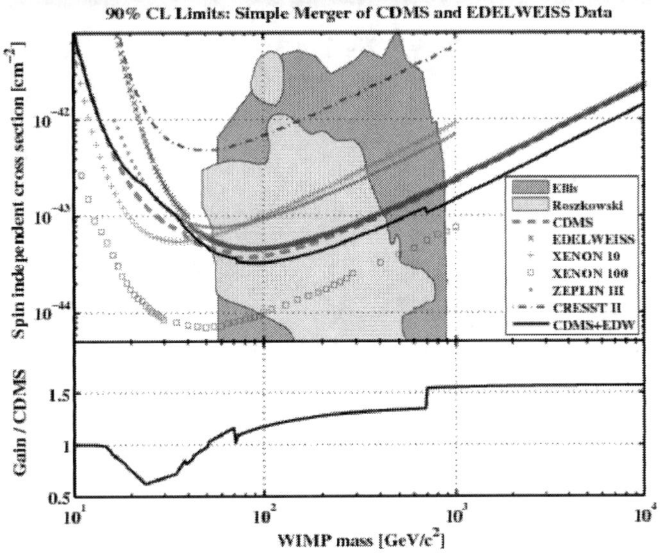

Figure 6. Limit on the WIMP-nucleon interaction cross-section obtained by a combined analysis of the EDELWEISS and CDMS data.

3.3. Combination with CDMS dataset

The EDELWEISS and CDMS collaborations have achieved similar sensitivities in WIMP search. In addition, the use of the same target material allows a simple combination of data. A common analysis between both collaboration was carried out in [3].

Though other methods have also been tested, a simple merger of datasets was chosen prior to analysis. The exposures of EDELWEISS (384 kg.d between 20 and 200 keV) and CDMS ($\sim$ 379 kg.d between $\sim$ 10 and 100 keV, see [8]) were added, as well as the observed WIMP candidate events (5 events for EDELWEISS and 4 for CDMS). Applying the optimal interval method [7] to set an upper limit to the WIMP-nucleon cross-section, the obtained sensitivity, shown on Fig. 6, represents a $\sim$ 50% gain with respect to CDMS for large WIMP masses. There is no gain for WIMP masses below 50 GeV due to the threshold and the presence of events at low energy in the EDELWEISS dataset.

4. Status of the EDELWEISS-III project

The EDELWEISS-III project consists in an upgrade of both the current EDELWEISS setup and detectors in order to reach a sensitivity to the WIMP-nucleon cross-section of the order of 5×10^{-9} pb in a short term. This will require to obtain an exposure of 3000 kg.d.

In order to reduce the beta and gamma backgrounds, we developed an improved detector design, named "Full InterDigit" (FID). These 800 g crystals are equipped with two NTD heat sensors, and are covered by interleaved electrodes on all their surface. There is therefore no "guard" region anymore inside the crystal volume. Both the increase of crystal mass and the removal of guard regions increase the fiducial mass for each individual detector by a factor $\sim$ 4. First calibrations of these detectors have also demonstrated an improvement of the gamma-ray rejection over the original ID detectors, as presented in Fig. 7. We plan to install 40 FID800 detectors, corresponding to $\sim$ 24 kg fiducial volume in 2012.

Infrastructure upgrades (cabling, cold electronics, cryogenics and acquisition) are necessary to host these new detectors, and will also reduce the neutron budget within the cryostat. In

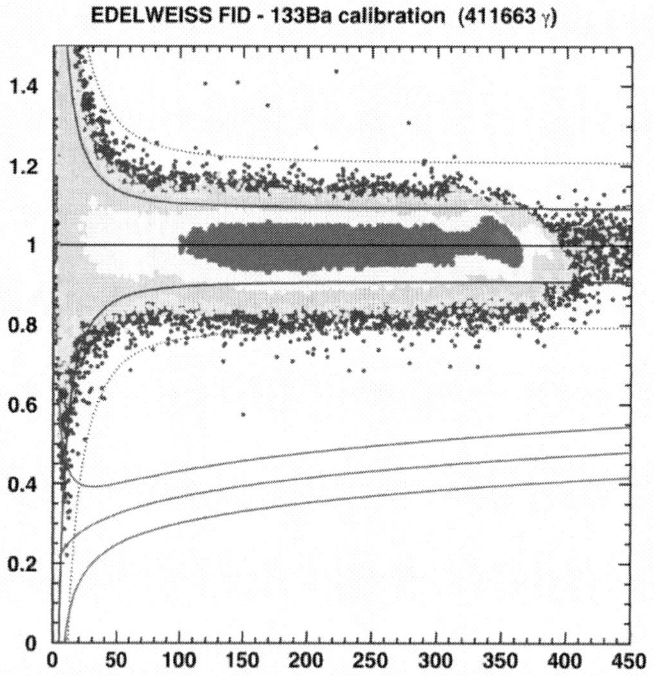

Figure 7. Ionization yield distribution as a function of recoil energy for the fiducial events recorded in two FID800 detectors during gamma-ray calibrations. Although the fiducial statistics is larger than what was collected on ID detectors, no anomalous event leaks in the NR region contrarily to what is observed on Fig. 3.

addition, an inner polyethylene shield will be installed to reduce the flux of fast neutrons coming from outside the cryostat.

Finally, let us mention that various efforts are also ongoing in order to improve the sensitivity of EDELWEISS detectors to low-mass WIMPs, of typical mass ~ 10 GeV.

References

[1] http://edelweiss.in2p3.fr/
[2] Armengaud E et al. 2011 *Phys. Lett.* B **702** 329
[3] Ahmed Z et al. 2011 *Phys. Rev.* D **84** 011102
[4] Broniatowski A et al. 2009 *Phys. Lett.* B **681** 305
[5] Loaiza P, these proceedings
[6] Eitel K, these proceedings
[7] Yellin S 2002 *Phys. Rev.* D **66** 032005
[8] Ahmed Z et al. 2010 *Science* **327** 1619

Journal of Physics: Conference Series **375** (2012) 012005 doi:10.1088/1742-6596/375/1/012005

CDMS and SuperCDMS

Wolfgang Rau
for the CDMS and SuperCDMS collaborations
Department of Physics, Queen's University Kingston ON, K7L 3N6, Canada
E-mail: rau@owl.phy.queensu.ca

Abstract. The CDMS experiment has operated cryogenic Ge and Si detectors for several years in a well shielded underground environment searching for signals from Weakly Interacting Massive dark matter Particles (WIMPs). Due to the low background and excellent background discrimination power, CDMS has provided the best sensitivity for WIMP-nucleon interactions for most of the past decade. New detectors with larger mass and further improved background discrimination, developed for the successor experiment SuperCDMS, have recently been deployed; WIMP search measurements are foreseen to resume before the end of 2011. In this paper we discuss the results from CDMS - in particular the constraints on low mass WIMPs recently proposed to explain results from other dark matter experiments - and present the status of and plans for SuperCDMS.

1. Introduction

Weakly Interacting Massive Particles (WIMPs) [1] have been proposed as one of the most probable solutions to the long standing dark matter problem [2]. Simple kinematic considerations lead to the conclusion that – due to the galactic escape velocity of about 600 km/s – elastic scattering of galactic WIMPs with electrons would in most cases lead to negligible energy transfer, while nuclear recoils will yield energies in the range of tens of keV. The very low expected interaction rate [3] stands in stark contrast with the high rate of interactions from environmental radiation in an unshielded detector. In direct dark matter detection experiments, shielding against cosmic radiation is achieved by operating in deep underground laboratories [4] and using veto detectors for the remaining cosmic ray muons, while environmental radioactivity is suppressed by elaborate shielding configurations and careful selection of construction materials. Suppression of remaining background is often achieved by detector specific methods to discriminate between electron recoil background and candidate nuclear recoil events.

The Cryogenic Dark Matter Search (CDMS) experiment has achieved an unprecedented control and understanding of the background in its cryogenic ionization detectors. Over much of the past decade CDMS has provided the most sensitive test of potential WIMP-nucleon interactions [5, 6, 7, 8, 9, 10] and was only recently surpassed by the XENON experiment [11]. SuperCDMS, the successor of CDMS, is about to start measurements with new and improved detectors [12] and an increased target mass with the aim to enhance the sensitivity by a factor of about 5-8, being limited by the residual cosmogenic background at the current location. The next phase of SuperCDMS with a further increase in target mass by an order of magnitude is planned to be installed at SNOLAB near Sudbury, ON (Canada), the deepest large underground laboratory worldwide.

Section 2 describes the CDMS technology, followed by a discussion of our main data analysis (section 3) and a discussion of the recent joint analysis of CDMS and EDELWEISS data (section 4). Section 5 is dedicated to a specific analysis aimed at lowering the analysis threshold at the cost of an increasing background, in order to gain sensitivity for WIMPs with masses below 10 GeV. Finally we will discuss the status of the SuperCDMS experiment at the Soudan Underground laboratory (section 6) and our plans to move to SNOLAB (section 7) which includes a discussion of a detector test facility to be installed at SNOLAB in 2012.

2. CDMS Technology

CDMS detectors are based on cylindrical germanium or silicon single crystals, 7.6 cm in diameter and 1 cm high, which are equipped with a phonon sensor to measure the deposited energy and a charge collecting electrode to identify the type of interaction based on the ionization efficiency or *yield*, which is the ratio of the amount of collected charges to the total energy deposited in the primary interaction.

The phonon sensor is a superconducting tungsten film evaporated onto one of the flat crystal surfaces. The tungsten is held in the transition from the superconducting to the normal state (therefore called *Transition Edge Sensor* or TES) where a small change in temperature leads to a large change in the measured resistance. The phonon sensor is divided into four quadrants, each of which consists of about 1000 individual TES connected in parallel. The detectors are operated at roughly 40 mK.

The aluminum grid on the opposite face of the detector through which the bias voltage (typically 3 V) is applied, acts as charge collecting electrode. The grid is separated into two sensors, an inner disk and an outer ring, allowing us to discard events with incomplete charge collection caused by the distorted electric field at the edge of the detector. Figure 1 shows a sketch of the sensor layout of the CDMS detectors.

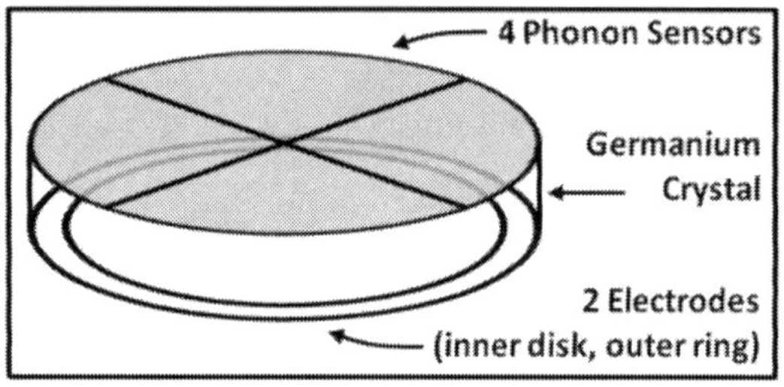

Figure 1. Sensor layout of CDMS detectors: one side is equipped with four phonon sensors (in quadrants), while the other side has two concentric charge electrodes. Events where the dominant charge signal occurs in the outer electrode are discarded. The phonon sensor layout allows event position reconstruction.

The phonon signal carries information about the event position, including the depth or z-position, giving rise to the name *Z-sensitive Ionization and Phonon detector* or ZIP. For more details about the detectors see e.g. [13].

The detectors are mounted in stacks of six, attached to a mechanical structure (the *tower*) with low heat conductivity, which brings the detector wiring up to 4 K and holds the first stage amplifier circuits (SQUID based for the low-impedance phonon sensors and FET based for the high-impedance charge readout). The towers are mounted in the innermost of a set of nested copper cans (*icebox*) which are attached to the different thermal stages of a dilution refrigerator by a horizontal set of concentric cold fingers. Thanks to this configuration the detectors can be fully surrounded by lead and polyethylene shielding against gamma and neutron radiation from the environment as well as radioactive trace contamination in the refrigerator. The outermost layer of the setup is a set of plastic scintillator detectors to identify residual cosmic ray muons (*muon veto*) which may interact inside the setup and generate WIMP like events [13]. The setup has been operated for WIMP search from 2003 until 2008 with up to five towers (30 detectors).

3. Main CDMS Analysis

The analysis is described in detail in [10]. Here we give a short summary of the main points.

Out of the 30 CDMS ZIP detectors 19 were from germanium and 11 from silicon; due to their better performance and much higher sensitivity for WIMP interactions, only germanium detectors were considered in the analysis discussed here. Several of the detectors suffered from insufficient performance during parts of the measurements. The data from these detectors were removed from the dark matter search for the respective periods; however they were included in the identification of multiple scattering which cannot happen for WIMP interactions. For all remaining detectors events in the outer part (identified by the outer charge electrode) were removed since the distorted electric field configuration at the edge of the detector can lead to reduced charge collection and thus to a potential misidentification of electron recoils as WIMP-candidate events.

A major part of the analysis is a careful analysis of the expected background. Two major types of background need to be considered: the first type are interactions from neutrons which produce nuclear recoil events and therefore are not distinguishable from WIMPs on an event-by-event basis; the second type are surface electron recoil events which suffer from reduced charge collection.

Neutrons from radioactive processes in the material surrounding the experiment (mostly the rock walls) are negligible due to the thick polyethylene shielding. Neutron background from construction material in the experiment is estimated based on contamination measurements of such materials. The event rate from this source is expected to be in the range of 0.03-0.06 for the latest data period comprising a total of 612 kg-days (raw exposure).

Neutrons can also be produced by muons interacting in material surrounding the detectors; those neutrons can have very high energies and thus penetrate the neutron shielding, leading to an expected background of $0.04^{+0.04}_{-0.03}$ events in the same data set. Muons interacting in the experimental setup can be identified thanks to the highly efficient (well above 99%) muon veto detectors (for details see e.g. [13]) and thus do not contribute to the background.

When produced close to the electrodes, the initially energetic charge carriers can move against the external electric field and enter the wrong electrode which reduces the amount of collected charges. This can lead to electron recoils with a reduced charge yield typical for nuclear recoils. Fortunately also the phonon signal of such events is changed: the fast thermalization due to the phonon-electron interaction in the metal layers at the nearby surface leads to a difference in pulse shape, identifying these surface events. This discrimination, however, is not perfect and comes at a significant cost of signal efficiency. In the latest data set the expected background from this source is 0.8±0.2 events and thus dominates our background estimate.

The complete analysis was performed based on calibration data (gamma and neutron calibrations), multiple scatter events and events outside the signal region to avoid bias (*blind analysis*). The effective exposure of the latest data period, after all the discussed data selection criteria were applied, was 194 kg-days. After finalizing the analysis, the signal region was inspected for the first time and 2 candidate events were identified. This is perfectly compatible with the expected background of roughly one event total and a careful inspection concluded that the parameters of the events allow no further conclusion about the true nature of the events (residual surface event background or actual nuclear recoil events). Combined with earlier data taken with the same setup this leads to an upper limit on the WIMP-nucleon cross-section as shown by the light (red) solid line in figure 2. [10] The limit is calculated based on Yellin's *optimum interval* method [14]. The two kinks in the limit curve (around 40 GeV and 70 GeV respectively) are a consequence of this method since different intervals are selected to determine the limit for different mass ranges.

4. Joint CDMS-EDELWEISS Analysis

The EDELWEISS experiment also uses cryogenic germanium detectors to search for potential WIMP-nucleon interactions [15]; the effective exposure of CDMS (energy dependent with a maximum of 379 kg-days) is slightly lower than that of EDELWEISS (384 kg-days), but the EDELWEISS threshold is higher (20 keV compared to ~10 keV for CDMS) with a few background events just above threshold. Overall CDMS is slightly more sensitive especially at lower WIMP masses. The similarity in both technology and sensitivity prompted us to pursue a joint analysis [16]. Since the same target material is used it is straight forward to just add the effective exposure and combine the list of candidate events in order to obtain a combined upper limit on the WIMP-nucleon cross-section. The result of this analysis is shown in figure 2 together with the individual CDMS and EDELWEISS limits. The slightly reduced sensitivity of the combined analysis when compared with CDMS data alone results from the events just above threshold in the EDELWEISS data set. For comparison also the most recent limit from the XENON100 experiment is plotted which is so far the only experiment that surpassed the CDMS sensitivity.

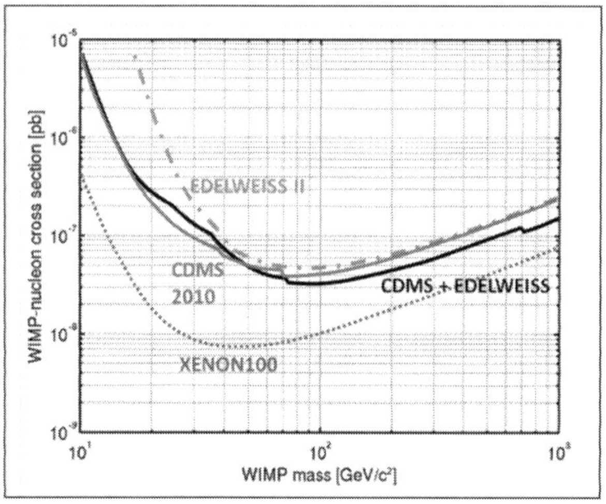

Figure 2. Exclusion limits on the spin-independent WIMP-nucleon cross-section under standard assumptions about the galactic dark matter halo (for details see the individual references). The light (red) solid line is the limit derived from the standard analysis of CDMS-II data [10]. Also shown are the more recent limits from EDELWEISS (dash-dotted) [15] and XENON100 (dotted) [11]. The black solid line is the result of the joint analysis of CDMS and EDELWEISS data [16].

There are in fact different possibilities to combine the data of different experiments. For this first joint analysis it was decided beforehand that the above method, just adding exposures and event lists, should be used; however we considered a set of different possibilities [17] which take advantage of the a-priori knowledge of the differences in expected background between the two experiments. More details about the advantages and disadvantages of some of these methods can be found in [16].

5. Low Threshold Analyses

Several experiments in the field of direct detection of dark matter have claimed evidence for a positive signal, most notably the DAMA/LIBRA collaboration, observing an annual modulation in the low energy region of their recoil spectrum in a large NaI target, which is proposed to be caused by the variable speed of the earth relative to the galactic WIMP distribution due to the combination of the earth's rotation about the sun and the sun's movement within the Milky Way [18]. A more recent claim comes from the CoGeNT experiment which employs low-background and low-threshold germanium detectors operated at 77 K. An exponential rise in the low-energy part of the spectrum is interpreted as evidence for WIMP scattering in the detector [19]. Hooper et al. [20] proposed a consistent interpretation of the results from DAMA and CoGeNT in terms of low-mass WIMPs. More recently also CRESST published data showing an excess of events consistent with a low-mass WIMP interpretation [21]

The energy transfer to a given target nucleus drops strongly with the WIMP mass as shown in figure 3. Lowering the threshold thus can drastically increase the expected interaction rate for low mass WIMPs. The threshold in the standard CDMS analysis is set with the goal to have a near background-free experiment which leads to the maximum discovery potential; however, even at low energy the event rate is rather low. Thus a lower threshold in CDMS leads to a significant improvement in the sensitivity for low-mass WIMPs.

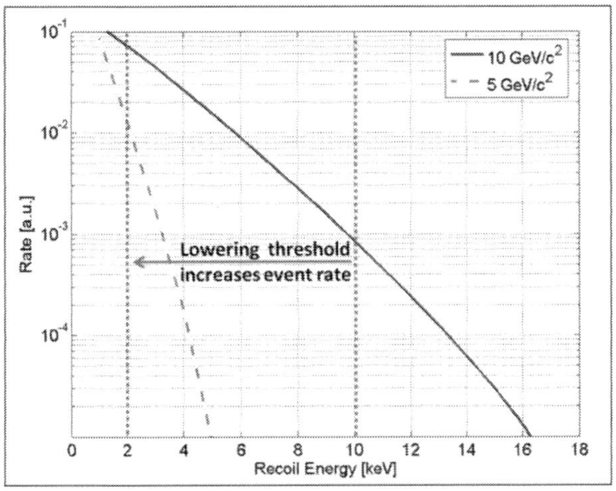

Figure 3. Expected energy spectra from elastic scattering of WIMPs with a mass of 5 GeV/c^2 (dashed line) and 10 GeV/c^2 (solid line) respectively off germanium nuclei, calculated according to [22]. Lowering the experimental threshold drastically increases the sensitivity for low-mass WIMPs (approximate thresholds for standard and low threshold CDMS analyses are indicated).

We have applied this philosophy in slightly different ways for two of our data sets. The first data set was taken at a shallow site at Stanford (*Stanford Underground Facility*, SUF) with only ~15 mwe of shielding against cosmic radiation leading to a significant background rate, but the good electronic noise environment allowed for a very low trigger threshold of <1 keV for some detectors [23]. The second data set is the same as discussed above for the CDMS standard analysis. The background here is lower, but the worse electronic noise led to a threshold of 2 keV [24].

Figure 4 shows the low-mass WIMP interpretation of CoGeNT and DAMA/LIBRA by Hooper et al. [20] as shaded regions. Also shown is the preferred region originally published by CoGeNT. This proposed evidence is contrasted with the limits obtained by our standard analysis as well as our two low-threshold analyses mentioned above. While the data set from Stanford still leaves a significant fraction of the overlap region found by Hooper et al. [19] untested, this interpretation is in strong tension with the low-threshold analysis of our data taken at Soudan.

It should be noted that part of the original favored region published by CoGeNT extends to below the CDMS limit, but more importantly: more recent discussions of the CoGeNT data [26] seem to indicate that the uncertainty on the background in CoGeNT is underestimated in the above publication. This allows the favored region to shift to considerably lower cross-sections removing the tension between CDMS and CoGeNT, but potentially also weakens the claim for a signal.

6. SuperCDMS at Soudan

To overcome the limitation from surface events discussed in section 3, we have developed a new type of detector with a different sensor layout. The most important changes are a larger mass per detector module achieved by an increased thickness of the crystals (one inch instead of 1 cm),

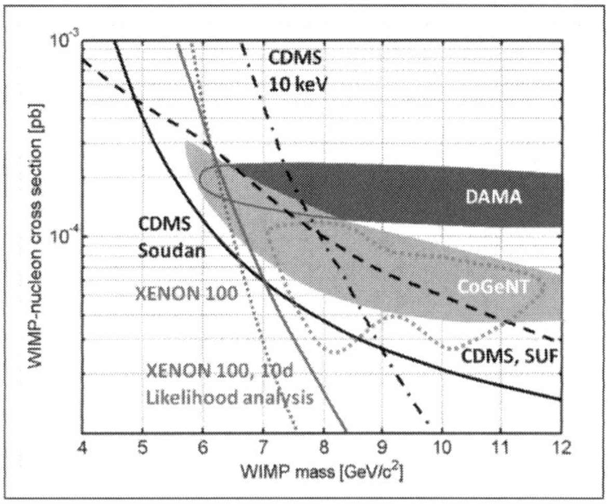

Figure 4. Exclusion limits for low-mass WIMPs. Dash-dotted: standard CDMS analysis (10 keV threshold); dashed: low-threshold analysis of early CDMS SUF data [23]; black solid: low-threshold analysis of recent CDMS Soudan data [24]. For comparison light dotted: XENON100 [11]; light solid: re-analysis of older XENON100 data [25]. Dark/light shaded regions: low-mass WIMP interpretations of DAMA and CoGeNT by Hooper et al. [20]; dotted contour: original CoGeNT publication [19].

phonon sensors on both sides of the detector and a new geometry of the charge sensors which now are interleaved with the phonon sensors. These new detectors are called *iZIP* where the "*i*" stands for *interleaved.* Figure 5 shows a sketch of the new geometry of the phonon sensors (three equal inner sensors and an outer circle rather than four equal quadrants, for a better radial position determination).

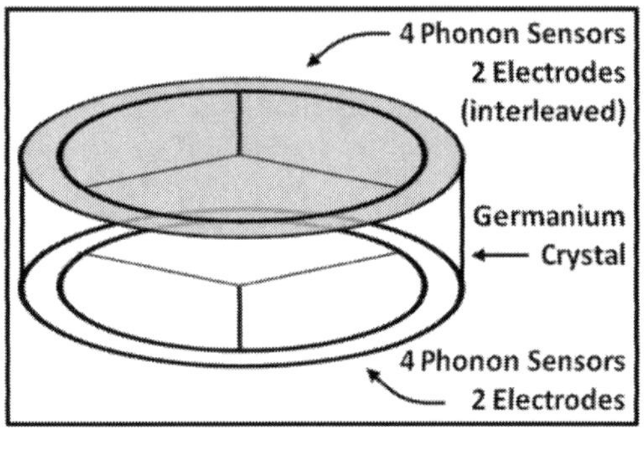

Figure 5. Sensor layout of the new iZIP detectors. There are four phonon sensors on each side arranged as three equal sensors in the center and one outer circle. The arrangement at the bottom is rotated by 180° with respect to the one at the top for better spatial sensitivity. The electrodes for charge collection are interleaved with the phonon sensors, generating an inhomogeneous electric field close to the surface.

Figure 6 demonstrates the effect of the interleaved electrodes: for events in the bulk the electrons and holes drift to the electrodes on opposite sides of the detector, while for events near the surface the charge carriers distribute between the electrode and the phonon sensors (acting as ground reference) on the same side of the detector. Thus, the ratio of the measured charge on both sides of the detector can be used to identify surface events with high efficiency. The surface event discrimination performance of these new detectors is considerably better than that of the CDMS ZIP detectors, while keeping a larger fraction of the actual nuclear recoil events. For more details on the iZIP detectors and their performance see [12] and [27].

In early fall 2011 SuperCDMS has implemented a full complement of 15 germanium iZIP detectors into the existing experimental setup at the Soudan Underground Laboratory - half as many as before since iZIP detectors have twice as many sensors as the CDMS ZIP detectors, but with an increased total mass of $sim10$ kg. After a commissioning period which is expected to last until end of 2011, dark matter search will restart and is expected to last for two to three years with a final sensitivity for the WIMP-nucleon cross-section of $\sim 5 \times 10^{-9}$ pb.

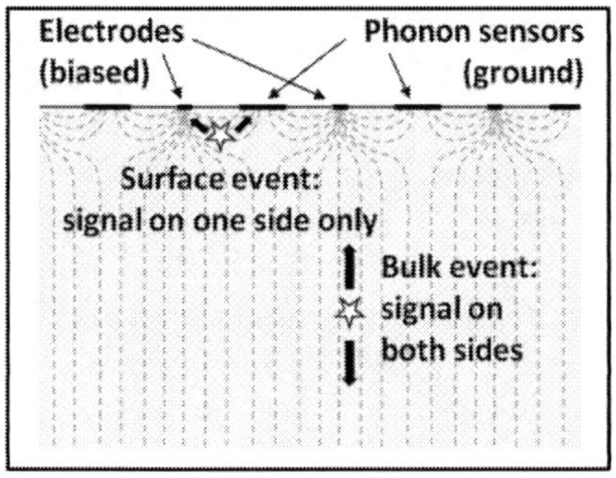

Figure 6. Surface electric field configuration (top part; bottom is analogous) in an iZIP detector. The charge collecting bias is applied to the narrow electrodes; the wider phonon sensors act as ground reference (bottom of detector is equivalent, but with opposite sign bias). The resulting field configuration draws charges from a bulk event to opposite sides of the detector; charges from surface events distribute between the charge and phonon sensors on the same side of the detector.

7. SuperCDMS at SNOLAB

The goal sensitivity for SuperCDMS at Soudan is given by the expected residual cosmogenic background at the depth of the Soudan Underground Laboratory (2100 mwe). This is the main motivation for the plans of SuperCDMS to build the next generation experiment with a total target mass in the 100 kg range at SNOLAB, the new underground laboratory near Sudbury, Ontario (Canada), which provides not only a several order of magnitude better shielding against cosmogenic radiation due to its larger depth of 6000 mwe, but is also operated as a whole as class 2000 cleanroom [4].

Work has started to develop detectors with larger diameters (100 mm) and further increased mass. The iZIP technology is expected to provide enough discrimination power between nulcear recoils and surface events to keep the background from this source negligible even for the total exposure aimed for with SuperCDMS at SNOLAB (order of hundred kg-years). However, in our existing test facilities the discrimination power cannot be tested, since it is impossible to recognize surface events that are misidenfitied as nuclear recoils due to the presence of actual nuclear recoil events from neutrons produced by cosmogenic radiation. Therefore we have started to construct a well shielded detector test facility underground at SNOLAB. The heart of this facility is the cryostat formerly used for the first CDMS measurements at Stanford. The cryostat has been refurbished so that the larger detectors can be mounted directly in the cold volume inside the cryostat. The cryostat will be installed in a dry-well surrounded by a cylindrical (passve) water shield of $\sim$3.7 m in diameter and height. Figure 7 shows a sketch of the setup.

The external background rate per detector is expected to meet the specifications of $<$1 neutron/day and $<$1 gamma/second [28]. This will allow us to test the discrimination against surface events to better than 1 in 10^6. In order to minimize the operational costs the cryostat will be automated and equipped with re-liquefiers for the cryogenic liquids. Installation at SNOLAB is expected in 2012.

8. Conclusion

The sophisticated detector technology employed by CDMS has led to a decade of leading results in direct dark matter search with unprecedented background control. Continued improvements in detector design and eventually the move to SNOLAB for a further reduction in cosmogenic background will guarantee that the successor experiment SuperCDMS continues to play a leading role in WIMP search for the coming years.

A first step towards SNOLAB will be the installation of a well shielded cryogenic detector test facility at SNOLAB. This facility will be commissioned in 2012 and will be used to qualify the detectors for SuperCDMS at SNOLAB.

With the experimental sensitivity starting to cut deep into the parameter range favored by theoretical considerations and cosmological observations we can expect an exciting time in dark matter search over the next few years.

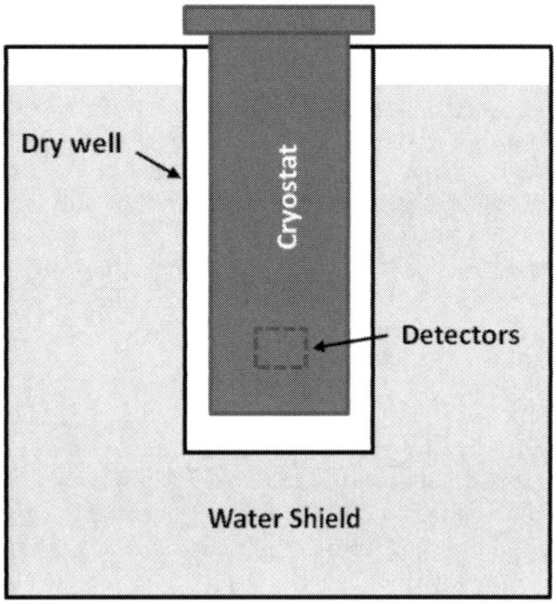

Figure 7. Sketch of the shielded test facility for SuperCDMS at SNOLAB. The detectors will be mounted in the enlarged sample volume inside the cryostat. The cryostat sits in a dry well inside the water shield. The cylindrical water tank has a diameter and height of about 3.7 m; the thickness of the water layer is ~1.5 m (slightly less at the bottom). The gashandling and temperature and pressure control and monitoring systems are connected to the top of the cryostat; to change the detectors or calibration source configuration, the water will have to be drained to gain access to the bottom of the cryostat.

References

[1] G. Steigman, M.S. Turner, 1985 *Nucl. Phys.* B **253** 375
[2] G. Bertone, D. Hooper, J. Silk, 2005 *Phys. Rep.* **405** 279;
J. Einasto, 2010 *e-print* arXiv:0901.0632v2 [astro-ph.CO]
[3] G. Jungman, M. Kamionkowski, K. Griest, 1996 *Phys. Rep.* **267** 195
[4] Nigel Smith, this volume
[5] CDMS Collaboration, R. Abusaidi et al., 2000 *Phys. Rev. Lett.* **84** 5699
[6] CDMS Collaboration, D.S. Akerib et al., 2003 *Phys. Rev.* D **68** 082002
[7] CDMS Collaboration, D.S. Akerib et al., 2004 *Phys. Rev. Lett.* **93** 211301
[8] CDMS Collaboration, D.S. Akerib et al., 2006 *Phys. Rev. Lett.* **96** 011302
[9] CDMS Collaboration, Z. Ahmed et al., 2009 *Phys. Rev. Lett.* **102** 011301
[10] CDMS Collaboration, Z. Ahmed et al., 2010 *Science* **327** 1619
[11] XENON Collaboration, E. Aprile et al., 2011 *Phys. Rev. Lett.* **107** 131302
[12] M. Pyle et al., 2009 *AIP Conf. Proc.* **1185** 223
[13] CDMS Collaboration, D.S. Akerib et al., 2005 *Phys. Rev.* D **72** 052009
[14] S. Yellin, 2002 *Phys. Rev.* D **66** 032005
[15] EDELWEISS Collaboration, E. Armengaud et al., 2011 *Phys. Lett.* B **702** 329
[16] CDMS and EDELWEISS Collaborations, Z. Ahmed et al., 2011 *Phys. Rev.* D **84** 011102(R)
[17] S. Yellin, 2011 *e-print* arXiv:1105.2928 [physics.data-an]
[18] DAMA/LIBRA Collaboration, R. Bernabei et al., 2008 *Eur. Phys. J.* C **56** 333;
DAMA/LIBRA Collaboration, R. Bernabei et al., 2010 *Eur. Phys. J.* C **67** 39
[19] CoGeNT Collaboration, C.E. Aalseth et al., 2011 *Phys. Rev. Lett.* **106** 131301
[20] D. Hooper et al., 2010 *Phys. Rev.* D **82** 123509
[21] CRESST Collaboration, G. Angloher et al., *e-print* arXiv:1109.0702v1 [astro.ph.CO]
[22] F. Donato, N. Fornengo, S. Scopel, 1998 *Astropart. Phys.* **9** 247
[23] CDMS Collaboration, D.S. Akerib et al., 2010 *Phys. Rev.* D **82** 122004
[24] CDMS Collaboration, Z. Ahmed et al., 2011 *Phys. Rev. Lett.* **106** 131302
[25] XENON Collaboration, E. Aprile et al., 2011 *Phys. Rev.* D **84** 052003
[26] Juan Collar, TAUP 2011 (Munich); slides available at `http://taup2011.mpp.mpg.de/?pg=Agenda`
[27] S. Scorza, this volume
[28] S. Liu, 2011 *Queen's University Kingston*, MSc thesis

Technical aspects in dark matter investigations

R. Cerulli[1], R. Bernabei[2], P. Belli[2], F. Cappella[3], C.J. Dai[4], A. d'Angelo[3], A. Di Marco[2], H.L. He[4], A. Incicchitti[3], X.H. Ma[4], F. Montecchia[2,5], X.D. Sheng[4], R.G. Wang[4], Z.P. Ye[4,6]

[1] Laboratori Nazionali del Gran Sasso, I.N.F.N., Assergi, Italy
[2] Dip. Fisica, Univ. di Roma "Tor Vergata" and INFN, sez. Roma Tor Vergata, Rome, Italy
[3] Dip. Fisica, Univ. di Roma "La Sapienza" and INFN, sez. Roma, Rome, Italy
[4] IHEP, Chinese Academy, P.O. Box 918/3, Beijing, China
[5] Lab. Sperim. Policentrico di Ingegneria Medica, Università di Roma "Tor Vergata"
[6] University of Jing Gangshan, Jiangxi, China

E-mail: `riccardo.cerulli@lngs.infn.it`

Abstract.
Some theoretical and experimental aspects regarding the direct dark matter field are mentioned. In particular some arguments, which play a relevant role in the evaluation of model dependent interpretations of experimental results and in comparisons, are shortly addressed.

Many Dark Matter (DM) candidate particles have been proposed in theories extending Particle Physics beyond the Standard Model (SM). A large number of possibilities is available in different scenarios, and model frameworks, for particles with very different phenomenology and interaction type with ordinary matter. A wide literature is available.

From the experimental point of view, many detection processes can be considered in order to study the DM particle interaction with the target materials. One of the many cases is the elastic scattering on target nuclei where the measured quantity is the nuclear recoil energy. However many DM candidates can give rise to signals that either have totally an electromagnetic nature (see e.g. [1]) or involve electromagnetic signals associated to nuclear recoils (see e.g. [2, 3]); obviously, approaches that are based on multiple subtraction procedure of the electromagnetic component of the experimental counting rate are blind to these latter scenarios. It is worth noting that also the neutralino in supersymmetric extensions of the SM could have, in some cases, interaction producing e.m. radiation rather than nuclear recoil. Moreover, well known side processes for nuclear recoils exist (such as recoils induced by neutrons, fission fragments, end-range alphas, surface electrons, etc.).

In order to interpret the experimental result of a direct search experiment many theoretical and experimental parameters and models appear in the evaluation and many hypotheses must also be assumed. Large possibilities are open considering the lack of knowledge about the real nature of the candidate, its distribution in the Galaxy, its coupling to target materials, etc. In addition, as regards calculations, one has to consider that: i) each model requires its parameters; ii) each parameter has an allowed range of values and not just a single value; iii) uncertainties in the models and in the parameters can play a relevant role in model dependent interpretations of results and in comparisons.

In the following just few of these quantities, models and parameters – mostly for the case of DM interactions on nuclei – are reminded as examples:

Published under licence by IOP Publishing Ltd

- cross section, mass, and other quantities describing the phenomenology and the space parameter of the considered DM candidate particle;
- Spin-Independent (SI) and/or Spin-Dependent (SD) interaction: elastic scatterings with electromagnetic contribution arising from excitation of bound electrons, inelastic scatterings on target nuclei with either SI and/or SD coupling in various scenarios, interaction of light DM either on electrons or on nuclei with production of a lighter particle, preferred interaction with electrons, conversion of DM particles into e.m. radiation, etc.;
- Effective couplings to nucleon: possible isospin dependence of the couplings can be considered; effective DM particle-nucleon coupling strengths for either SI and/or SD interaction can be defined, and this is important in the comparison of results obtained with experiments using different target nuclei depending on the unpaired nucleon (compare e.g. odd spin isotopes of Xe, Te, Ge, Si, W with ^{23}Na and ^{127}I).
- Form Factor: it depends on the target nucleus; there is not a universal formulation for it, many profiles are available in literature; in these profiles some parameters – whose value is not fixed – appear. In case of SD interaction there is no decoupling between nuclear and DM particles degrees of freedom and it depends on adopted nuclear potential; the form factor profiles can differ one by another by order of magnitude; the value strongly affects the expected signal and the model dependent interpretation of the results;
- Spin Factor: in the SD interaction it is a crucial quantity. It depends on the nuclear potential; large differences in the measured counting rate can be expected when you consider different target nuclei, different modelizations and when they have different unpaired nucleon (compare e.g. odd spin isotopes of Xe, Te, Ge, Si, W with ^{23}Na and ^{127}I);
- Scaling Law: the experimental observable in direct detection when searching for scatterings of DM particles on target nuclei is the nucleus cross section of the interaction; in order to compare the results obtained by using different target nuclei a scaling law for cross section is required. For example, it has been proposed that two-nucleon currents from pion exchange in the nucleus can give different contribution for nuclei with different atomic number [4]; as a consequence the cross section for some nuclei can be enhanced with respect to others in the SI interactions even within the framework of the MSSM [4]; also this argument has a great relevance in the comparisons;
- Halo model and DM velocity distribution: the Dark Halo of the Galaxy is an open problem of the field; the expected counting rate for the DM signal depends on the local DM density and on its velocity distribution at Earths position. The experimental observations regarding the dark halo of our Galaxy do not allow us to get information on them without introducing models. The uncertainties existing in these models, and in their parameters affect the expected counting rate and they must be taken into account. The effect of these uncertainties is very important considering also that each experiment generally use different target material, energy threshold, quenching factor, form factor, etc.. and also the fact that the expectation for different DM candidate particle is more sensitive to some of these parameters rather than others. In addition, as regard the velocity distribution, the possible existence of non-thermalized component of DM particles in the halo, due e.g. to the tidal stream of satellite galaxies of Milky Way, can give relevant contribution and impact in results and comparisons.

For some of the many other complementary aspects regarding uncertainties in the experimental results and in the comparison see e.g. ref. [5].

The DM direct experiments can be classified in two classes: experiments investigating a peculiar model independent DM signature able to point out the presence of a DM contribution in their measured rate, and experiments based on the comparison of their rate or of their recoil-like candidates surviving many subtractions with an expectation calculated in one assumed scenario (which implies to adopt many assumptions and approximations).

As regard the first class of experiments, at present the only feasible model independent signature is the DM annual modulation exploited by the DAMA experiments at the Gran Sasso National Laboratory [6, 7, 8]. This experiment with the present DAMA/LIBRA and the former DAMA/NaI (exposed masses: $\simeq$ 250 kg and $\simeq$ 100 kg of highly radiopure NaI(Tl), respectively) set-ups has released so far a total exposure of 1.17 ton $\times$ yr over 13 annual cycles, obtaining a positive model independent evidence for the presence of DM particles in the galactic halo at 8.9 σ C.L.. No systematics or side reactions able to mimic the signature (that is, able to account for the measured modulation amplitude and simultaneously satisfy all the requirements of the signature) has been found or suggested [6, 9, 7, 8, 10]. In the last years CoGeNT experiment with a P-type Point Contact Germanium detector reported a preliminary positive hint [11].

As regard the second class, these experiments, in order to reduce their experimental counting rate, generally perform huge data selections and many subtraction procedures; after all, they derive a set of recoil-like candidates. It is worth noting that not only uncertainties in the many applied large selection/subtraction procedures and in the related efficiencies are present, but well known side reactions exist giving recoil-like candidates surviving all the applied subtractions.

The use of different techniques and experimental approaches, different target materials and experimental conditions must be considered in the comparisons. Some general requirements for DM experiments must be satisfied. In particular, it is mandatory for an experiment to accurately know the energy scale and energy resolution in the keV energy range of interest; this is crucial especially for dis-uniform detectors (as the liquid noble gas one; see their and other literature). In addition, it is mandatory to know the response of the detector at threshold validated by careful calibrations performed in the same condition as in the production runs. In some experiments, as e.g. DAMA/LIBRA, the energy scale and resolution are continuously measured by external/internal known sources from MeV down to the energy threshold [9] while in other experiments these quantities, in the low energy region of interest for DM investigation, are extrapolated from calibrations at much higher energy (see e.g. [12]).

Another important quantity – in the case of elastic scattering on target nuclei – in results and comparisons is the quenching factor, q_f and, in particular for double phase experiments, the L_{Eff} parameter, and their behavior at low energy. As regard L_{Eff}, tensions exist in the available measurements and also in theoretical estimation (see e.g. [13]). The sensitivity for DM particle of double phase experiments strongly depends on this quantity and a cautious approach in its extrapolation is mandatory in order to not overestimate experimental sensitivity. The quenching factor is a specific property of the employed detector(s) and not a general quantity universal for a given material. For example, in liquid noble-gas detectors, it depends, among other things, on the level of trace contaminants which can vary in time and from one liquefaction process to another, on the cryogenic microscopic conditions, etc.. In bolometers the quenching factor depends for instance on specific properties, trace contaminants, cryogenic conditions, etc. of each specific detector, while generally it is assumed exactly equal to unity. In scintillators, the quenching factor depends, for example, on the dopant concentration, on the growing method/ procedures, on residual trace contaminants, etc., and is expected to have some energy dependence. The uncertainty of q_f value is not easy to be deeply evaluated in neutron measurement and it can be large. In the last years it has been proposed a general empirical formula to derive the quenching factor values for each used detector [14]. For example it derives quenching factors values for NaI(Tl) larger at low energy than those usually adopted; in addition, the possible effect of channeling has also to be considered [3]. These two effects can play an important role in the interpretation of results and in comparisons.

As regard problems related to the application of multiple subtraction procedures of the measured counting rate, as pursued by experiments trying to identify the presence of recoil-like events, some cases will be mentioned.

In particular, for the rate claimed by XENON-100 in an inner "fiducial" volume some arguments can be pointed out: i) the energy calibration of the underground detector has been carried out in significantly higher energy region than the claimed keV one, and thus the energy threshold, the energy scale, the energy resolution and all the necessary efficiencies for the many applied subtraction procedures are unproved; ii) the small number of available photoelectrons/keV in the underground detector, the huge detector disuniformity, the far UV scintillation light emission, etc. do not support that energy threshold; iii) the stability levels of all the cuts windows applied in the many subtraction procedures and of the related efficiencies have not been suitably proved; iv) the reproducibility of identical detector performances in different Xe liquefaction processes in the relatively large underground volume (as well as along production runs) and the needed level of stability of all the detector features during each data taking period have not been suitably proved as well; etc. Some other general aspects have been addressed in [12, 5, 13].

As regards recent results by CDMS-II [15], a huge selection of the available detectors and data is performed and many subtraction procedures are applied as well, further reducing to marginal the already very limited available exposure; the stabilities of all the applied cuts windows are unproved, the quenching factor is always assumed equal to unity, all the efficiencies values and their stability are not suitably proved as well, etc.

Finally it is worth noting that the limits achieved by experiments searching for DM particle giving rise to nuclear recoils, are not robust reference points. In fact as discussed before similar results are quite uncertain not only because of possible underestimated or unknown systematics in the huge data subtractions and in some experimental aspects, but also because the results refer only to a certain (generally largely arbitrary) set of assumptions. When one consider, for example, the energy threshold dependence of the exclusion plots, up to several orders of magnitude differences can be present between claimed and realistic evaluation of the experimental sensitivity.

In conclusion, all the arguments mentioned above allow one to understand that the uncertainties existing in the models and in their parameters, the sometimes not suitably determined experimental quantities, etc., can drastically affect model dependent results and comparisons. Thus the comparison of the results achieved by different experiments must be handled with very cautious actitude without neglecting the many sources of uncertainties. This holds in particular for model dependent exclusion plot, that cannot be considered as universal limit. Model independent approaches are mandatory.

References

[1] R. Bernabei et al., Int. J. Mod. Phys. A 21 (2006) 1445; R. Bernabei et al., Phys. Rev. D 77 (2008) 023506; R. Bernabei et al., Mod. Phys. Lett. A 23 (2008) 2125.
[2] R. Bernabei et al., Int. J. Mod. Phys. A 22 (2007) 3155.
[3] R. Bernabei et al., Eur. Phys. J. C 53 (2008) 205.
[4] G. Prezeau et al., Phys. Rev. Lett. 91 (2003) 231301.
[5] R. Bernabei et. al, J. Phys.: Conf. Ser. 203 (2010) 012040 (arXiv:0912.4200);
[6] R. Bernabei el al., La Rivista del Nuovo Cimento 26 n.1, 1 (2003).
[7] R. Bernabei et al., Eur. Phys. J. C 56, 333 (2008).
[8] R. Bernabei et al., Eur. Phys. J. C 67, 39 (2010).
[9] R. Bernabei et al., Nucl. Instr. & Meth. A 592 (2008) 297.
[10] R. Bernabei et al., arXiv:0912.0660; Can. J. Phys. 89 (2011) 11; S.I.F.Atti Conf.103(2011) (arXiv:1007.0595); pre-print ROM2F/2011/12, TIPP2011 Conf., Chicago, USA (2011).
[11] C.E. Aalseth et al., Phys. Rev. Lett. 106 (2011) 131301; Phys. Rev. Lett. 107 (2011) 141301.
[12] R. Bernabei et al., ISBN 978-88-95688-12-1, pages 1-53 (2009) Exorma ed. (arXiv:0806.0011[astro-ph]).
[13] J.I. Collar and D.N. McKinsey, arXiv:1005.0838; arXiv:1005.3723; J.I. Collar, arXiv:1006.2031; arXiv:1010.5187; arXiv:1103.3481; arXiv:1106.0653; arXiv:1106.3559.
[14] V.I. Tretyak, Astrop. Phys. 33 (2010) 40
[15] CDMS-II Coll., in this Proc. and references therein.

Background effects on reconstructed WIMP couplings

Chung-Lin Shan

Institute of Physics, Academia Sinica
No. 128, Sec. 2, Academia Road, Nankang, Taipei 11529, Taiwan, R.O.C.

E-mail: clshan@phys.sinica.edu.tw

Abstract. In this talk, I presented effects of small, but non–negligible unrejected background events on the determinations of WIMP couplings/cross sections.

1. Introduction

Residue background events which pass all discrimination criteria and then mix with other real signals in data sets are one of the most important issues in all underground experiments. Hence, as a more realistic study, we reexame our model–independent data analysis methods for determining (ratios between) different couplings/cross sections of Weakly Interacting Massive Particles (WIMPs) on nucleons in direct Dark Matter detection experiments [1, 2] by taking into account small fractions of unrejected background events [3, 4].

2. Background effects on reconstructed WIMP–nucleon couplings/cross sections

In this section I present simulation results of the reconstructed (ratios between) different WIMP–nucleon couplings/cross sections with mixed data sets from WIMP–induced and background events. 2 (3) $\times$ 5,000 experiments have been simulated by the Monte Carlo method. Each experiment contains 50 total events on average. The experimental threshold energies of all experiments have been assumed to be negligible and the maximal cut–off energies are set the same as 100 keV. Note here that both signal and background events are treated as WIMP signals in the analyses.

2.1. Reconstructed $|f_{\rm p}|^2$

In Fig. 1 I show the reconstructed squared SI WIMP–nucleon couplings $|f_{\rm p}|^2$ as functions of the input WIMP mass. The SI WIMP–nucleon cross section is set as 10^{-8} pb, the commonly used value for the local WIMP density $\rho_0 = 0.3\ {\rm GeV/cm^3}$ has been used for both data generating and analyzing. A ^{76}Ge target has been used for reconstructing $|f_{\rm p}|^2$, whereas another *independent* data set of ^{76}Ge and one of ^{28}Si have been used for determining the needed WIMP mass. The background ratios shown here are no background (dashed green), 10% (long–dotted blue) and 20% (solid red).

It can be seen here that, not surprisingly, due to extra unexpected background events, the larger the background ratio in the analyzed data set, the more strongly overestimated the reconstructed SI WIMP–nucleon coupling for all input WIMP masses. Moreover, for a heavy target, e.g., ^{76}Ge, it has been found that this overestimate could be at the strongest for WIMP masses between 30 GeV and 100 GeV, once the background ratio rises to $\gtrsim$ 20% [3].

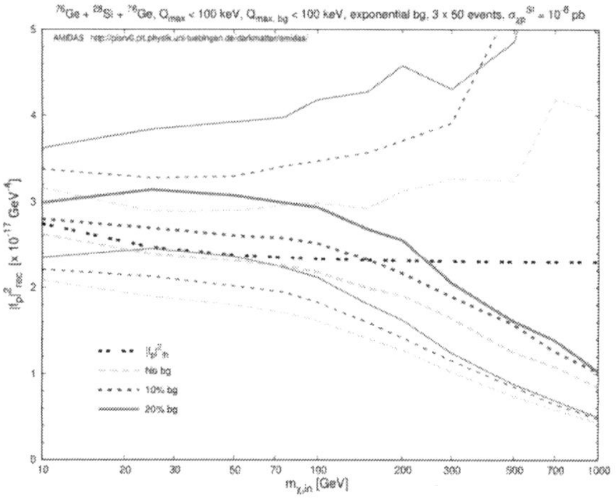

Figure 1. The reconstructed squared SI WIMP–nucleon couplings $|f_{\rm p}|^2$ and the lower and upper bounds of their 1σ statistical uncertainties as functions of the input WIMP mass reconstructed with a ^{76}Ge target. Another *independent* data set of ^{76}Ge and one of ^{28}Si have been used for determining the needed WIMP mass. The double–dotted black curve is the theoretical value of $|f_{\rm p}|^2$ corresponding to the fixed SI WIMP–nucleon cross section $\sigma_{\chi{\rm p}}^{\rm SI} = 10^{-8}$ pb. The background ratios shown here are no background (dashed green), 10% (long–dotted blue) and 20% (solid red) (plot from Ref. [3]).

Nevertheless, Fig. 1 shows that one could in principle estimate the SI WIMP–nucleon coupling with an uncertainty of a factor $\lesssim 2$ by using data sets with maximal 20% background events.

2.2. Reconstructed $a_{\rm n}/a_{\rm p}$

The left frame of Figs. 2 shows the reconstructed $a_{\rm n}/a_{\rm p}$ ratios with $n = -1$ (dashed blue), 1 (solid red), and 2 (dash–dotted cyan) as functions of the input $a_{\rm n}/a_{\rm p}$ ratio for the case of a dominant SD WIMP–nucleus interaction. A combination of ^{19}F + ^{127}I targets has been used. The background ratio in the analyzed data sets is 20%. The mass of incident WIMPs has been set as $m_\chi = 100$ GeV.

It can be seen that, due to the non–negligible background events, the reconstructed $a_{\rm n}/a_{\rm p}$ ratios (for this ^{19}F + ^{127}I target combination) become now underestimated; in fact, the larger the background ratio the larger this systematic deviation of the reconstructed $a_{\rm n}/a_{\rm p}$ ratios [4]. Meanwhile, for the same data sets, the larger the n value (or, equivalently, the larger the used moment of the WIMP velocity distribution), the smaller this systematic deviation [4]. This (in)compatibility between the reconstructed $a_{\rm n}/a_{\rm p}$ ratios with different n could thus offer us a simple check for the purity/availability of our data sets.

On the other hand, in the right frame of Figs. 2 I show the reconstructed $a_{\rm n}/a_{\rm p}$ ratios for the case of a general combination of both SI and SD WIMP interactions. Here we analyzed the same data sets with (solid red) and without (dashed blue, $n = 1$) considering the extra SI interaction term.

It can be seen that the (in)compatibility between the reconstructed $a_{\rm n}/a_{\rm p}$ ratios under different assumptions about the WIMP–nucleus interactions becomes larger. Moreover it has been found that, firstly, with an increased background ratio the systematic deviations and the statistical uncertainties of the reconstructed $a_{\rm n}/a_{\rm p}$ ratio for considering the general combination of the SI and SD WIMP interactions (solid red) grow only (very) slightly [4]. Secondly, once residue background events exist regularly between the experimental minimal and maximal cut–off energies or (even better) (mostly) in high energy ranges, one can in principle estimate the

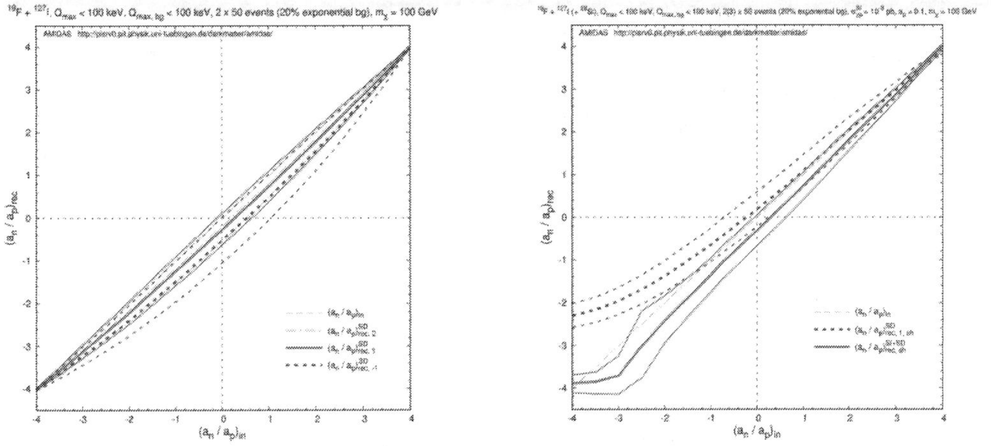

Figure 2. The reconstructed $a_{\rm n}/a_{\rm p}$ ratios and the lower and upper bounds of their 1σ statistical uncertainties as functions of the input $a_{\rm n}/a_{\rm p}$ ratio. Left: a SD–dominant WIMP–nucleus interaction with $n = -1$ (dashed blue), 1 (solid red), and 2 (dash–dotted cyan); right: adding a non–zero SI WIMP–nucleus interaction, analyzed with (solid red) and without (dashed blue, $n = 1$) taking into account this term. The mass of incident WIMPs has been set as $m_\chi = 100$ GeV. The input SI WIMP–nucleon cross section and the input SD WIMP–proton coupling have been set as $\sigma_{\chi\rm p}^{\rm SI} = 10^{-8}$ pb and $a_{\rm p} = 0.1$, respectively. The background ratio in the analyzed data sets is 20% (plots from Ref. [4]).

ratio between two SD WIMP–nucleon couplings (pretty) precisely *without* worrying about the non–negligible backgrounds [4].

Moreover, by considering different input WIMP masses, a WIMP–mass independence of the reconstructed $a_{\rm n}/a_{\rm p}$ ratios as well as of their statistical uncertainties with even non–negligible background events has been observed [4]. With data sets of $\lesssim$ 20% background ratios, the reconstructed 1σ statistical uncertainty intervals could in principle always cover the input (true) $a_{\rm n}/a_{\rm p}$ ratios pretty well for WIMP masses $\gtrsim$ 25 GeV [4].

2.3. Reconstructed $\sigma_{\chi(\rm p,n)}^{\rm SD}/\sigma_{\chi\rm p}^{\rm SI}$

Figs. 3 show the reconstructed $\sigma_{\chi\rm p}^{\rm SD}/\sigma_{\chi\rm p}^{\rm SI}$ (left) and $\sigma_{\chi\rm n}^{\rm SD}/\sigma_{\chi\rm p}^{\rm SI}$ (right) ratios as functions of the input $a_{\rm n}/a_{\rm p}$ ratio. It can be seen that, interestingly, while the $\sigma_{\chi\rm p}^{\rm SD}/\sigma_{\chi\rm p}^{\rm SI}$ ratios reconstructed with F + I + Si targets become more and more strongly underestimated with an increased background ratio, those reconstructed with Na + Ge targets become in contrast more and more strongly overestimated [4]. Meanwhile, more interestingly, while the $\sigma_{\chi\rm n}^{\rm SD}/\sigma_{\chi\rm p}^{\rm SI}$ ratios reconstructed with F + I + Si targets become more and more strongly underestimated with an increased background ratio for all input $a_{\rm n}/a_{\rm p}$ values, those reconstructed with Xe + Ge targets become more and more strongly *underestimated* for $a_{\rm n}/a_{\rm p} \gtrsim 0$ and more and more strongly *overestimated* for $a_{\rm n}/a_{\rm p} \lesssim 0$ [4]. Nevertheless, with data sets of $\lesssim$ 20% background ratios, the reconstructed 1σ statistical uncertainty intervals could in principle always cover the input (true) $\sigma_{\chi(\rm p,n)}^{\rm SD}/\sigma_{\chi\rm p}^{\rm SI}$ ratios pretty well [4].

3. Conclusions

In this article we reexamine the model–independent data analysis methods introduced in Refs. [1, 2] for determining (ratios between) different WIMP–nucleon couplings/cross sections from measured recoil energies of direct detection experiments directly by taking into account

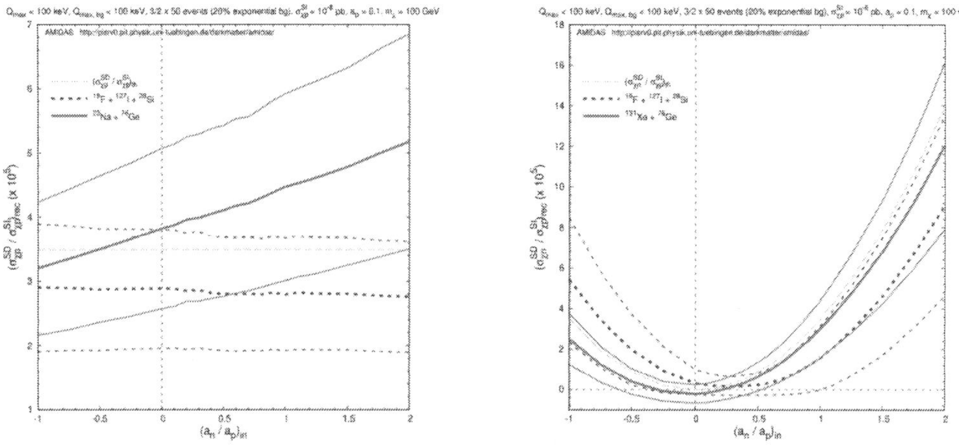

Figure 3. The reconstructed $\sigma_{\chi\mathrm{p}}^{\mathrm{SD}}/\sigma_{\chi\mathrm{p}}^{\mathrm{SI}}$ (left) and $\sigma_{\chi\mathrm{n}}^{\mathrm{SD}}/\sigma_{\chi\mathrm{p}}^{\mathrm{SI}}$ (right) ratios and the lower and upper bounds of their 1σ statistical uncertainties as functions of the input $a_{\mathrm{n}}/a_{\mathrm{p}}$ ratio. The dashed blue curves indicate the ratios reconstructed with F+I+Si, whereas the solid red curves indicate the ratios with ^{76}Ge and combined with ^{23}Na (left) and ^{131}Xe (right). Parameters are as in the right frame of Figs. 2 (plots from Ref. [4]).

small fractions of residue background events.

Our simulations show that, due to extra unexpected background events, the SI WIMP–nucleon coupling would be overestimated; nevertheless, the maximal acceptable background ratio in the analyzed data set of $\mathcal{O}(50)$ total events could be $\sim 20\%$.

On the other hand, the maximal acceptable background ratio for determining the ratio of the SD WIMP coupling on neutrons to that on protons is also $\sim 20\%$. However, the larger the relative strength between the SD WIMP–nucleus interaction to the SI one, the smaller the systematic deviations as well as the statistical uncertainties. Moreover, it has also been found that, by taking different assumptions about the relative strength between the SI and SD WIMP–nucleus interactions, and/or using different moments of the one–dimensional WIMP velocity distribution, there would be an (in)compatibility between different reconstructed $a_{\mathrm{n}}/a_{\mathrm{p}}$ ratios, which could allow us to check the purity/availability of the analyzed data sets.

Furthermore, we found also that only background events in the lowest energy ranges could affect the reconstructions (significantly); those in high energy ranges would almost not change the reconstructed ratios or only very slightly.

Acknowledgments
The author appreciate the Physikalisches Institut der Universität Tübingen for the technical support of the computational work presented in this article. This work was partially supported by the National Science Council of R.O.C. under contract no. NSC-99-2811-M-006-031.

References
[1] Shan C-L 2011 Estimating the spin–independent WIMP–nucleon coupling from direct dark matter detection data (*Preprint* 1103.0481)
[2] Shan C-L 2011 Determining ratios of WIMP–nucleon cross sections from direct dark matter detection data *J. Cosmol. Astropart. Phys.* JCAP07(2011)005 (*Preprint* 1103.0482)
[3] Shan C-L 2011 Effects of residue background events in direct dark matter detection experiments on the estimation of the spin–independent WIMP–nucleon coupling (*Preprint* 1103.4049)
[4] Shan C-L 2011 Effects of residue background events in direct dark matter detection experiments on the determinations of ratios of WIMP–nucleon cross sections (*Preprint* 1104.5305)

Directional Detection of Dark Matter with MIMAC

J. Billard, F. Mayet and D. Santos

Laboratoire de Physique Subatomique et de Cosmologie, Université Joseph Fourier Grenoble 1, CNRS/IN2P3, Institut Polytechnique de Grenoble, Grenoble, France

E-mail: billard@lpsc.in2p3.fr

Abstract. Directional detection is a promising search strategy to discover galactic Dark Matter. We present a Bayesian analysis framework dedicated to Dark Matter phenomenology using directional detection. The interest of directional detection as a powerful tool to set exclusion limits, to authentify a Dark Matter detection or to constrain the Dark Matter properties, both from particle physics and galactic halo physics, will be demonstrated. However, such results need highly accurate track reconstruction which should be reachable by the MIMAC detector using a dedicated readout combined with a likelihood analysis of recoiling nuclei.

1. Introduction

Taking advantage of the astrophysical framework, directional detection of Dark Matter is an interesting strategy to distinguish WIMP events from background ones [1]. Indeed, like most spiral galaxies, the Milky Way is supposed to be immersed in a halo of WIMPs which outweighs the luminous component by at least one order of magnitude. As the Solar System rotates around the galactic center through this Dark Matter halo, WIMP events should mainly come from the direction to which points the Sun velocity vector and which happens to be roughly in the direction of the constellation Cygnus ($\ell_\odot = 90°, b_\odot = 0°$). Hence, we argue that a clear and unambigous signature of a Dark Matter detection could be achieved by showing the correlation of the measured signal with the direction of the solar motion.

However, the simultaneous measurement of the recoil energy and the recoil direction of a nucleus target of the detector material is a major experimental challenge. In the case of a Fluorine target at 50 mbar, the mean track length of a 10 keV recoil is of the order of 300 μm and about 3 mm at 100 keV. Then, directional detection of Dark Matter requires very sensitive experiment combined with highly performant technology. In this context, the MIMAC project has been proposed [2] and the track reconstruction strategy and some expected resolutions will presented hereafter.

2. Directional Detection Phenomenology

The first step when analysing directional data should be to look for a signal pointing toward the constellation Cygnus with a sufficiently high significance [3]. If no evidence in favor of a Galactic origin of the signal is deduced from the previous analysis, then an exclusion limit should be derived [4]. On the left panel of figure 1 we present the projected discovery regions and exclusion limits for a forthcoming directional detector proposed by the MIMAC collaboration. We consider a 10 kg CF_4 detector operated during $\sim$ 3 years, allowing 3D track reconstruction, with a 10°

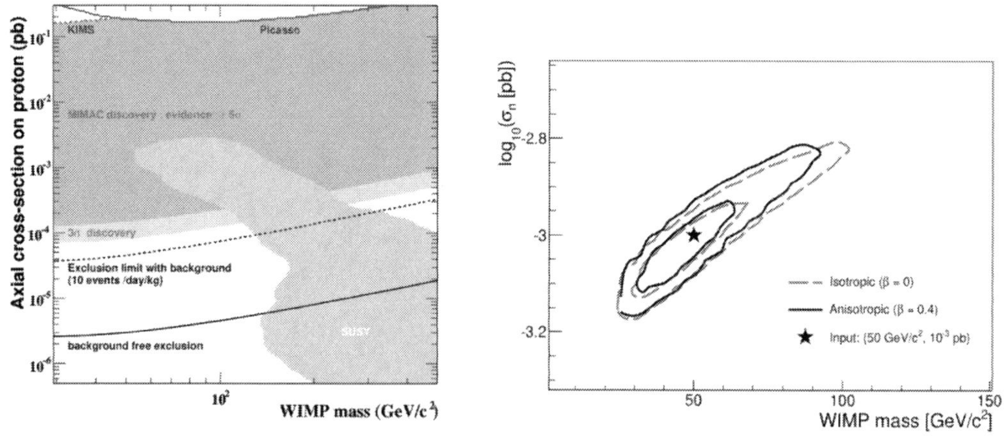

Figure 1. Left: Spin dependent cross-section on proton (pb) as a function of the WIMP mass (GeV/c^2). Exclusion limits from some direct detection experiments are presented, KIMS [5] and Picasso [6] as well as the theoretical region, obtained within the framework of the constrained minimal supersymmetric model from [7]. Contours corresponding to a significance greater than 3σ and 5σ are presented in dark and light grey. The exclusion limit corresponding to pure background data is presented as the black dashed line and the detector sensitivity as the black solid line. Right: 68% and 95% contour level in the (m_χ, σ_n) plan, for a 50 GeV/c^2 WIMP and for two input models : isotropic ($\beta = 0$) and triaxial ($\beta = 0.4$), and considering a background event rate of 10 events/kg/year.

angular resolution, a recoil energy range of 5-50 keV and with a conservative background rate of 10 evts/kg/year. From the left panel of figure 1, two different scenarii may be distinguished:

- The dark and light-grey shaded areas represent the contours where a directional detection of Dark Matter would have an expected significance greater than 3σ and 5σ. These contours are deduced from the map-based likelihood method presented in [3]. As an illustration, if the WIMP-nucleon cross-section is about 10^{-4} pb with a WIMP mass of 100 GeV.c^{-2}, the detector would have a Dark Matter detection with a significance greater than 3σ.
- If the WIMP-nucleon cross-section is lower than 10^{-5} pb, then an exclusion limit is deduced using the extended likelihood method (black dashed line) presented in [4]. As a benchmark and to illustrate the effect of background on exclusion limits, the detector sensitivity (no event) is presented (black solid line).

Left panel of figure 1 also presents exclusion limits from direct detection experiments, KIMS [5] and Picasso [6] as well as the theoretical region, obtained within the framework of the constrained minimal supersymmetric model taken from [7]. We can conclude that a directional detector like MIMAC will cover an important region of interest worth being investigated.

As shown in [8], it is possible to go further by exploiting all the information from a directional detector, *i.e.* the energy and the direction of each event. We will then consider the most optimistic scenario where the WIMP nucleon cross-section is sufficiently large ($\sigma_p = 10^{-3}$ pb) to get a high significance Dark Matter detection. In such case, we have shown the possibility to constrain the WIMP properties, both from particle physics (m_χ, σ_n) and galactic Dark Matter halo physics (velocity dispersions) [8] with a single directional experiment. To do so, we used a Markov Chain Monte Carlo algorithm dedicated to the sampling of the likelihood function characterized by 8 free parameters which are $\{m_\chi, \log_{10}(\sigma_n), l_\odot, b_\odot, \sigma_x, \sigma_y, \sigma_z, R_b\}$, where the direction $(l_\odot, b_\odot)$ refers to the main direction of the recorded events, σ_p is the WIMP-nucleon cross section, m_χ the WIMP mass, $\sigma_{x,y,z}$ the velocity dispersions of the WIMP velocity distribution and R_b is the background event rate. As a result, one can see from the right panel of figure 1 that strong and consistent constraints on the $(m_\chi, \log_{10}(\sigma_n))$ plane can be obtained within the framework of a multivariate Gaussian WIMP velocity distribution, where the dispersions, and hence the anisotropy, are set as free parameters. Indeed, even in the case of an extremely anisotropic halo model, the WIMP mass and cross section are consistent with the input values as the velocity distribution is being constrained simultaneously. This is the first step toward a Dark Matter halo model independent analysis that could be done using data from upcoming directional experiment.

3. Track reconstruction with MIMAC

3.1. MIMAC prototype

The MIMAC prototype is the elementary chamber of a future large matrix of chambers. It allows us to perform a track measurement with the required performance to achieve directional detection of Dark Matter. The primary electron-ion pairs produced by a nuclear recoil in one chamber of the matrix are detected by drifting the electrons to the grid of a bulk micromegas [9] and by producing the avalanche in a very thin gap (128 or 256μm). The electrons move towards the grid in the drift space and are projected on the anode thus allowing to get information on X and Y coordinates. To access the X and Y dimensions, a bulk micromegas with a 10 by 10 cm^2 active area segmented in pixels with a pitch of 424 μm is used as a 2D readout. In order to reconstruct the third dimension Z of the recoil, a self-triggered electronics has been developed. It allows to perform the anode sampling at a frequency of 50 MHz, leading to time slices of 20 ns width along the Z dimension. This electronic also includes a dedicated 16 channels ASIC [10] associated to a DAQ [11]. The total recoil energy is deduced from the measured ionization quenching factor (IQF) [2].

3.2. Likelihood approach for track reconstruction

A straightforward track reconstruction algorithm consists in using a three dimensional linear fit of the center of gravity of each time slice. However, we found that this method failed at recovering the correct θ angle due to the longitudinal diffusion of the electrons. In order to recover correctly the track properties, two strategies can be used:

- If one is able to get the Z position of the track in the detector, *e.g.* by using a PMT sensitive to the primary scintillation, the electron diffusion can be correctly estimated and substracted
- If no additional readout is used, the only way to recover without bias the track properties, is to fit all the parameters at the same time using a likelihood approach. This is the strategy that we have developed and which is presented hereafter.

To perform this likelihood approach dedicated to track reconstruction within the MIMAC experiment, we combined the use of three different softwares: SRIM [12] for the track simulation, MagBoltz [13] for the estimation of the electron drift properties (velocity and diffusion), and the MIMAC DAQ simulation software. This way, we are able to simulate tracks as measured by the detector. Of course, this highlights the need for the detector commissioning as any discrepancy between the simulation and the measurement will result in bias in the track reconstruction.

In order to estimate the expected performance of the detector, in terms of spatial and angular resolution, systematical studies have been done. To do so, we have simulated about one thousand tracks generated isotropically in the lower half sphere (going downward) at a fixed Z coordinate and energy. In the case of a Fluorine recoil in our gas mixture (70% CF_4 + 30%CHF_3), we found that $\sigma_{x,y}$ is about 1 mm at 20 keV and 0.4 mm at 100 keV and that it is independent from the Z coordinate. About σ_z, we found a resolution of about 1.5 mm at 40 keV and 1 mm at 100 keV, with a weak dependence with the Z coordinate. The fact that the spatial resolutions are of the order of a millimeter will allow us to perform an accurate three dimensional fiducialization of the detector volume to prevent from surface events. The expected angular resolution σ_γ depends strongly on the energy of the recoiling nucleus. Indeed, the angular resolution is about 55° at 20 keV while it is about 30° at 100 keV. However, it does not depend on the Z coordinate of the track in the 3 cm to 7 cm range.

We used a Boosted Decision Tree [14] as a sense recognition discriminant by first finding the maximum Likelihood according to the two hypothesis: Up and Down. Then, we generate a large sample of tracks (around 10,000) corresponding to the best fits of the two hypothesis. If the track is going upward or downward, the BDT value should be positive or negative respectively. As a working example, we have simulated 1000 tracks generated at 100 keV, in the direction ($\theta = 45°, \phi = 0°$), going downward and at 1.3 cm of the anode. We found that 77% of the tracks have a negative BDT value, thus corresponding to a 77% sense recognition efficiency for this given set of parameters (position, direction and energy).

4. Conclusion

In this paper, we have shown that directional detection of Dark Matter could be sensitive to very low cross-sections $\sim 10^{-5}$ pb, in the case of a 10 kg CF_4 detector running during three years, and could hence be very competitive with existing experiments. Moreover, using this unique kind of direct search of Dark Matter could lead to unambiguous arguments in favor of a positive detection of Dark Matter. Also, we have shown that a directional detection of Dark Matter will allow to constrain the Dark Matter properties both from particle physics and galactic halo physics. However, directional detection represents a major experimental challenge. In this context, the MIMAC experiment has been proposed and combined with a dedicated likelihood analysis, we have shown that MIMAC should be able to perform a highly performant directional detection of Dark Matter.

References

[1] Spergel D N, Phys. Rev. D **37** (1988) 1353.
[2] Santos D *et al.*, arXiv:1012.1166.
[3] Billard J, Mayet F, Macias-Perez J F and Santos D, Phys. Lett. B **691** (2010) 156-162.
[4] Billard J, Mayet F and Santos D, Phys. Rev. D **82** (2010) 055011.
[5] Lee H S *et al.*, Phys. Rev. Lett. **99** (2007) 091301.
[6] Archambault S *et al.*, Phys. Lett. B **682** (2009) 185.
[7] Trotta R *et al.*, JHEP **0812** (2008) 024.
[8] Billard J, Mayet F and Santos D, Phys. Rev. D **83** (2011) 075002.
[9] Giomataris I *et al.*, Nucl. Instrum. Methods A **560** (2006) 405.
[10] Richer J P *et al.*, Nucl. Instrum. Methods A **620** (2010) 470.
[11] Bourrion O *et al.*, Nucl. Instrum. and Meth. A**662**(2010) 207.
[12] Ziegler J F, Biersack J P and Littmark U, Pergamon Press New York (1985), www.srim.org.
[13] Biagi S F, Nucl. Instrum. and Meth. A**421** (1999) 234-240.
[14] Hocker, A. *et al.*, 2007, PoS **ACAT**, 040.

12th International Conference on Topics in Astroparticle and Underground Physics (TAUP 2011) IOP Publishing
Journal of Physics: Conference Series **375** (2012) 012009 doi:10.1088/1742-6596/375/1/012009

Bayes and present dark matter direct search status

Chiara Arina

Institut für Theoretische Teilchenphysik und Kosmologie, RWTH Aachen, 52056 Aachen, Germany

E-mail: chiara.arina@physik.rwth-aachen.de

Abstract. Recently there has been a huge activity in the dark matter direct detection field, with the report of an excess from CoGeNT and CRESST along with the annual modulated signal of DAMA/LIBRA and the strong exclusion bound from XENON100. We analyse these results within the framework of Bayesian inference and evidence. Indeed Bayesian methods are well suited for marginalizing over experimental systematics and background. We present the results for spin-independent interaction on nuclei with particular attention to the low dark matter mass region and the compatibility between experiments. In the same vein we also investigate the impact of astrophysical uncertainties on the WIMP preferred parameter space within the class of isotropic dark matter velocity distributions.

1. Introduction

Dark Matter (DM) direct searches experience a fervent activity with the report of an annual modulation by the CoGeNT collaboration [1] along with the established DAMA/LIBRA (DAMA hereafter) signal [2]. The exclusion bound of the XENON100 collaboration [3] (Xe100 hereafter), obtained via a profile likelihood approach [4], puts to test those excesses. The aim of this proceeding is to update the Bayesian analysis of [5] with the CoGeNT data 2011 and to investigate the mutual agreement/disagreement between those experiments for spin-independent interaction. As a novelty, we present the assessment of the need of including astrophysical uncertainties in the direct detection (DD) analysis, with the use of the Bayesian evidence.

2. Bayesian framework and experimental likelihoods

The Bayesian statistics is a well defined framework that allows the inclusion of experimental systematics and backgrounds, as well as astrophysical uncertainties common to all experiments. Indeed the posterior probability density function (pdf) is given by Bayes' theorem:

$$\mathcal{P}(\theta \mid X) = \pi(\theta)\, \frac{\mathcal{L}(X|\theta)}{\mathcal{Z}(X)}\,, \tag{1}$$

where the likelihood $\mathcal{L}$ describes how the theoretical model, with free parameters θ, describes the data X. The prior probability density π encodes the state of knowledge on θ before observing the data and is therefore independent of X. The posterior pdf is sampled via MCMC (Markov Chain Monte Carlo) techniques using the package `CosmoMC` [6, 7] and then marginalized over nuisance parameters. For DD purposes, the important quantity is $\mathcal{P}_{\rm marg}(m_{\rm DM}, \sigma_n^{\rm SI})$, where $m_{\rm DM}$ is the DM mass and $\sigma_n^{\rm SI}$ is the DM-nucleus cross-section.

We take as nuisance parameters the mean velocity of the DM at the sun position v_0, the escape velocity from the DM halo v_{esc} and the DM density at the sun position $\rho_\odot$ (altogether named astrophysical variables). We consider two models for the DM velocity distribution: SMH (Standard Model Halo) – that is a simple Maxwellian halo with astrophysical variables fixed at their mean value – and NFW (Navarro-Frenk-White) – namely we derive the velocity distribution from this DM density profile and marginalize over all possible values of v_0, v_{esc} and $\rho_\odot$ as well as profile parameters. For all technical details about the analysis, prior choices and likelihood definitions, we refer to [5]. Here we just recall that the main systematic for Xe100 is the scintillation efficiency, while for DAMA the quenching factor on Sodium q_{Na} and on Iodine q_I are marginalized over all their experimental range. Regarding the new CoGeNT data, the likelihood is given by the product of two gaussian distributions, one for the total rate and one for the modulated signal. The data binning, analysis and subtraction of cosmogenic peaks follow closely [1] and [8]. A flat and exponential background, which does not modulate in time and which is described by 3 nuisance parameters, is added on top of the DM signal. The exponential background accounts for the effect of a bad rejection of surface events near threshold.

In Eq. 1 the denominator $\mathcal{Z}(X)$, called Bayesian evidence, is defined as the average of the likelihood over the prior for a specific model $\mathcal{M}$

$$\mathcal{Z} = \int \mathcal{L}(X|\theta)\pi(\theta)d^D\theta\,, \tag{2}$$

where D is the dimensionality of the parameter space. This quantity is of crucial importance in model comparison, which is defined as the ratio of posterior probabilities. A model $\mathcal{M}_0$ can be compared with $\mathcal{M}_1$, which has extra parameters, through the Bayes factor B, see *e.g.* [9, 10]:

$$\frac{\mathcal{P}(\mathcal{M}_0|X)}{\mathcal{P}(\mathcal{M}_1|X)} = \frac{\mathcal{Z}_0\,\pi(\mathcal{M}_0)}{\mathcal{Z}_1\,\pi(\mathcal{M}_1)} = B_{01}\frac{\pi(\mathcal{M}_0)}{\pi(\mathcal{M}_1)}\,, \tag{3}$$

where $\pi(\mathcal{M}_i)$ is the prior probability of each model. The Bayes factor automatically favours simpler models unless the data justify the complexity of more complicated alternatives, because of the marginalization procedure used to calculate the evidence, Eq. 2. In the following we will compare the model SMH with NFW, the evidence for each model being computed with `MULTINEST` [11, 12].

3. Results

Figure 1 summarizes the 2D inference in the $\{m_{\rm DM}, \sigma_n^{\rm SI}\}$ plane for individual experiments (blue dashed curve for Xe100, gray dashed and solid black contours for CoGeNT and DAMA respectively) and for the combined fit (solid red lines), in the SMH and NFW model (left and right panel respectively). Table 1 summarizes the preferred values for the DM and astrophysical observables in each model (recalling that v_0, v_{esc} and $\rho_\odot$ are fixed for SMH model). Comparing the left and right panel one notes that the closed regions become wider, while the Xe100 exclusion bound moves slightly to the right, improving the agreement between various search results. This is a volume effect due to the marginalization procedure over the astrophysical variables. In both models, the inclusion of uncertainties on the scintillation energy leads to compatibility between CoGeNT, the combined fit and Xe100 exclusion bounds at 90_S% C.L.[1], while leaving a marginal compatibility with DAMA at 99% C.L. It is always possible to combine experiments and find a common region, but actually the question is whether the preferred point for $\{m_{\rm DM}, \sigma_n^{\rm SI}\}$ is a good fit for both experiments and what are the values for the nuisance parameters that are selected. Notice from table 1 that the combined fit chooses values which are in line with the

[1] This confidence level is based on the WIMP signal S and not on the volume of the posterior pdf, see [5] for an in-depth discussion.

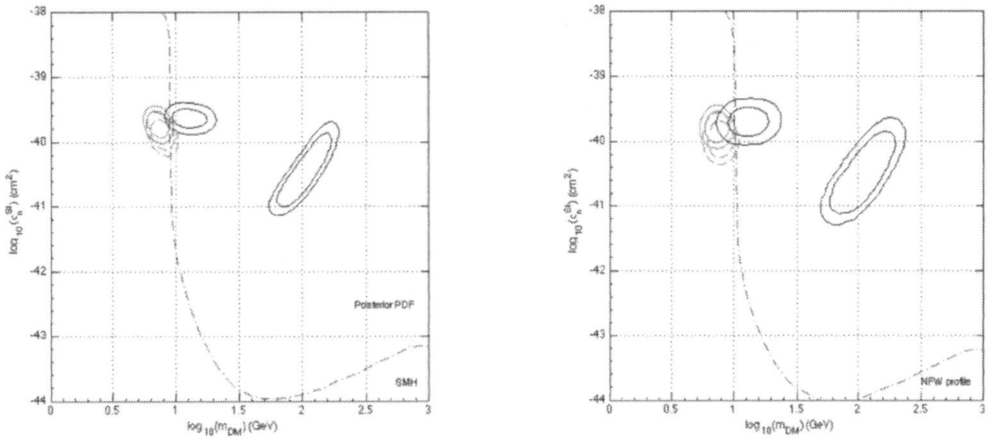

Figure 1. *Left*: 2D credible regions for the individual experimental bounds and regions assuming the SMH, combined in a single plot. For DAMA (black solid), CoGeNT (gray dashed) and the combined fit (red solid) we show the 90% and 99% contours. The blue dashed line represents the 90_S% C.L. for Xe100. *Right*: Same as left for the NFW model.

one of CoGeNT and DAMA alone, within the statistical errors. The large deviation from the DAMA alone fit concerns the quenching factor on Sodium: as it has been shown in figure 2 of [5] the 1D posterior pdf for $q_{\rm Na}$ is flat all along its prior range, while the combined fit is strongly peaked at the value $q_{\rm Na} = 0.57 \pm 0.03$. This can be understood thinking that the larger $q_{\rm Na}$ the lower the DM mass should be to account for the DM modulated signal. The selected value for the combined fit therefore pushes the DAMA region towards the direction of the CoGeNT region.

Table 2 resumes the results for model comparison in the case of individual experiments and for the DAMA+CoGeNT combined fit. A positive (negative) value for $\ln B$ represents an increase (decrease) of the support in favour of the simplest model given the observed data. The strength of evidence is given by the empirical 'Jeffreys scale' (see table 1 in [10]), where threshold values are set for inconclusive, moderate or strong evidence: *e.g.* $\ln B = 5$ states that the odds against the most complicated models are 150:1 and corresponds to a probability of 0.993. This holds in the case where the probability of each model is the same, namely $\pi(\mathcal{M}_0) = \pi(\mathcal{M}_1) = 1/2$. Looking at $\ln B$ in table 2 it is clear that there is strong or moderate evidence against the NFW model for single experiments. Therefore data at present time do not allow a determination of astrophysical observables. On the contrary there is a very strong evidence for NFW for the combined fit, namely experiments need to adjust astrophysical parameters to find a common agreement. Similar results hold for the whole class of spherically symmetric DM halos.

4. Conclusion

We have employed Bayesian inference to study the concordance between various DD experiments and model comparison to infer the necessity of a more complicated model than the SMH at present time. The outcome of the analysis may be summarized in a few main points:

- the combined fit of DAMA and CoGeNT requires parameters closely in line with those of individual fits, the only exception being a large quenching factor on Sodium;
- the CoGeNT region and the combined fit are compatible at 90_S% C.L. with the Xe100 exclusion bounds, while DAMA is only marginally allowed at 99% C.L.;

Table 1. 1D posterior pdf modes and 90% credible intervals for the circular velocity v_0, escape velocity v_{esc}, and the local DM density $\rho_\odot$ for the two DM halo models considered in this work.

Model		m_{DM} (GeV)	$\sigma_n^{SI}(cm^2)$	v_0 (km/s)	v_{esc} (km/s)	$\rho_\odot$ (GeV/cm^3)
SMH	DAMA	12	2.2×10^{-40}	230	544	0.4
	CoGeNT	7.5	1.9×10^{-40}	230	544	0.4
	Combined	6.9	1.8×10^{-40}	230	544	0.4
	Xe100	–	–	230	544	0.4
NFW	DAMA	12	1.7×10^{-40}	220^{+40}_{-20}	558^{+19}_{-16}	$0.39^{+0.15}_{-0.09}$
	CoGeNT	7.4	9.7×10^{-41}	219^{+39}_{-22}	557^{+18}_{-15}	$0.37^{+0.15}_{-0.10}$
	Combined	7.	1.5×10^{-40}	214^{+32}_{-21}	556^{+14}_{-15}	$0.35^{+0.14}_{-0.09}$
	Xe100	–	–	219^{+43}_{-20}	559 ± 18	$0.37^{+0.16}_{-0.08}$

Table 2. The odds against the NFW model for individual and combined experiments.

	DAMA	CoGeNT	Combined
$\ln B$	5.3	3.9	−32

- a combination of at least two experiments demonstrates the ability of constraining astrophysical variables, contrary to a single DD experiment.

Remarkably the excess region falls in the same ballpark of $\{m_{DM}, \sigma_n^{SI}\}$ values that could explain the excess of events reported by the CRESST experiment [13]. The CoGeNT collaboration has presented an ongoing analysis on the rejection of surface events near threshold [14], which could account for at least 50% of the counts at low energies: the total rate would therefore be accommodated by smaller DM cross-sections than those in figure 1.

Acknowledgments

It's a pleasure to thank R. Trotta for discussions on model comparison and comments on the manuscript and the CoGeNT collaboration for making data available for public analysis. This work acknowledges use of the COSMO computing resource at CP3 of Louvain University.

References

[1] Aalseth C *et al.* (CoGeNT Collaboration) 2011 *Phys. Rev. Lett.* **107** 141301 (*Preprint* 1106.0650)
[2] Bernabei R *et al.* (DAMA/LIBRA Collaboration) 2008 *Eur. Phys. J.* **C56** 333–355 (*Preprint* 0804.2741)
[3] Aprile E *et al.* (XENON100 Collaboration) 2011 *Phys. Rev. Lett.* **107** 131302 (*Preprint* 1104.2549)
[4] Aprile E *et al.* (XENON100 Collaboration) 2011 *Phys. Rev.* **D84** 052003 (*Preprint* 1103.0303)
[5] Arina C, Hamann J and Wong Y Y Y 2011 *JCAP* **1109** 022 (*Preprint* 1105.5121)
[6] Lewis A and Bridle S 2002 *Phys. Rev.* **D66** 103511 (*Preprint* astro-ph/0205436)
[7] Lewis A and Bridle S *CosmoMC Notes* Http://cosmologist.info/notes/CosmoMC.pdf
[8] Collar J *How to use time-stamped CoGeNT data.* Private communication.
[9] Kunz M, Trotta R and Parkinson D 2006 *Phys. Rev.* **D74** 023503 (*Preprint* astro-ph/0602378)
[10] Trotta R 2008 *Contemp. Phys.* **49** 71–104 (*Preprint* 0803.4089)
[11] Feroz F and Hobson M 2008 *Mon. Not. Roy. Astron. Soc.* **384** 449 (*Preprint* 0704.3704)
[12] Feroz F, Hobson M and Bridges M 2009 *Mon. Not. Roy. Astron. Soc.* **398** 1601–1614 (*Preprint* 0809.3437)
[13] Angloher G *et al.* 2011 (*Preprint* 1109.0702)
[14] Collar J *Talk at TAUP 2011 conference*

12th International Conference on Topics in Astroparticle and Underground Physics (TAUP 2011) IOP Publishing
Journal of Physics: Conference Series **375** (2012) 012010 doi:10.1088/1742-6596/375/1/012010

Direct detection of hot dark matter including light sterile neutrinos

Y.F. Li and Zhi-zhong Xing

Institute of High Energy Physics and Theoretical Physics Center for Science Facilities, Chinese Academy of Sciences, Beijing 100049, China

E-mail: xingzz@ihep.ac.cn

Abstract. Both active and sterile sub-eV neutrinos can serve for hot dark matter (DM) in the Universe. We consider the beta-decaying (e.g., ^{3}H) and EC-decaying (e.g., ^{163}Ho) nuclei as the most promising targets to capture hot DM in the laboratory. We calculate the capture rates of relic electron neutrinos and antineutrinos against the corresponding beta decay or electron capture (EC) decay backgrounds in different flavor mixing schemes. We stress that such direct measurements of hot DM might not be hopeless in the long term.

1. Introduction

Although the existence of dark matter (DM) in the Universe has been established, what it is made of remains a fundamental puzzle [1]. Within the standard model (SM) three kinds of active neutrinos and their antiparticles, whose masses lie in the sub-eV range, may constitute hot DM. Beyond the SM one or more species of sterile neutrinos and antineutrinos at a similar mass scale may also form hot DM, if they were thermalized in the early Universe as their active counterparts. Such light sterile particles are hypothetical, but their existence is more or less implied by current experimental and cosmological data [2, 3]. We are therefore open-minded to conjecture that hot DM might in general consist of both active and sterile components.

How to detect hot DM is a great challenge to the present experimental techniques. Among several possible ways [4], the most promising one is the relic neutrino capture experiment by means of radioactive beta-decaying nuclei [5]–[10]. The signal of this method is measured by the monoenergetic electron's kinetic energy for each neutrino mass eigenstate, well beyond that of the corresponding beta-decay background. A measurement of the gap between the capture and decay processes will directly probe hot DM neutrinos and determine or constrain their masses and mixing angles. However, this method does not directly apply to the capture of hot DM antineutrinos, simply because it is ν_e (instead of $\overline{\nu}_e$) that is involved in the capture reaction. A possible way out for relic antineutrino detection is to make use of some radioactive nuclei which can decay via electron capture (EC) [11]–[13].

2. Relic neutrino captures

In the presence of $3 + N_s$ species of active and sterile neutrinos, the flavor eigenstates of three active neutrinos can be written as [1]

$$|\nu_\alpha\rangle = \sum_i V^*_{\alpha i} \, |\nu_i\rangle \; , \tag{1}$$

where α runs over e, μ and τ, ν_i is a mass eigenstate of active (for $1 \leq i \leq 3$) or sterile (for $4 \leq i \leq 3 + N_s$) neutrinos, and $V_{\alpha i}$ stands for an element of the $3 \times (3 + N_s)$ lepton flavor mixing matrix V in the weak charged-current interactions.

The relic neutrino capture on radioactive beta-decaying nuclei (i.e., $\nu_e + N \to N' + e^-$) can happen for any kinetic energy of the incident neutrino, because the corresponding beta decay $N \to N' + e^- + \overline{\nu}_e$ always releases some energies ($Q_\beta = m_N - m_{N'} - m_e > 0$). So it has a unique advantage in detecting the cosmic neutrino background with both $m_i \ll Q_\beta$ and extremely low energies [5]–[10]. In the low-energy limit, the differential neutrino capture rate reads [7]

$$\frac{\mathrm{d}\lambda_\nu}{\mathrm{d}T_e} = \sum_i |V_{ei}|^2 \sigma_{\nu_i} v_{\nu_i} n_{\nu_i} \, R(T_e,\, T_e^i) \; , \tag{2}$$

where n_{ν_i} denotes the number density of relic ν_i neutrinos around the Earth. The standard Big Bang model predicts $\langle n_{\nu_i} \rangle \approx \langle n_{\overline{\nu}_i} \rangle \approx 56 \text{ cm}^{-3}$ today for each species of active neutrinos and antineutrinos [1], and this prediction is also expected to hold for each species of light sterile neutrinos and antineutrinos if they could be completely thermalized in the early Universe. The number density of hot neutrino and antineutrino DM around the Earth may be more or less enhanced by the gravitational clustering effect when m_i is larger than 0.1 eV [14]. In Eq. (2), $R(T_e,\, T_e^i)$ is a Gaussian energy resolution function defined by [9]

$$R(T_e,\, T_e^i) = \frac{1}{\sqrt{2\pi}\,\sigma} \exp\left[-\frac{(T_e - T_e^i)^2}{2\sigma^2}\right] \; , \tag{3}$$

in which T_e is the overall kinetic energy of the electrons detected in the experiment, and $T_e^i = Q_\beta + E_{\nu_i}$ is the kinetic energy of the outgoing electron for each incoming mass eigenstate ν_i. Using Δ to denote the experimental energy resolution, we have $\Delta = 2\sqrt{2\ln 2}\,\sigma \approx 2.35482\,\sigma$.

The main background of a neutrino capture process is its corresponding beta decay. The energy spectrum (i.e., $\mathrm{d}\lambda_\beta/\mathrm{d}T_e$) of a beta decay can be found in [15] and be convolved with the same energy resolution in Eq. (3). Note that the numerical results of λ_ν and λ_β can be properly normalized by using the half-life of the mother nucleus via the relation ($\lambda_\beta T_{1/2} = \ln 2$). Then the distributions of the numbers of signal and background events are expressed, respectively, as

$$\begin{aligned} \frac{\mathrm{d}N_\mathrm{S}}{\mathrm{d}T_e} &= \frac{1}{\lambda_\beta} \cdot \frac{\mathrm{d}\lambda_\nu}{\mathrm{d}T_e} \cdot \frac{\ln 2}{T_{1/2}} N_\mathrm{T}\, t \; , \\ \frac{\mathrm{d}N_\mathrm{B}}{\mathrm{d}T_e} &= \frac{1}{\lambda_\beta} \cdot \frac{\mathrm{d}\lambda_\beta}{\mathrm{d}T_e} \cdot \frac{\ln 2}{T_{1/2}} N_\mathrm{T}\, t \end{aligned} \tag{4}$$

for a given target factor N_T (i.e., the number of target atoms) and for a given exposure time t in the experiment.

3. Numerical illustration

We illustrate the relic neutrino capture signals against the beta-decay backgrounds by considering two neutrino mixing schemes: (a) the standard scheme with three sub-eV active neutrinos; (b) the (3 + 2) scheme with three sub-eV active neutrinos and two sub-eV sterile neutrinos. In our numerical calculations we typically take 100 g ^{3}H as the target.

In the standard scheme we shall use $\Delta m^2_{21} \approx 7.6 \times 10^{-5}$ eV^2 and $|\Delta m^2_{31}| \approx 2.4 \times 10^{-3}$ eV^2 together with $\theta_{12} \approx 34^\circ$ and $\theta_{13} \approx 10^\circ$ as typical inputs in our numerical estimates and only take the case of $\Delta m^2_{31} > 0$ for illustration. Fig. 1 shows the relic neutrino capture rate as a function of the kinetic energy T_e of electrons for different values of m_1. The finite energy resolution Δ

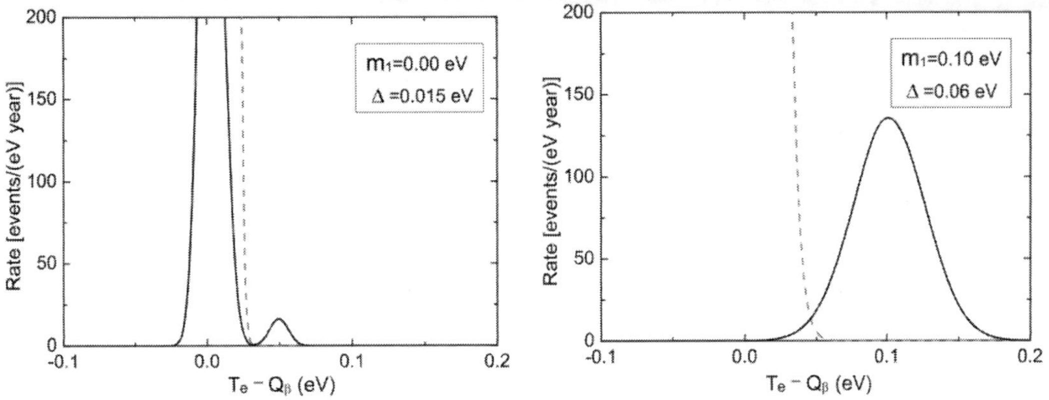

Figure 1. The relic neutrino capture rate as a function of the kinetic energy of electrons in the standard neutrino mixing scheme with $\Delta m^2_{31} > 0$.

is taken in such a way that only one single peak can be observed beyond the background. As the smallest neutrino mass m_1 increases from 0 to 0.1 eV, the capture signal moves towards the larger $T_e - Q_\beta$ region. In comparison, the shift of the corresponding background is less obvious because m_1 and Δ have the opposite effects on the location of the spectral endpoint of the beta decay. Hence the distance between the peak of the signal and the background becomes larger for a larger value of m_1, and accordingly the required energy resolution Δ becomes less stringent.

Next we consider the $(3 + 2)$ neutrino mixing scheme with two sub-eV sterile neutrinos. In view of the preliminary hints of sub-eV sterile neutrinos [2, 3], we simply assume $m_4 = 0.2$ eV and $m_5 = 0.4$ eV together with $|V_{e1}| \approx 0.792$, $|V_{e2}| \approx 0.534$, $|V_{e3}| \approx 0.168$, $|V_{e4}| \approx 0.171$ and $|V_{e5}| \approx 0.174$ in our numerical estimates. We calculate the relic neutrino capture rate as a function of the kinetic energy T_e of electrons against the corresponding background for both $\Delta m^2_{31} > 0$ and $\Delta m^2_{31} < 0$ cases in Fig 2, where $m_1 = 0$ or $m_3 = 0$ is taken for simplicity. To illustrate possible gravitational clustering effects, we assume $\zeta_1 = \zeta_2 = \zeta_3 = 1$ (without clustering effects for three active neutrinos because their maximal mass is only about 0.05 eV in the scenario under discussion) and $\zeta_5 = 2\zeta_4 = 10$ (with mild clustering effects for two sterile neutrinos because their masses are 0.2 eV and 0.4 eV, respectively). As shown in Fig. 2, the signals of two sterile neutrinos are obviously enhanced due to $\zeta_4 > 1$ and $\zeta_5 > 1$. If the gravitational clustering of non-relativistic neutrinos is very significant around the Earth, it will be very helpful for us to detect the cosmic neutrino background.

As mentioned in section 1, the beta-decaying nuclei can only be used as the targets of relic neutrino captures and one should employ the EC-decaying nuclei to detect relic antineutrinos. The rather stable isotope ^{163}Ho [11] is expected to be one of the most promising targets for relic antineutrino detection. We stress that the fine structure near the spectral endpoint of the EC decay of ^{163}Ho nuclei could be used to study the lepton flavor effects in a similar manner as that of the beta decays, and the properties of the antineutrino capture rate against its background are more or less the same as those discussed above. More discussions can be found in [12, 13].

4. Conclusion

To pin down what DM is really made of has been one of the most important and most challenging problems in particle physics and cosmology. Both the active and sterile species of sub-eV neutrinos and antineutrinos might be a part of hot DM. Here we have addressed ourselves to the direct laboratory detection of relic neutrinos and antineutrinos. The beta-decaying (e.g.,^{3}H) and EC-decaying (e.g., ^{163}Ho) nuclei have been considered as the most promising targets to

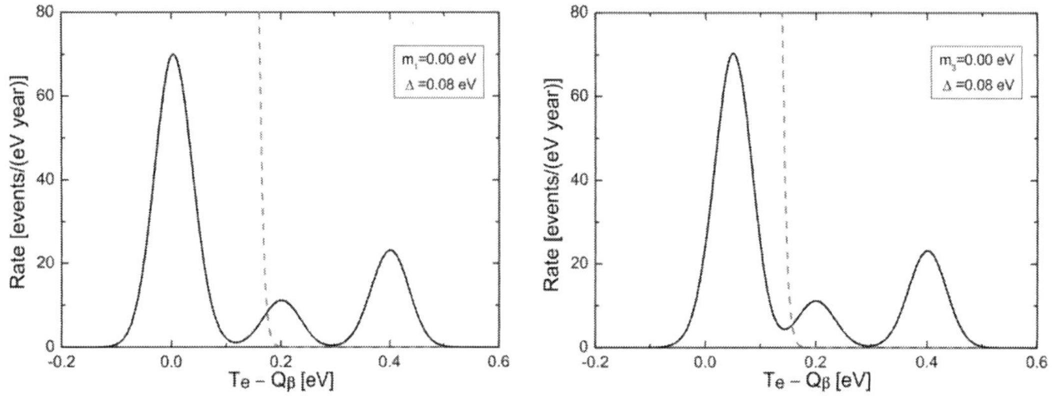

Figure 2. The relic neutrino capture rate as a function of the kinetic energy of electrons in the (3 + 2) neutrino mixing scheme with $\Delta m^2_{31} > 0$ (left panel) or $\Delta m^2_{31} < 0$ (right panel). The gravitational clustering of relic sterile neutrinos around the Earth is illustrated by taking $\zeta_1 = \zeta_2 = \zeta_3 = 1$ and $\zeta_5 = 2\zeta_4 = 10$ for example.

capture such extremely low energy objects. Although the present experimental techniques are unable to lead us to a guaranteed measurement of hot DM in the foreseeable future, we might have a chance to make a success of this great exploration in the long term.

Acknowledgements

This work was supported in part by the China Postdoctoral Science Foundation under grant No. 20100480025 (Y.F.L.) and in part by the National Natural Science Foundation of China under grant No. 11135009 (Z.Z.X.).

References

[1] See, e.g., Z.Z. Xing and S. Zhou, *Neutrinos in Particle Physics, Astronomy and Cosmology* (Zhejiang University Press and Springer-Verlag, 2011).
[2] J. Kopp, M. Maltoni, and T. Schwetz, Phys. Rev. Lett. **107**, 091801 (2011).
[3] J. Hamann *et al.*, Phys. Rev. Lett. **105**, 181301 (2010).
[4] A. Ringwald, Nucl. Phys. A **827**, 501c (2009).
[5] S. Weinberg, Phys. Rev. **128**, 1457 (1962).
[6] J.M. Irvine and R. Humphreys, J. Phys. G **9**, 847 (1983).
[7] A. Cocco, G. Mangano, and M. Messina, JCAP **0706**, 015 (2007).
[8] Y.F. Li and Z.Z. Xing, Phys. Lett. B **695**, 205 (2011).
[9] Y.F. Li, Z.Z. Xing, and S. Luo, Phys. Lett. B **692**, 261 (2010).
[10] Y.F. Li and Z.Z. Xing, arXiv: 1110.2293 [hep-ph].
[11] M. Lusignoli and M. Vignati, Phys. Lett. B **697**, 11 (2011).
[12] Y.F. Li and Z.Z. Xing, Phys. Lett. B **698**, 430 (2011).
[13] Y.F. Li and Z.Z. Xing, JCAP **1108**, 006 (2011).
[14] A. Ringwald and Y.Y.Y. Wong, JCAP **0412**, 005 (2004).
[15] E.W. Otten and C. Weinheimer, Rept. Prog. Phys. **71**, 086201 (2008).

Recent results from the SIMPLE dark matter search

M Felizardo,[1, 2,3,§] **TA Girard,**[2, 4] **T Morlat,**[2, 4] **A C Fernandes,**[3,2] **A R Ramos,**[3,2] **J G Marques**[3 ,2]

[1] Department of Physics, Universidade Nova de Lisboa, 2829-516 Monte da Caparica, Portugal

[2] Centro de Física Nuclear, Universidade de Lisboa, 1649–003 Lisbon, Portugal

[3] Instituto Tecnológico e Nuclear, Estrada Nacional 10, 2686-953 Sacavém, Portugal

[4] Department of Physics, Universidade de Lisboa, 1749-016 Lisboa, Portugal

[§] felizardo@itn.pt

Abstract. We present the results of the recent Phase II SIMPLE search effort, comprising two stages each of ~14 kgd exposure of 15 Superheated Droplet Detectors with ~0.2 kg active mass and recoil energy threshold of 8 keV. In the second stage, the neutron shielding was increased to reduce the on-detector rate by a factor 3.5. Combined with an improved nucleation efficiency and analysis of the detector pressure evolution during the measurements, the results yield improved results in the phase space of both spin -dependent and -independent WIMP interactions.

1. Introduction

The SIMPLE [1,2] measurements are based on high concentration superheated droplet detectors (SDDs), 1-2% suspensions of superheated liquid C_2ClF_5 droplets (<r>~30 μm radius) in a viscoelastic 900ml gel matrix, which undergo transitions to the gas phase upon energy deposition by incident radiation. The physics of the superheated liquid response to incident radiation is generic: nucleation of the gas phase in the liquid requires [3]: (i) that the energy deposited be greater than a thermodynamically defined minimum energy, and (ii) that this energy be deposited within a thermodynamically defined minimum distance inside the droplet. Together, energy depositions of order ~ 150 keV/μm are required for a bubble nucleation, which renders the technique effectively insensitive to the majority of traditional light particle backgrounds which complicate more conventional dark matter search detectors (including electrons, γ's and cosmic muons).

The SDDs are operated in the 1500 mwe deep, 60 m^3 GESA facility of the Laboratoire Souterrain à Bas Bruit [4] in southern France. The cavern is shielded from the rock environment by a 30-100 cm thickness of concrete, which is internally sheathed by a 1 cm thickness of iron. The SDDs were immersed to a depth of 20 cm in a temperature-controlling, 700 liter water pool within the cavern, which currently rests on a dual vibration absorber placed atop a 30 cm thick wood platform and 10 cm PE, resting on a 50 cm thick concrete floor. The pool was surrounded by layers of sound and thermal insulation. An additional 50-75 cm thick water shielding surrounded the pool and platform, with a 75 cm water thickness overhead; 50 cm of water separated the pool bottom from the detector bases. Monte-Carlo simulations of the on-detector neutron field [5], based on radio-assays of the shielding materials and which account for spontaneous fission, decay-induced (α,n) reactions and (μ,n) reactions in the rock, show negligible variations for concrete thicknesses ≥ 20 cm, and yielded an expected neutron background of 0.253 ± 0.002 (stat) ± 0.003 (syst) evt/kgd.

The ambient radon level varies seasonally between 28–1000 Bq.m^{-3} as a result of water circulation in the mountain, which is reduced to 50-80 Bq.m^{-3} during measurements by circulating the cavern air at ~ 0.2 m^{2}s^{-1}; circulation of the water-pool at 25 liter/min (equivalent to replacing the top 1 cm water layer each minute) through the temperature-controlling external cryothermostat, as well as the short radon diffusion lengths of the SDD construction materials (glass, plastic, metal), the N_2 over-pressuring, and the glycerin layer covering the gel reduces the overall α contribution to the measurement, including the radon progeny and detector contribution to an estimated 5.72 ± 0.12 (stat) ± 0.29 (syst) evt/kgd.

Each SDD was fabricated in an underground (210 mwe) clean room near GESA, following procedures previously described elsewhere [6]; the gel ingredients, all biologically-clean food products, were α-purified using actinide-specific ion-exchanging resins. The freon was single distilled; the water, double distilled. Each was capped with an electret microphone [7] and immersed in the GESA water pool; each was pressurized to 2.00±0.05 bar and maintained at 9.0±0.1°C by the water pool. The presence of U/Th contaminations in the gel, measured at ~ 0.1 ppb by low-level α and γ spectroscopy of the production gel, yielded an overall α-background level of < 0.5 evt/kg/d; a similar level is measured for the detector containment materials.

The characterization of the SDDs has been extensively studied [8-10] through fabrication variations, the use of different refrigerants, SDD lifetime, spatial localization of events, calibrations and neutron tomography.

2. Experimental detail and analysis

Phase II of SIMPLE consisted of 2 science runs, the first between October 2009 and February 2010, and the second (following a rebuilding of the neutron shielding, to include addition of the PE, paraffin and wood) later in May-July 2010.

The signals of the 15 SDDs and respective pressure readings in each measurement stage, temperature, plus the acoustic veto, were monitored continuously, as also the radon level. Data losses during the two periods, resulting from the detectors being introduced at one device per day over the three week installation periods, from weather-induced power failures, and from detector failure, yielded a total net exposure of 14.1 kgd and 10.1 kgd, respectively.

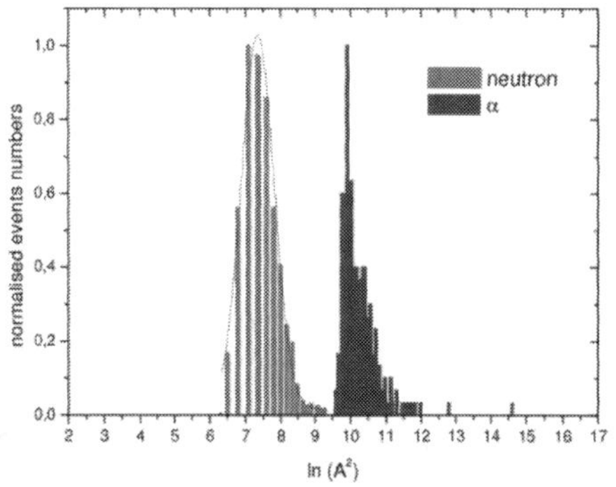

Fig. 1 neutron and α calibration results, obtained from irradiations of a single C_2ClF_5 detector.

Each detector was first inspected for its raw signal rate and pressure evolution over the measurement period. These were analyzed according to their signal characteristics for origin as described in [11], yielding 60 and 45 particle-induced events respectively. The event and pressure records were next correlated, and reduced data sets extracted for pressures below 2.20±0.05 bar corresponding to a recoil energy threshold of 8 keV, with reduced exposures of 13.5 and 6.71 kgd and 56 and 39 particle-induced events, respectively.

At 9°C, the reduced superheat of the devices is 0.3, and the probability of events from backgrounds such as electrons, γ's and MIP's negligible [1] over the exposures. The reduced signal events were compared with similarly-analyzed neutron and α calibration signals [1,11], obtained with weak sources of Am/Be and U_3O_8 respectively, and recently re-confirmed using the same detector module for both calibrations as shown in Fig. 1. These yielded 10 and 1 neutron-like events for Stage I and II, respectively, most likely of neutron origin, with an acceptance of $\geq 97\%$.

The Stage I data above 2.20 bar was used to precise the C_2ClF_5 nucleation parameter at Λ = 1.40±0.05. Monochromatic (54 and 149 keV) neutron irradiation data, taken at 1 and 2 bar as a function of temperature [12], were re-analyzed to yield a new efficiency of $\eta = 1 - \exp[-\Gamma(1 - E_{thr}/E_n)$ with Γ=4.2±0.3, independent of pressure.

3. Results and conclusions

Upper limits on the number of WIMP events was estimated for each measurement Stage by applying the Feldman-Cousins method [13], based on the observation of j neutron-like events with a background of k events one σ_{sys} below the central value of the expected neutron background. The resulting spin-dependent (SD) exclusion contours are presented in Fig. 2 (a), together with leading direct and indirect search results. Aside from providing the most restrictive limits on a possible spin-dependent WIMP-proton coupling to date, they for the first time intersect with result from the indirect search experiments.

The impact of the results in the spin-independent (SI) sector are similarly shown in Fig 2 (b), in comparison with results from the leading search efforts. The contours differ slightly from [2] as a result of an arithmetic error correction in application of Feldman-Cousins, and relaxation of the previously-reported neutron identification based on single detector double-scatterings pending further investigation. Nevertheless, the results remain in tension with a large part of the phase space recently claimed as indicative of a light mass WIMP discovery.

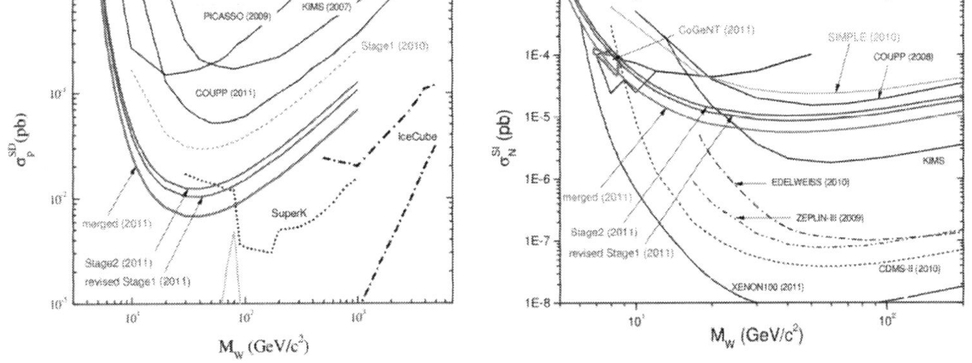

(a) (b)

Fig. 2 (a) various SD WIMP-proton SIMPLE exclusion contours, together with leading direct and indirect search results; **(b)** various SI SIMPLE contours, together with those of leading spin-independent search results; shown in both are previous and reanalyzed Stage 1 results, Stage 2, and a merging of the two.

The improved restrictions are a direct result of the more detailed signal analysis permitted by the use of a true microphone-based data acquisition, improved radio-assays of the shielding materials, and the revised nucleation efficiency in the analysis: Stage 2, with the additional benefit of its improved neutron shielding, provides an almost identical sensitivity with about half the Stage 1 exposure.

There remain some concerns regarding these results which require further investigation. First, a full characterization of the GESA measurement site, using a Bonner sphere spectrometer system of the IRSN LSBB partner is being conducted. Second, some of the materials used in the experiment are still under radio-assay, which may impact on the estimates of the neutron backgrounds as used in the contour extraction. Last, the identification of the apparent neutron events by a single detector double scattering is being fully re-examined.

While these issues are in progress, improvement on the above results nevertheless would require an increase in the experiment active mass to provide for increased exposures over shorter measurement times. A new device [14] re-prototyped in 2010, based on a 5 kg superheated C_3F_8 droplet contained within a gel-sheathed vessel (effectively a bubble chamber), is currently completing development; it will permit a factor 25 increase in the active mass with reduced space requirements. An additional all-around 60 cm of purified water shielding increases the neutron suppression by more than 10^3, giving the possibility to achieve exposures of 10^2 kgd in measurements of a few weeks duration.

Acknowledgments

This work was funded, in part, by the Nuclear Physics Center of the University of Lisbon, and by grants PDTC/FIS/83424/2006 and PTDC/FIS/115733/2009 of the Portuguese Foundation for Science and Technology (FCT). The activity of M. Felizardo was supported by grant SFRH/BD/46545/2008 of FCT.

References

[1] Felizardo M, et. al. 2010 *Phys. Rev. Lett.* **105** 211301
[2] Felizardo M, et. al. 2011 *Preprint* arxiv:1106.3014v1
[3] Seitz F 1958 *Phys Fluids 1*, **1**
[4] LSBB http://lsbb.oca.eu
[5] Fernandes A C, et.al. 2010 *Nucl. Instr. Meth.* A **623** 960
[6] Morlat T, et. al. 2008 *Astrop. Phys.* **30** 159
[7] Felizardo M, et. al. 2008 *Nucl. Instr. Meth.* A **589** 72
[8] Felizardo M, et. al. 2010 *Nucl. Instr. Meth.* A **614** 278
[9] Felizardo M, et. al. 2009 *Nucl. Instr. Meth.* A **599** 93
[10] Morlat T, et. al. 2006 *Nucl. Instr. Meth.* A **560** 339
[11] Felizardo M 2010 *Proceedings of Identification of Dark Matter PoS(IDM2010)* **006**
[12] Giuliani F, et. al. 2004 *Nucl. Instr. & Meth.* A **526** 348
[13] Feldman G J and Cousins R D 1998 *Phys. Rev.* D **57** 3873
[14] Girard TA et. al. 2011 *these Proceedings*

Recent Progress from the MiniCLEAN Dark Matter Experiment

Jocelyn Monroe

Royal Holloway University of London, Department of Physics, Egham Hill, Surrey TW20 0EX, UK

E-mail: `Jocelyn.Monroe@rhul.ac.uk`

Abstract. The MiniCLEAN dark matter direct detection experiment is a single-phase liquid argon detector, instrumented with photomultiplier tubes to observe scintillation light from a 150kg fiducial mass. This detector design strategy emphasizes scalability to target masses of order 10 tons or more. The projected light yield is 6 photo-electrons per keV, which allows pulse shape discrimination to separate the electron background from a WIMP-induced nuclear recoil signal. MiniCLEAN is also designed for a liquid neon target which, in the event of a positive signal in argon, will enable an independent verification of backgrounds and provide a unique test of the expected A^2 dependence of the WIMP interaction rate. This talk will review the experimental technique and current status of MiniCLEAN.

1. Introduction

The nature of dark matter is one of the fundamental questions about the universe today, at the forefront of experimental and theoretical physics research. From a variety of astrophysical measurements we now know that dark matter makes up approximately 23% of the energy density of the universe [1]. However, the properties of dark matter particles are unknown. A leading dark matter particle candidate is the hypothesized lightest super-symmetric particle, which is predicted to interact weakly with atomic nuclei, with a mass in the range of 10 to 10^4 GeV/c^2, and a cross section in the range of 10^{-42} to 10^{-48} cm^2 [2]. Current direct detection dark matter searches have excluded spin-independent cross sections above 4×10^{-44} cm^2 [3]. In the past few years there have also been several experimental results that are consistent with a dark matter signal, from both direct and indirect searches, although none are yet confirmed independently [4] [5] [6]. The field may be on the brink of discovery, or, finding new sources of background.

Current experimental limits indicate that the dark matter interaction cross section is at or below the scale of the solar neutrino interaction cross section. Solar neutrino experiments like SNO have 10 kilo-tonne fiducial mass, and observe roughly 1 solar neutrino interaction per tonne of detector fiducial mass [7]. Dark matter direct detection experiments will need to be at similar scales to measure appreciable numbers of events. Current experiments are at the 10-100 kg scale, and therefore demonstrating scalability of detector technology to the multi-tonne scale is of crucial importance. This is the primary goal of the MiniCLEAN experiment.

Figure 1. Left: MiniCLEAN detector assembly drawing, showing the water-filled veto tank instrumented with PMTs surrounding the outer cryostat containing the LAr detector and calibration source access port. Right: veto tanks for MiniCLEAN (right) and DEAP3600 (left) in the experiment hall at SNOLab.

2. Experiment Overview

The DEAP/CLEAN collaboration is engaged in a staged program to demonstrate the feasibility of very large single-phase LAr detectors with low energy thresholds and very low background levels. The detector design approach follows that of many successful, large neutrino experiments, emphasizing scale and simplicity.

The LAr target is chosen because of the potential for scalability to kilo-tonne masses while preserving excellent discrimination between signal and background. The dark matter detection concept is that WIMP-argon scattering produces recoiling argon nuclei, which deposit their kinetic energy in the detector. This produces single and triplet excited states of the ar detector atoms, which have 2 ns and 1.6μs lifetimes respectively [8]. The ratio of singlet to triplet excitation, which determines the number of scintillation photons produced vs. time, depends on the ionization density along the recoil track, and therefore is different for nuclear vs. electron recoils [8]. Exploiting this difference in the fraction of prompt light very effectively differentiates between these, demonstrated at the level of 3×10^{-8} at 90% confidence level [9]. A simple statistical model fit to this data projects that this rejection is sufficient to eliminate gamma and electron backgrounds for tonne-scale LAr dark matter detectors [10]. Other appealing properties of LAr include the relative ease of purification and resulting high light yield of 5-8 detected photo-electrons (p.e.) per keV of energy deposited through ionization channels [11], and relatively low cost.

The drawback of LAr as a dark matter target is the ^{39}Ar contamination, which has a beta decay with a 565 keV endpoint, at a rate of 1/s/kg [12]. A few percent of the decay electrons populate the energy window of interest for dark matter searches. This high rate imposes stringent requirements for distinguishing electron backgrounds from nuclear recoil signals, and poses a limitation to time-projection chamber readout of LAr dark matter detectors because of the ms-timescales of drifting ionization electrons leading to pile-up. However, recent progress on extracting sources of ^{39}Ar depleted-argon from underground has demonstrated reduction of the ^{39}Ar by a factor of 20, and further gains may be possible [13]. With the factor of 20 reduction, the projected maximum detector mass for dual-phase detectors is several tons. To defeat the pile-up limitation development of LAr scintillation-only dark matter detectors, several efforts using liquid targets only with no electric field ("single phase") are underway [10, 17, 14].

Since 2001, the DEAP/CLEAN collaboration has built a number of single-phase prototypes, miniDEAP (1 kg), microCLEAN (4 kg), DEAP-1 (7 kg), and the current detectors are MiniCLEAN (360 kg) (Mini Cryogenic Low Energy Astrophysics with Noble liquids) and DEAP-3600 (3600 kg). The MiniCLEAN detector is under construction now, and installation in SNOLab will begin in 2012, with the dark matter run beginning early in 2013. With an energy threshold of 50 keVr, fiducial mass of 150 kg, and <1 background event in 1 year of operation, MiniCLEAN's projected sensitivity to the WIMP-nucleon scattering cross section is 2×20^{-45} cm^2, at 90% confidence level for a 100 GeV WIMP mass [15]. DEAP-3600 construction will begin in 2012, and the dark matter run will begin in mid-2013, with dark matter sensitivity of 10^{-46} cm^2 [9]. The goal of the DEAP/CLEAN program is a 100 tonne-scale detector [17]. This scale would enable detailed study of the properties of WIMP dark matter, investigating predicted properties of neutrinos and the existing paradigm of neutrino oscillations, and testing the basic tenets of stellar evolution in our nearest star [16]. In addition, it would be sensitive to neutrinos from core-collapse supernovae and, in particular, to ν_μ and ν_τ flavors. This would be highly complementary to water detectors (such as Super-Kamiokande or future very larget detectors for proton decay and neutron oscillations) that will measure the $\overline{\nu_e}$ flux [17, 16].

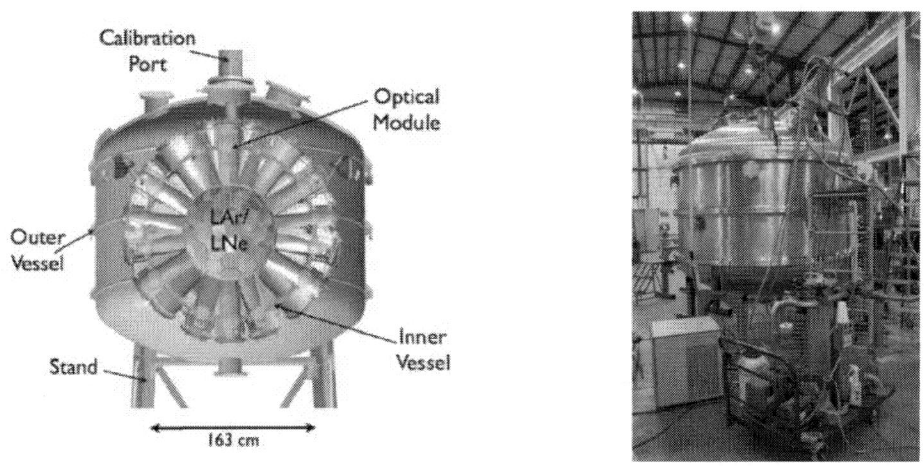

Figure 2. Left: MiniCLEAN LAr detector assembly drawing, showing the cryostat containing inner vacuum vessel, which is composed of 93 optical cassettes surrounding the LAr target volume. Right: the outer cryostat under vacuum testing at the manufacturer.

The MiniCLEAN detector design is an open target volume of 360 kg of LAr, with no electric field, viewed by a sphere of PMTs, inside a water-filled veto insttumented with PMTs. The detector assembly and current status of the experimental hall at SNOLab are shown in Figure 1. Scintillation light from dark-matter induced recoils is detected at the PMT sphere at 85 cm radius. The central target of liquid cryogen is contained in a low-background stainless steel vessel, with ports to 92 optical cassettes viewing the LAr through light guides filled with LAr and acrylic. A schematic of the LAr detector assembly and the outer cryostat are shown in Figure 2. The front faces of these light guides form a 4π wavelength-shifting sphere viewing the target volume at a radius of 45 cm. The wavelength shifting sphere is coated with TPB to shift the UV LAr scintillation light to 420 nm, where it can be detected by the array of 92 Hamamatsu R5912mod 8" PMTs. Each PMT is optically isolated from its neighbors in an optical cassette, and the fiducial volume is shielded from radioactivity from the PMT by a 10 cm acrylic plug. The optical cassette design and prototype test stand are shown in Figure 3. The PMT signals are read out at 250 MHz via CAEN V1720 waveform digitizing electronics.

The The process systems of the MiniCLEAN detector are designed for operation with a liquid neon (LNe) target as well, which gives unique sensitivity to verify the expected A^2 dependence of the WIMP-nucleon cross-section in the event that a candidate dark matter signal is found.

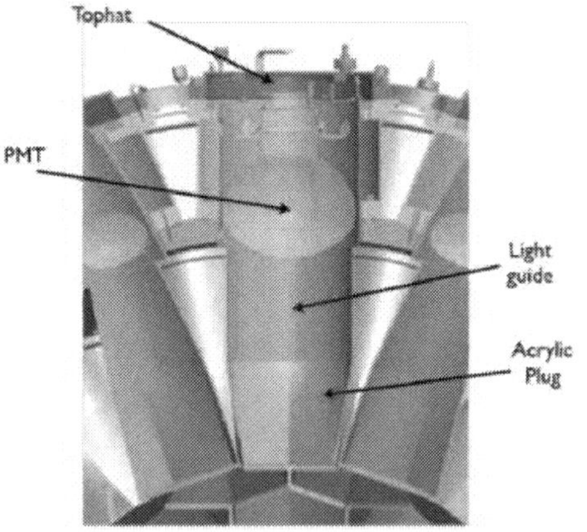

Figure 3. Left: MiniCLEAN optical casette assembly drawing, showing the R5912mod PMT, reflector, light guide, acrylic plug, and TPB-coated acrylic plate forming the wavelength-shifting sphere. Right: optical casette prototype test stand at LANL, which is used to characterize the optical casette components.

3. Conclusions and Outlook

The nature of dark matter is one of the fundamental questions about the universe today. Conclusive discovery of dark matter will require multiple technologies and targets. MiniCLEAN is unique in the global program with the ability to exchange the LAr for LNe target, in order to verify the expected A^2 dependence of a candidate dark matter signal. The ultimate goal of dark matter experiments is to measure particle properties, e.g. the mass and the interaction cross section, which requires hundreds to thousands of events. LAr has great promise as the target for future multi-tonne detectors, and the DEAP/CLEAN program will explore the scalability of the single-phase approach. The MiniCLEAN and DEAP-3600 detectors are under construction, and will begin operations at SNOLab from 2013.

References

[1] Spergel D N *et al.* *2003 Astrophys. J. Suppl.* **148** *175*
[2] *Ellis J R, Olive K A, Santoso Y, and Spanos V C 2005 Phys. Rev. D* **71** *095007*
[3] *Aprile E et al. 2011 Phys. Rev. Lett.* **107** *131302*
[4] *Bernabei R et al. 2008 Eur. Phys. J. C* **56** *333*
[5] *Alseth C E et al. 2011 Phys. Rev. Lett.* **107** *141301*
[6] *Angloher G et al. 2011 proceedings of this conference arXiv:1109.0702*
[7] *Aharmim B et al. 2007 Phys. Rev. C* **75** *045502*
[8] *Lippincott W H et al. 2008 Phys. Rev. C* **78** *035801*
[9] *Boulay M 2011 J. Phys.: Conf. Series proceedings of this conference*
[10] *Boulay M and Hime A 2006 Astropart. Phys.* **25** *179*
[11] *Lippincott W H et al. 2010 Phys. Rev. C* **81** *045803*
[12] *Benetti et al. 2007 Nucl. Instrum. Meth. A* **574** *83*
[13] *Wright A 2011 proceedings of DPF2011 arXiv:1109.2979*
[14] *Ueshima K et al. 2011 Nucl. Instrum. Meth. A* **659** *161*
[15] *Hime A 2011 proceedings of DPF2011 arXiv:1110.1005*
[16] *Horowitz C J, Coakley K J, and McKinsey D N 2003 Phys. Rev. C* **68** *023005*
[17] *McKinsey D N and Coakley K J 2005 Astropart. Phys.* **22** *355*

12th International Conference on Topics in Astroparticle and Underground Physics (TAUP 2011) IOP Publishing
Journal of Physics: Conference Series **375** (2012) 012013 doi:10.1088/1742-6596/375/1/012013

NEWAGE

Kiseki Nakamura[1], Kentaro Miuchi[1], Satoru Iwaki[1], Hidetoshi Kubo[1], Tetsuya Mizumoto[1], Hironobu Nishimura[1], Joseph D. Parker[1], Tatsuya Sawano[1], Atsushi Takada[1], Toru Tanimori[1], Hiroyuki Sekiya[2] and Atsushi Takeda[2]

[1] Kyoto University, Kitashirakawaoiwakecho Sakyoku Kyotoshi Kyoto, 606-8502, Japan
[2] Kamioka Observatory, ICRR, The University of Tokyo, 456 Higashimozumi Kamiokacho Hidashi Gifu, 506-1205, Japan

E-mail: nakamura@cr.scphys.kyoto-u.ac.jp

Abstract. The NEWAGE project (NEw generation WIMP search with an Advanced Gaseous tracking device Experiment) is a direction-sensitive dark matter search experiment, searching for WIMPs (Weakly Interacting Massive Particle) via nuclear recoil. We adopted μ-PIC (micro PIxel Chamber) to our detector. Then, we first obtained and updated the direction-sensitive exclusion limit. In this paper, we introduce NEWAGE experiment.

1. Direction sensitive dark matter search

Non-baryonic dark matter is widely believed to account for a large fraction of the mass in the universe. But the nature of this dark matter is still unknown. The Weakly Interacting Massive Particle (WIMP) is one dark matter candidate which is well motivated by cosmology and supersymmetry. Among the particles of the superysmmetric standard model, the lightest neutral supersymmetric particle, called the neutralino, is thought to be a good WIMP candidate. Direct detection of the dark matter is achieved via WIMP-nucleus elastic scattering. The expected event rate is very low because the cross section for a WIMP interacting with ordinal matter is very small. So, if we are to discriminate dark matter signals from background signals, an additional signal other than the simple energy spectrum is needed.

An annual modulation of the event rate is one of the well studied distinct signals. Dark matter is considered to be distributed throughout our milky-way galaxy, and since the earth goes around the sun, the relative velocity of the dark matter with respect to the earth would have an annual variation. In this scenario, the event rate is expected to show an annual modulation of a few %[1]. The DAMA group reported that they detected dark matter by observing such an annual modulation[2][3]. Recently, CoGeNT group also reported the observation of the annual modulation[4]. But some other groups have reported inconsistent results[5][6]. From these recent results, a new method is required to obtain more precise evidence

A direction-sensitive method is capable of providing another distinct signal[7][8]. Owing to the motion of the solar system within the galaxy, the dark matter seems to come from one direction like a wind on the earth. The direction-sensitive method detects this incoming direction by measuring the recoil nuclear tracks. The recoil angle relative to the expected wind direction is expected to have a large asymmetry between the forward scattering and back scattering. We can thus obtain a signal of a similar significance with 1000 times fewer statistics with this method compared to the conventional method.

2. NEWAGE experiment

We have been promoting the NEWAGE project (NEw generation WIMP search with an Advanced Gaseous tracking device experiment). We use a μ-TPC (micro Time-Projection-Chamber)[9] to detect three-dimensional nuclear tracks. We applied the μ-PIC[10] for the readout of the μ-TPC. Though the typical length of nuclear track scattered by dark matter is less than 1cm, an μ-PIC with a 400μm pitch can detect these short tracks with a certain angular resolution. Additionally, such detailed track information can be used to discriminate from the gamma-ray background. The length of a nuclear event is much shorter than those of background electron tracks scattered by gamma-rays of the same observed energy. Thus, we can discriminate background electron tracks from nuclear tracks by the track length with a good discrimination factor ($< 1 \times 10^{-5}$).

2.1. Detector

A schematic drawing of μ-TPC is shown in Figure 1. We have two large-size μ-TPCs named NEWAGE-0.3a (at Kamioka underground laboratory) and NEWAGE-0.3b (at Kyoto surface laboratory). We use NEWAGE-0.3a as the dark matter serarch, on the other hand, NEWAGE-0.3b as R&Ds.

The μ-PIC, the readout of the μ-TPC, is made by PCB (Printed Circuit Board) technology. PCB technology is suitable to develop a large area detector at a low cost, and thus, the μ-PIC can be a good tool for large-volume dark matter detectors. The largest available size is $30 \times 30\text{cm}^2$[11]. Figure 2 shows the structure of the μ-PIC. Gas multiplication occurs around each anode electrode. The charge multiplied at each electrode is read out through orthogonally-formed anode strips and cathode strips. The size of the μ-PIC is 30cm $\times$ 30cm, and the number of pixels is 768 $\times$ 768. We supply high voltage to the anode electrodes of the μ-PIC.

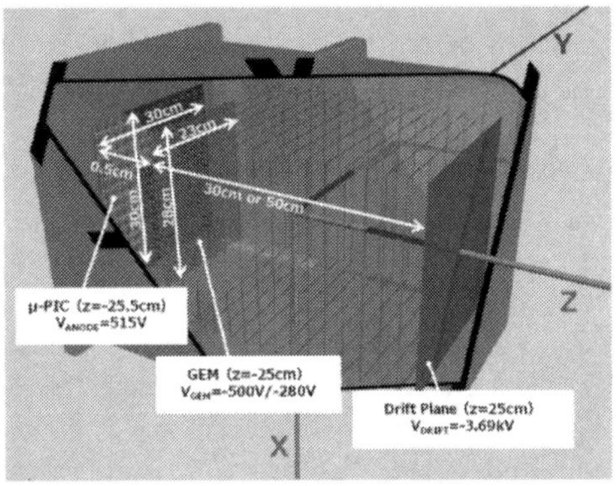

Figure 1. Schematic drawing of NEWAGE-0.3b detector. The size of the μ-PIC is 30cm$\times$30cm, the size of the GEM is 28cm$\times$23cm, and drift length is 30cm (at Kamioka underground laboratory) or 50cm (Kyoto surface laboratory). Induction gap, the length between μ-PIC and GEM, is 5mm.

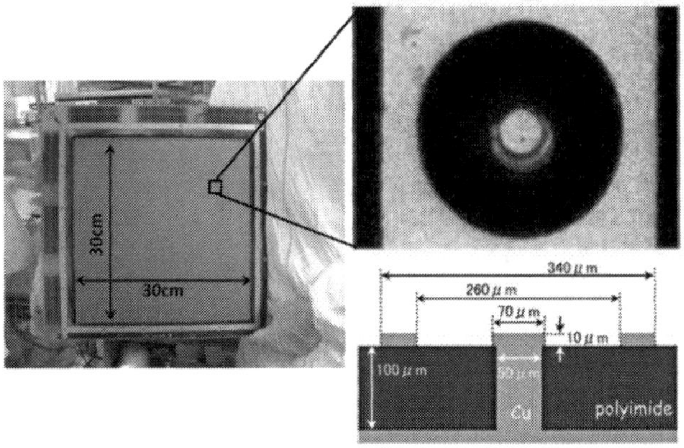

Figure 2. Left figure is the photograph of the μ-PIC (30cm$\times$30cm). Right figure is the enlarged schematic drawing of the μ-PIC. Anode electrodes with a diameter of 50μm are formed with a pitch of 400μm

Also, We put a GEM (Gas Electron Multiplier)[12], consisting of a 50μm-thick polyimide insulator sandwiched between 5μm-thick copper electrodes, 5mm above the μ-PIC to obtain sufficient gain. The size of the GEM is 23cm × 28cm, and this GEM is segmented into 8 parts to reduce the damage by discharges. We supply negative high voltages at the top and bottom of the GEM.

The drift length is 30cm (NEWAGE-0.3a) or 50cm (NEWAGE-0.3b). The detection volume is surrounded by a drift cage to form the electric field. Copper electrodes are formed on the inner surface of the drift cage by every 1cm. We supply high voltage to the drift plane. These detector components are set in a chamber made of SUS304 stainless steel. The inner surface of the chamber was electrolycally-polished to reduce out-gassing. We use CF_4 as the μ-TPC gas because fluorine has a relatively large spin-dependent cross section for the neutralino. The gas pressure is 152torr (NEWAGE-0.3a : dark matter search up to 2011) or 76torr (NEWAGE-0.3b : low pressure R&D).

A dedicated electronics system was used for the measurement. Details of the electronics are described in [13] and the references therein. Three-dimensional tracks were detected by the discriminated "hit" signals of 768 anode strips and 768 cathode strips and the anode-cathode coincidence-timing. Energy of each track was determined by the summed waveform.

2.2. Current status

We performed our first underground dark matter search in 2010 at the Kamioka underground laboratory, at a depth of 2700 m water equivalent.[13]. The exposure was 0.524 kg · days. The gas pressure was 152torr, and the energy threshold was 100keV. Figure 3 shows the direction-sensitive exclusion limit obtained by measuring the directions of nuclear tracks. We set a limit of 5600pb for 150GeV WIMPs.

Now, we aim to reach the DAMA-region[3] as the next milestone. To do this, we need to improve the sensitivity by 3 orders of magnitude (Figure 3). We plan to reach this sensitivity through the following improvements.

- Lowering the energy threshold by using a lower pressure gas. By reducing the energy threshold by one-half, the sensitivity is expected to improve by 1 order of magnitude. To check the performance of lower pressure gas, we measured the angular resolution with 76torr gas. The result is shown in Figure 4. The results with 76torr gas is shown by the red histogram, while our previous work with 152torr gas is shown by the blue histogram for comparison. By using the lower pressure gas, we were able to measure the angular

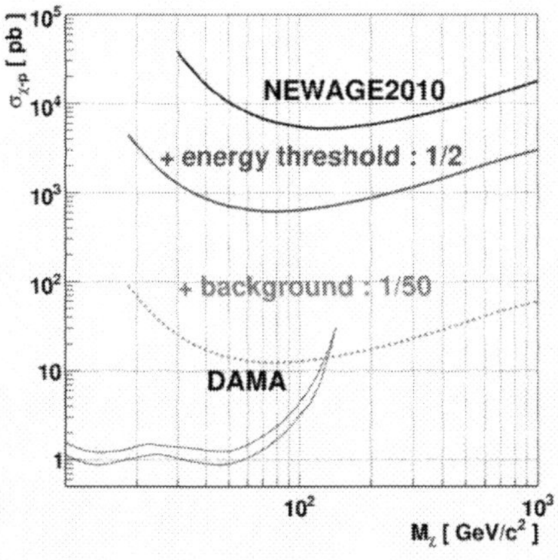

Figure 3. Exclusion limits and expected sensitivities. Horizontal axis is the WIMP mass, and vertical axis is the spin-dependent WIMP-proton cross section. Thick black line represents the limit set by NEWAGE in [13]. Thin black line represents allowed region by DAMA[3]. Blue line represents expected limit when we decrease the energy threshold by one half. Red line represents expected limit when we additionally decrease the background to 1/50.

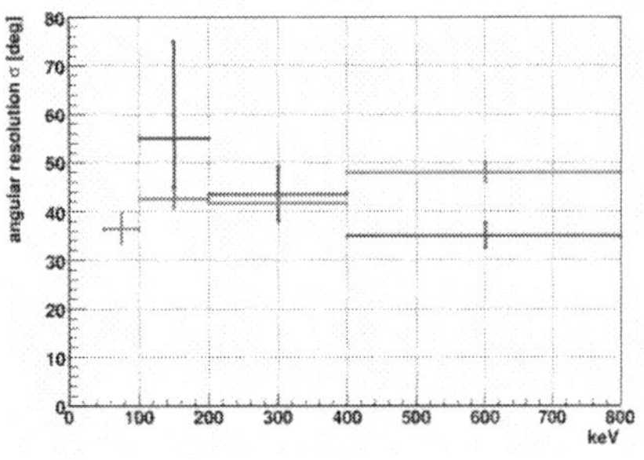

Figure 4. Energy dependence of measured angular resolutions. Red and blue histograms show the results using 76torr and 152torr gas, respectively.

resolution in the 50-100keV energy range in addition to other energy ranges. We are working on understanding the observed energy dependence.

- Lowering background by using a radon elimination system and replacing the detector components. We plan to reduce the background by at least 1 order of magnitude.
- Recognizing the *sense* of the nuclear tracks (i.e., head-tail recognition). With this recognition, the sensitivity is expected to improve by several factors.
- Taking higher statistics data by constructing a large size detector. We plan to construct a detector with 10-times larger volume.

3. Conclusions

Direction sensitive method is strong tool to reveal the dark matter problem. NEWAGE is the direction sensitive dark matter search experiment using a μ-PIC. We performed our first underground direction-sensitive dark matter search in 2010 at the Kamioka underground laboratory, and obtained 5600pb limit for 150GeV WIMPs. To reach DAMA region, we now improve the detector on lowering the energy threshold, lowering background, head-tail recognition and large size detector.

Acknowledgments

This work was partially supported by KAKENHI Grant-in-Aids for Young Scientist(A) (19684005 and 23684014); KAKENHI Grant-in-Aids for Scientific Research(B) (21340063); KAKENHI Grant-in-Aids for Challenging Exploratory Research (23654084); KAKENHI Grant-in-Aids for JSPS Research Fellow; and Global COE Program "The Next Generation of Physics, Spun from Universality and Emergence" from the Ministry of Education, Culture, Sports, Science and Technology (MEXT) of Japan.

References

[1] Lewin J D and Smith P F 1996 *Astropart. Phys.* **6** 87
[2] Bernabei R *et al.* 2003 *Riv. N. Cim.* **26 n.1** 1
[3] Savage C, Gondolo P and Freese K 2004 *Phys. Rev. D* **70** 123513
[4] Aalseth C E *et al.* 2011 *Phys. Rev. Lett.* **107** 141301
[5] Behnke E *et al.* 2011 *Phys. Rev. Lett* **106** 021303
[6] Aprile E and others (XENON Collaboration) 2010 *Phys. Rev. Lett.* **105** 131302
[7] Masek G, Buckland K and Mojaver M 1989 *Proceedings of the Workshop on Particle Astrophysics* 43
[8] Gerbier G, Rich J, Spiro M and Tao C 1990 *Nuclear Phys. B Proc. Sup.* **13** 207
[9] Miuchi K *et al.* 2007 *Nucl. Instrm. Methods Phys. Res. Sect. A* **576** 43
[10] Ochi A *et al.* 2001 *Nucl. Instrm. Methods Phys. Res. Sect. A* **471** 264
[11] Takada A *et al.* 2007 *Nucl. Instrm. Methods Phys. Res. Sect. A* **573** 195
[12] Sauli F and Sharma A 1999 *Annu. Rev. Nucl. Part. Sci* **49** 341
[13] Miuchi K *et al.* 2010 *Phys. Lett. B* **686** 11

Journal of Physics: Conference Series **375** (2012) 012014 doi:10.1088/1742-6596/375/1/012014

Dark matter sensitivities of the MAJORANA DEMONSTRATOR

G K Giovanetti[1,2], E Aguayo[10], F T Avignone III[16,8], H O Back[7,12], A S Barabash[3], J R Beene[8], M Bergevin[6], F E Bertrand[8], M Boswell[5], V Brudanin[4], M Busch[2,12], Y-D Chan[6], C D Christofferson[11], J I Collar[14], D C Combs[7,12], R J Cooper[8], J A Detwiler[6], P J Doe[19], Yu Efremenko[18], V Egorov[4], H Ejiri[9], S R Elliott[5], J Esterline[2,12], J E Fast[10], N Fields[14], P Finnerty[15,12], F M Fraenkle[15,12], V M Gehman[5], M P Green[15,12], V E Guiseppe[17], K Gusey[4], A L Hallin[13], R Hazama[9], R Henning[15,12], E W Hoppe[10], M Horton[11], S Howard[11], M A Howe[15,12], R A Johnson[19], K J Keeter[1], C Keller[17], M F Kidd[5], A Knecht[19], O Kochetov[4], S I Konovalov[3], R T Kouzes[10], B D LaFerriere[10], B H LaRoque[5], J Leon[19], L E Leviner[7,12], J C Loach[6], S MacMullin[15,12], M G Marino[19], R D Martin[6], D-M Mei[17], J H Merriman[10], M L Miller[19], L Mizouni[16,10], M Nomachi[9], J L Orrell[10], N R Overman[10], D G Phillips II[15,12], A W P Poon[6], G Perumpilly[17], G Prior[6], D C Radford[8], K Rielage[5], R G H Robertson[19], M C Ronquest[5], A G Schubert[19], T Shima[9], M Shirchenko[4], K J Snavely[15,12], D Steele[5], J Strain[15,12], K Thomas[17], V Timkin[4], W Tornow[2,12], I Vanyushin[3], R L Varner[8], K Vetter[6], K Vorren,[5,12], J F Wilkerson[15,12,8], E Yakushev[4], A R Young[7,12], C-H Yu[8], V Yumatov[3] and C Zhang[17]

[1] Department of Physics, Black Hills State University, Spearfish, SD, USA
[2] Department of Physics, Duke University, Durham, NC, USA
[3] Institute for Theoretical and Experimental Physics, Moscow, Russia
[4] Joint Institute for Nuclear Research, Dubna, Russia
[5] Los Alamos National Laboratory, Los Alamos, NM, USA
[6] Nuclear Science Division, Lawrence Berkeley National Laboratory, Berkeley, CA, USA
[7] Department of Physics, North Carolina State University, Raleigh, NC, USA
[8] Oak Ridge National Laboratory, Oak Ridge, TN, USA
[9] Research Center for Nuclear Physics and Department of Physics, Osaka University, Ibaraki, Osaka, Japan
[10] Pacific Northwest National Laboratory, Richland, WA, USA
[11] South Dakota School of Mines and Technology, Rapid City, SD, USA
[12] Triangle Universities Nuclear Laboratory, Durham, NC, USA
[13] Centre for Particle Physics, University of Alberta, Edmonton, AB, Canada
[14] Department of Physics, University of Chicago, Chicago, IL, USA
[15] Department of Physics and Astronomy, University of North Carolina, Chapel Hill, NC, USA
[16] Department of Physics and Astronomy, University of South Carolina, Columbia, SC, USA
[17] Department of Physics, University of South Dakota, Vermillion, SD, USA
[18] Department of Physics and Astronomy, University of Tennessee, Knoxville, TN, USA
[19] Center for Experimental Nuclear Physics and Astrophysics, and Department of Physics, University of Washington, Seattle, WA, USA

E-mail: gkg@unc.edu

Abstract. The MAJORANA DEMONSTRATOR is an array of natural and enriched high purity germanium detectors that will search for the neutrinoless double-beta decay of Germanium-76 and perform a search for weakly interacting massive particles with masses below 10 GeV. To reach the background rate goal in the neutrinoless double-beta decay region of interest of 4 counts/keV/t/y, the DEMONSTRATOR will utilize a number of background reduction strategies, including a time-correlated event cut for ^{68}Ge that requires a sub-keV energy threshold. This low energy threshold allows the DEMONSTRATOR to extend its physics reach to include a search for light WIMPs. We will discuss the detector systems and data analysis techniques required to achieve sub-keV thresholds as well as present the projected dark matter sensitivities of the MAJORANA DEMONSTRATOR.

1. The MAJORANA DEMONSTRATOR

The MAJORANA collaboration is currently building the MAJORANA DEMONSTRATOR, an ultra-low background array of 40 kg of high purity germanium (HPGe) detectors. Between 20 kg and 30 kg of these detectors will be constructed from material enriched to 86% in ^{76}Ge, allowing the DEMONSTRATOR to establish the feasibility of constructing a future tonne-scale neutrinoless double-beta decay germanium experiment [1, 2] while also testing the Klapdor-Kleingrothaus claim of neutrinoless double-beta decay in ^{76}Ge [3, 4]. The HPGe detectors will be mounted in holders constructed of ultra-low background electroformed copper [5] and arranged in closely packed arrays inside two vacuum cryostats. Both cryostats will be surrounded by a compact shield of electroformed copper, commercial copper, lead, plastic neutron moderator, and an active muon veto. The array will located at the 4850' level of the Sanford Underground Research Facility (SURF) in Lead, SD. The background goal for the DEMONSTRATOR is 4 background counts/tonne/year in the 4 keV region of interest (ROI) around the 2039 keV ^{76}Ge endpoint energy. This scales to a rate of 1 count/tonne/year for a DEMONSTRATOR style tonne-scale ^{76}Ge experiment, the desired background level for sensitivity to the inverted hierarchy mass spectrum region. The first underground cryostat, containing a mixed array of natural and enriched germanium, is expected to begin taking data in mid-2013.

To achieve the background goal of 4 counts/tonne/year/ROI, the DEMONSTRATOR must be able to identify background multisite events, resulting from the interaction of gamma rays within an HPGe detector, from the neutrinoless double-beta decay signal, which is a localized, effectively single-site event. Initially, segmented N-type coaxial detectors were considered for their ability to tag events occurring across multiple segments. However, these detectors are more difficult to manufacture than P-type detectors and, because each segment requires its own readout channel, require an increase in the amount of material with potential radio-impurities inside the cryostat. An alternative to segmented detectors are P-type point contact (PPC) detectors, a relatively new HPGe detector technology with a cylindrical geometry and a small, ~6 mm diameter signal contact. These detectors have an even weighting potential throughout the bulk of the crystal that rapidly increases around the point contact, resulting in characteristically different signal shapes for multi-site and single-site events. This feature allows for multi-site event discrimination comparable to or better than a segmented detector [6]. PPC detectors are simpler to fabricate than segmented detectors and require one set of readout electronics per detector, reducing the amount of material within the detector cryostat. For these reasons, the MAJORANA collaboration elected to use PPC detectors in the DEMONSTRATOR.

PPC detectors have other desirable properties that increase the background rejection capabilities and the physics reach of the DEMONSTRATOR. Because of their small point contact size, PPC detectors have a capacitance on the order of 1 pF, resulting in low intrinsic noise and the possibility of sub-keV energy thresholds [7]. This low threshold allows for the reduction

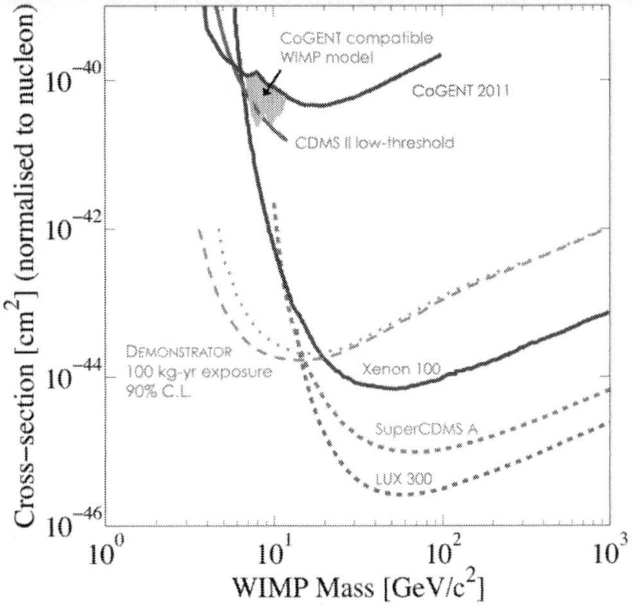

Figure 1. The projected sensitivity of the MAJORANA DEMONSTRATOR for 100 kg-yr of exposure with a 0.3 keV energy threshold (dashed blue line) and a 0.5 keV threshold (dotted blue line). A selection of current and proposed dark matter experiments are show for reference. Plot generated with dmtools.brown.edu.

of backgrounds from cosmogenically produced ^{68}Ge using a time correlated analysis cut. Long lived ^{68}Ge (Q_{ec} = 106 keV, $T^{1/2}$ = 270.8 days) decays via electron capture to ^{68}Ga (Q_{ec} = 2921.1 keV, $T^{1/2}$ = 67.7 min) , which can contribute background events to the $0\nu\beta\beta$ ROI. By tagging the 1.3 keV and 10.3 keV K and L-shell X-rays emitted during the ^{68}Ge decay and vetoing for several ^{68}Ga half-lives, this source of background can be reduced by 98%.

The low energy threshold of PPC detectors also makes them sensitive to low energy nuclear recoils from weakly interacting massive particles (WIMPs) [7]. The MAJORANA collaboration is currently investigating the requirements for maximizing the sensitivity of the DEMONSTRATOR to WIMPs. As part of this effort, we have deployed a modified, low-background broad energy germanium (BEGe) detector at the Kimballton Underground Research Facility (KURF) in Ripplemeade, VA (MALBEK) [8]. The cryostat and internal components of the MALBEK detector were constructed of radio-pure materials and assembled in a clean environment to achieve low intrinsic radioactive backgrounds. This detector is investigating point-contact, modified BEGe detector performance and serving as a test-bed for electronics, software, and data analysis techniques for the MAJORANA DEMONSTRATOR.

At its underground, shielded location the dominant backgrounds in the low-energy region of the MALBEK detector are gamma events that occur in the transition region between the bulk germanium and the lithium drifted n+ contact. Events occurring in this region are energy degraded and have significantly slower rise times than events that occur within the fully active region of the crystal. In order to detect a signal from nuclear recoils due to WIMPs, these transition region events must be removed. We are currently investigating the contribution of these slow rise time pulses to the low energy continuum, examining various pulse shape analysis techniques to cut the events, and characterizing our efficiencies in this part of the PPC spectrum.

2. Projected dark matter sensitivity

The DEMONSTRATOR will have at least 20 kg of Ge detectors constructed of material enriched in ^{76}Ge. A simulation of the low energy spectrum of these 20 kg of detectors was done to determine their sensitivity to low energy nuclear recoils from WIMPs [9]. Several assumptions were made about the contributions of various backgrounds to the low energy spectrum for this simulation.

Due to the expected high radio-purity of the DEMONSTRATOR, the major background at low energies should arise from the decay of cosmogenically produced tritium. A conservative estimate of 200 ^{3}H atoms/kg/day was used as the tritium activation rate [10], and it was assumed that detectors would spend a total of 15 days at the surface for fabrication after the initial crystal pulling. The second contribution to the background at low energy is a continuum from higher energy processes. The IGEX experiment was a low-background ^{76}Ge based experiment operated at the Canfranc Underground Laboratory. They measured a continuum of 0.1 counts/keV/kg/day in the region from 4-10 keV [11]. The goal is that the DEMONSTRATOR will be 100 times cleaner than the IGEX experiment, resulting in 0.001 counts/keV/kg/day in the region from 0-10 keV. The expected neutron background can also be estimated from IGEX, who estimated a contribution from cosmic-ray induced neutrons and neutrons from spontaneous fission events in the rock surrounding the detector as 0.01 counts/keV/kg/day [12]. Due to the increased external shielding of the DEMONSTRATOR and its deeper location, it is assumed that the contribution of neutrons will be negligible relative to the tritium beta-decay spectrum.

The projected sensitivity of the DEMONSTRATOR for a 0.3 eV and a 0.5 eV threshold after 100 kg-yr of exposure is shown in Figure 1 alongside the sensitivities of several existing and proposed dark matter experiments. The DEMONSTRATOR is particularly sensitive to masses smaller than 10 GeV and will be complementary to the next generation of dark matter experiments.

Acknowledgments

We acknowledge support from the Office of Nuclear Physics in the DOE Office of Science under grant numbers DE-AC02-05CH11231, DE-FG02-97ER41041, DE- FG02-97ER41033, DE-FG02-97ER4104, DE-FG02-97ER41042, DE-SCOO05054, DE-FG02-10ER41715, and DE-FG02-97ER41020. We acknowledge support from the Particle and Nuclear Astrophysics Program of the National Science Foundation through grant numbers PHY-0919270, PHY-1003940, 0855314, and 1003399. We gratefully ac- knowledge support from the Russian Federal Agency for Atomic Energy. We gratefully acknowledge the support of the U.S. Department of Energy through the LANL/LDRD Program. N. Fields is supported by the DOE/NNSA SSGF program. G. K. Giovanetti is supported by the Department of Energy Office of Science Graduate Fellowship Program (DOE SCGF), made possible in part by the American Recovery and Reinvestment Act of 2009, administered by ORISE-ORAU under contract no. DE-AC05-06OR23100.

References

[1] C E Aalseth et al. 2009 *AIP Conf. Proc.* **1182** 88-91
[2] V E Guiseppe 2011 *Nuclear Physics* B*(Proceedings Supplements)* **217** 44
[3] H V Klapdor-Kleingrothaus, A Deitz, I V Krivosheina and O Chkvorets 2004 *Phys. Lett.* B **586** 198
[4] H Klapdor-Kleingrothaus and I Krivosheina 2006 *Mod.Phys.Lett.* **A21** 1547-66
[5] E Hoppe et al. 2008 *J. Radioanal. Nucl. Chem* **277** 103-110
[6] R J Cooper et al. 2011 *Nucl. Instrum. Meth.* A **629** 303-310
[7] P S Barbeau, J I Collar and O Tench 2007 *JCAP* **0709** 009
[8] C E Aalseth et al. 2010 *Nucl. Instrum. Meth.* A **652** 692-695
[9] M G Marino 2010 doctoral dissertation University of Washington
[10] F T Avignone et al. 1992 *Nucl. Phys.* B **28** 280-285
[11] I G Irastorza et al. 2002 *Nucl. Phys.* B *(Proc. Suppl.)* **110** 55-57
[12] J M Carmona et al. 2004 *Astroparticle Phys.* **21** 523-533

Gamma and neutron background in the Edelweiss-II dark matter experiment

P. Loaiza for the Edelweiss Collaboration

Laboratoire Souterrain de Modane, CEA-CNRS, 1125 route de Bardonnèche, 73500 Modane, France

E-mail: ploaiza@lsm.in2p3.fr

Abstract. The Edelweiss-II collaboration has completed a search for WIMP dark matter using 400-g cryogenic Ge detectors and 384 kg·d of effective exposure. The next phase aims to probe the range of spin-independent WIMP-nucleon cross-sections between 10^{-8} and 10^{-9} pb. We present the study of gamma and neutron background coming from radioactive decays in the set-up and shield materials based on GEANT4 Monte Carlo simulations combined with measured radioactivity levels of all components. The results of these background studies have been useful to design the changes in the Edelweiss set-up to further reduce the background events in the next phase of the experiment.

1. Introduction

The reduction and discrimination of the background is one of the most important tasks in any dark matter experiment, as the signal rate expected from dark matter is extremely low. Edelweiss-II is a direct dark matter search experiment based on cryogenic Ge detectors. The combined measurement of the ionization and heat deposit in a particle interaction allows the rejection of the gamma background at the level of $(3\pm1)\times10^{-5}$ [1]. Near-surface interactions are efficiently rejected, 6×10^{-5} [3], thanks to the interleaved electrodes [2]. Using a total mass of 4 kg Edelweiss-II has recently published the final WIMP search result [1]. A cross-section of 4.4 x 10^{-8} pb is excluded at 90%CL for a WIMP mass of 85 GeV.
To reach the sensitivity goal of WIMP-nucleon cross-sections between 10^{-8} and 10^{-9} pb in the next phase of the experiment, the background reduction and discrimination is crucial. The sources of background in Edelweiss-II are neutrons, gamma-rays and surface beta contaminants. We present in this paper studies of the gamma-ray and neutron background coming from radioactive decays in the set-up and shield materials. Extensive Monte Carlo simulations have been performed and combined with radiopurity measurements of all components. These background studies are used for optimization of the next phase experiment, Edelweiss III.

2. Experimental set-up and simulations

Edelweiss-II is located in the Laboratoire Souterrain de Modane at 4800 m.w.e (4 $\mu/\mathrm{m}^2/\mathrm{day}$). The environmental gamma-ray flux dominated by natural radioactivity in the cavern walls has been reported in [4] and [5]. The neutron flux above 1 MeV is 10^{-6} $\mathrm{n/cm^2/s}$ [6].
The Edelweiss-II experiment uses Ge detectors installed in the 10 mK chamber. Each detector is supported by PTFE holders inside individual casings made of copper of type CuC2. The

detectors are arranged on plates supported by three vertical bars. The plates, the bars, the 10 mK thermal screen and the 10 mK plate (10 mK parts) are made of copper CuC1.
To simulate the response of the detectors to various type of particles, the complete set-up has been implemented in the GEANT4 package [7] as shown in Fig. 1.
Below the 10 mK plate, at 1K, a 14 cm thick lead plate shields the detectors from the gamma-rays induced by the radioactivity in the cold electronics, the dilution unit and other cryogenic parts. Five thermal screens at 1K, 4.2K, 40K, 100K and 300K, made of copper which has not been specially selected for its ultra-low radioactivity, complete the cryostat. Edelweiss is using coaxial cables from the detectors to room temperature. Passive resistances together with electrical connectors are installed at the 1K stage below the lead shielding. The warm electronics is all integrated in a single room-temperature module which is directly screwed on the cryostat. A 18 cm-thick outer layer of modern lead with a 2 cm-thick inner roman lead layer shields the cryostat against ambient gamma background. An outer 50-cm thick polyethylene shielding is used against ambient neutrons. The shields are mounted on a mild steel structure with rails. In addition a 100 m^2 plastic scintillator active muon veto [8] surrounds the polytehyelene.

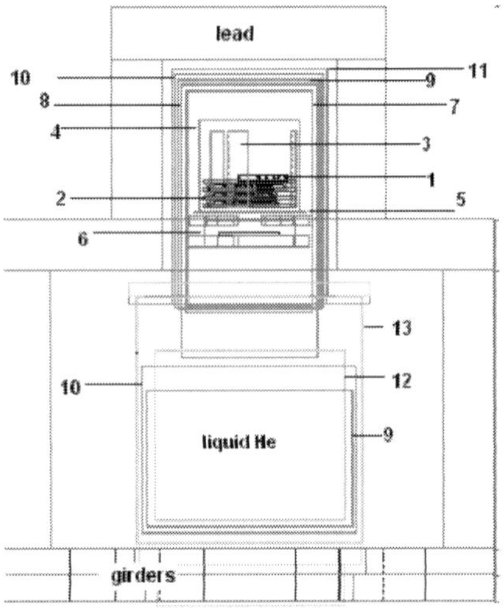

Figure 1. The GEANT4 geometry of the Edelweiss-II set-up. 1- germanium detectors with copper casings, 2- 10 mK copper plates, 3- support bars for the copper plates (10 mK) 4- 10 mK thermal screen, 5- 10 mK plate, 6- internal roman lead shielding, 7- 1K thermal screen, 8- 4.2K thermal screen, 9- 40K thermal screen, 10- 100K thermal screen, 11- 300K thermal screen, 12- stainless steel liquid He reservoir, 13- stainless steel pot. The outer polyethylene shielding and the muon veto are not shown.

All materials used in the construction have been measured to assess their radioactive contaminations. The results will be presented in [9].

3. Gamma-background

Figure 2 shows the gamma-background in the fiducial volume of the Edelweiss-II detectors compared to the GEANT4 MC simulation results. Details on the MC model will be given in [9]. The decays of ^{226}Ra, ^{228}Ra, ^{60}Co, ^{40}K, ^{54}Mn and ^{210}Pb were simulated in the most important set-up materials, including all components shown in Fig. 1. In addition, the cosmogenically induced isotopes ^{68}Ge and ^{65}Zn in germanium crystals were considered. Radiopurity measurements are used to calculate the gamma contributions of each component. As only upper limits were obtained for CuC1 copper and the copper of the screens, a χ^2 minimization with 10 free parameters was used to determine those contaminations, where the upper limits in the radioactive levels are taken as upper bounds for the fitting procedure.

The contributions to the gamma background in the low-energy region are presented in Table 1 for two fitting results, corresponding to the minimum and maximum contributions of the cryostat screens. The components which are not shown in Table 1 produce a negligeable number of gamma events. The primary source of gamma background are the U/Th descendants and ^{60}Co

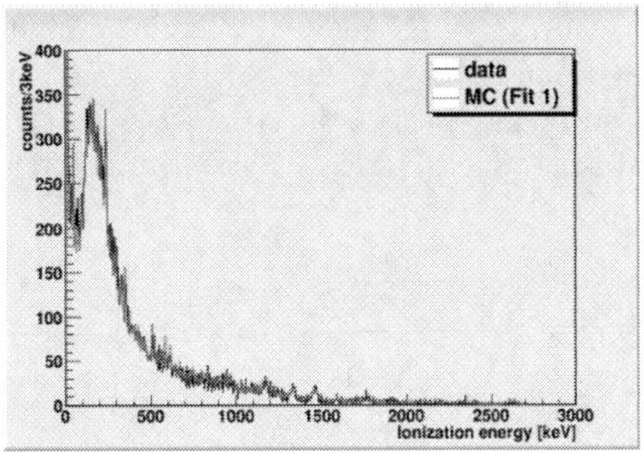

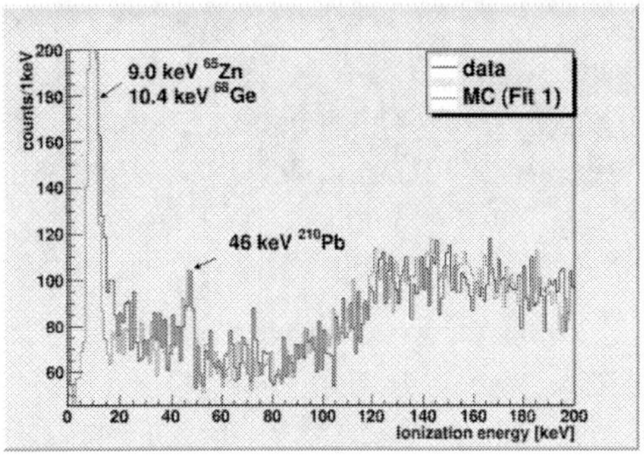

Figure 2. The background ionization energy spectrum in the fiducial volume of the Edelweiss-II detectors from measured data (black line) and Monte Carlo simulation (red line) for 185 kg d.

in the copper from the cryostat screens and 10mK parts, which represent between 39% and 52% of the total gamma-background. The second most important source (between 27% and 37%) is a pollution at 300K which must be introduced to match the data, which might be due to radioactivity in cryogenic pipes, warm electronics, radon or uncontrolled impurities on the 300K screen. The third most important gamma background source is the ^{210}Pb surface pollution on the detector casings (17%).

Table 1. Gamma-background sources in [20-200] keV. Rate in fiducial volume.

Material	Gamma event rate (evts/kg day) [20-200] keV						
	Fit 1						Fit 2
	^{226}Ra	^{228}Ra	^{60}Co	^{40}K	Others	Total (%)	Total (%)
Ge crystals	0	0	0	0	^{68}Ge: 1.6	1.6 (2)	1.6 (2)
Detector casings (CuC2 copper)	1.2	1.	1.	0	^{210}Pb: 11	14 (17)	14 (18)
10 mK parts (CuC1 copper)	0.2	1	5	0.3	^{57}Co: 0.7,^{54}Mn: 2.3	9.5 (12)	13.5 (17)
Cryostat screens (Cu)	12	15	3.	2	^{57}Co: 0.2,^{54}Mn: 0.3	32.5(40)	17 (22)
Pollution 300K	8	14	0	0	0	22 (27)	29 (37)
Pb boliden	0	2.6	0	0	0	2.6 (3)	4 (5)
Total MC	21	33.6	9	2.3	16	82	79
Total data						82	82

4. Neutron background

Details on the GEANT4 model used for the simulation of the neutron background will be given in [9]. To check the accuracy of the model, simulations of neutrons from an Am-Be source placed inside the lead shielding have been compared to the measured rate and energy spectrum of nuclear recoils. The ratio of measured-to-simulated event rates above 20 keV after all cuts, corrected for the dead time and averaged over all crystals, has been found to be 0.94±0.20, where the error is given at 68% CL. The ratio is consistent with 1 within errors, proving the validity of the geometrical model of the detector and neutron physcis in GEANT4.

To estimate the event rate due to neutrons in the data reported in [1] representing a total exposure of 384 kg d, the components detailed in the first column in Table 2 have been considered as potential sources of neutrons. The radiopurity data for the materials have been taken from radioactivity measurements. The maximum number of neutron events from the walls given in Table 2 takes into account the uncertainties in the U/Th concentration, in the neutron transport in polyethylene and in the environmental neutron flux. For the components located inside the lead shielding, the uncertainty due to neutron transport and geometry model should not exceed 20% (as follows from the agreement between simulations and data with a neutron source positioned within the shielding), and is taken into account in the values in Table 2. The neutron background is potentially dominated by neutrons from materials inside the shields, specially cables and electronics.

Table 2. Number of background events due to neutrons in Edelweiss-II in the run detailed in [1]. The column "Material" refers to the material in each source which contributes most to neutron production. The same cuts as for data have been applied to the simulated events.

Source	Material	Neutron events (384 kg d) [20-200] keV
Hall walls	Rock	<0.01
Hall walls	Concrete	<0.1
Shielding	Polyethylene	<0.05
Shielding	Lead	<0.08
Support	Stainless steel	<0.01
Support	Mild steel	<0.04
Thermal screens, 10mK plates,10mK bars (cryostat)	Copper	<0.03
1K connectors	Al, plastics	<0.4
Coaxial cables	PTFE	<0.65
Crystal holders	PTFE	<0.01
Electrodes	Aluminium	<0.01
Total		<1.4

Simply summing up the upper values from the materials considered in Table 2 yield less than 1.4 neutron events expected above 20 keV in the data in [1] where 5 nuclear recoil candidates are observed. Current investigations indicate that the number of neutron events from warm electronics is (1.0 ± 0.5) adding up to the sources detailed in Table 2.

5. Conclusion and future improvements

An extensive study of the gamma and neutron background have been performed, based on MC simulations combined with radiopurity data. The primary source of gamma background in Edelweiss-II is the copper from the cryostat screens and 10 mK parts. The neutron background is potentially dominated by neutrons produced by α-n reactions in materials inside the shields, specially cables and electronics The background studies performed in the present work have fostered progresses for the next stage of the experiment. New cryostat screens and 10mK parts will be built from ultra-pure copper and an inner polyethylene shield against neutrons from materials inside the shields will be installed. With these improvements and the coming increase by an order of magnitude of the detector mass, the goal is to soon probe the physically interesting range of spin-independent WIMP-nucleon cross-sections between 10^{-8} and 10^{-9} pb.

References

[1] Armengaud E. et al, 2011 *Physics Letters B*, Vol.**702** Issue 5, 329-335.
[2] Broniatowski A. et al 2008 *J. Low Tem. Phys.* **151** 830.
[3] Broniatowski A. et al 2009 *J. Physics Letters B* **681** 305.
[4] Ohsumi H. et al, 2002 *NIM A* **482** 832-839.
[5] Chazal V. et al, 1998 *Astrop. Phys.* **9** 163-172.
[6] Lemrani R. et al, 2006 *J. Phys. Conf. Ser.* **39** 145.
[7] Agostinelli et al, 2003 *NIM A* **506** 250-303.
[8] Eitel K. et al, these proceedings.
[9] Edelweiss Collaboration, in preparation.

12th International Conference on Topics in Astroparticle and Underground Physics (TAUP 2011) IOP Publishing
Journal of Physics: Conference Series **375** (2012) 012016 doi:10.1088/1742-6596/375/1/012016

Measurements of neutron fluxes in the LSM underground laboratory

Klaus Eitel for the EDELWEISS collaboration
Karlsruhe Institute of Technology, Institute of Nuclear Physics, 76021 Karlsruhe, Germany
E-mail: klaus.eitel@kit.edu

Abstract. A good particle candidate for Cold Dark Matter (CDM) is the supersymmetric neutralino or more generally a weakly interacting massive particle (WIMP). The expected interaction rate of WIMPs with the detector medium in direct detection experiments is below 0.01 events/kg/day. This makes a good knowledge of the background conditions highly important, especially with ever increasing sensitivity of the detectors. One major component of the background is related to energetic neutrons induced by cosmic muons. However, the flux of low energy and thermal neutrons typically from spontaneous fission and (α, n) reactions in the rock surrounding underground experiments gives also valuable information about the overall environmental activity. Detailed studies carried out in the framework of the EDELWEISS experiment are presented.

1. Introduction

The EDELWEISS experiment is a direct Dark Matter search located in the Modane underground laboratory LSM in the French Alps which provides a shielding factor of ~4800 m.w.e. The detection principle is based on measuring energy of the recoil nucleus originating from the WIMP elastic scattering. Bolometers of pure natural Ge are used both as the detectors and the target material. Latest results with a newly established readout technology have been presented recently [1, 2]. Despite the excellent suppression of β and γ background [3, 4], neutrons coming from the natural radioactivity [5] or induced by the remaining cosmic muons can still mimic the nuclear recoil signal of WIMP events. Therefore, this type of background requires special careful investigation. Its knowledge also becomes highly important in view of large 1-tonne-scale experiments of the next generation like EURECA [6].

2. Ambient neutron flux measurements

To monitor the ambient neutron flux at LSM, a setup of 4 proportional counter tubes of 3 cm diameter and 1 m length filled with ^{3}He gas [7] to measure neutrons via the capture reaction n+^{3}He$\rightarrow$T+p (Q =0.764 MeV) was installed. In an initial measuring campaign, these tubes were surrounded by a massive block of polyethylene (PE) (12×20 cm^2) to moderate energetic neutrons. With an internal background of 0.2 counts per day in the region of interest, the average count rate in 2008/2009 amounted to 266 ± 2 counts/day and in a more recent measurement (293 days in 2010/2011) to 258.1 ± 0.9 counts/day. To extract a total neutron flux from these count rates, detailed MC simulations have been performed, leading to a preliminary value of $\Phi_{\text{ambient}} = 9.6 \times 10^{-6}$ neutrons/cm^2/s. This is consistent with an earlier measurement $\Phi_{E>1\,\text{MeV}} = (1.06 \pm 0.1_{\text{stat}} \pm 0.6_{\text{syst}}) \times 10^{-6}$ neutrons/cm^2/s taking into account that, from the same simulations, only $\sim 16\,\%$ of all neutrons are expected to have energies above 1 MeV.

To be sensitive to thermal neutrons only, the PE block was removed for a second series of measurements starting in 2008, with one ^{3}He counter having a sensitivity of ≈ 250 counts/second for a thermal neutron flux of 1 neutron/cm^2/s [9]. The thermal neutron flux was then measured in 7 different locations of LSM near the concrete wall of the lab as well as near the existing setups to account for shielding effects of the large experiments EDELWEISS and NEMO. Preliminary results show a variation of the neutron flux by a factor of up to 3 between different positions [9]. Figure 1 shows the most recent long term measurements (207.4 days from July 2010 to

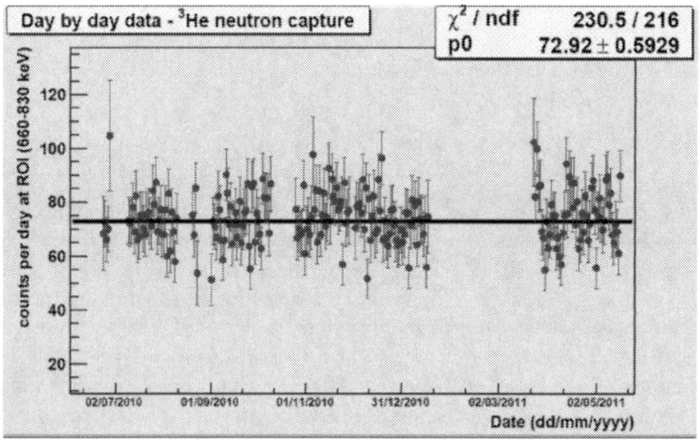

Figure 1. Count rate of thermal neutrons (events in the region of interest of energy deposits between 660 and 830 keV) as a function of time and fit parameters χ^2 and p0 for a constant rate.

May 2011) in the reference position near the installation of the EDELWEISS setup, giving no indication for an annual modulation. With an extracted count rate of 74.0 ± 0.6 counts/day, in excellent agreement with earlier measurements of 75.3 ± 0.9 this leads to a total flux of thermal neutrons of $\Phi_{\text{thermal}} = (3.57 \pm 0.05_{\text{stat}} \pm 0.27_{\text{syst}}) \times 10^{-6}$ neutrons/cm^2/s [9]. This value is higher than an earlier measurement performed in 1992 [10] obtaining a thermal neutron flux of $\Phi_{\text{thermal}} = (1.6 \pm 0.1) \times 10^{-6}$ n/cm^2/s. However, this measurement was performed before the large experimental installations in the LSM. We re-measured the thermal neutron flux in the location of the 1992 setup to $\Phi_{\text{thermal}} = (2.0 \pm 0.2_{\text{stat}} \pm 0.15_{\text{syst}}) \times 10^{-6}$ n/cm^2/s, in reasonable agreement taking into account the different environment.

3. Measurement of muon-induced neutrons in LSM

Cosmic muons with a flux of $\approx$4-5 μ/m^2/d at LSM and consecutive μ-induced neutrons constitute an important background component for direct DM search. While such μ-induced neutrons have been identified and rejected during the EDELWEISS DM search [1], the rate of these events (due to the total mass of a few kg for the Ge detectors) remains rather small. To study this background source in more detail, a special neutron detector has been installed. Details of the setup, its integration into the existing EDELWEISS readout scheme and calibration of the system are described in detail in [11]. Figure 2 shows a schematic view of the system with its central detector, a 1 m^3 Gd-loaded liquid scintillator to moderate and capture neutrons induced by muons. The neutron capture process on Gd results in several gammas with 8 MeV sum energy. The scintillator volume is viewed from each of the two module ends by 8

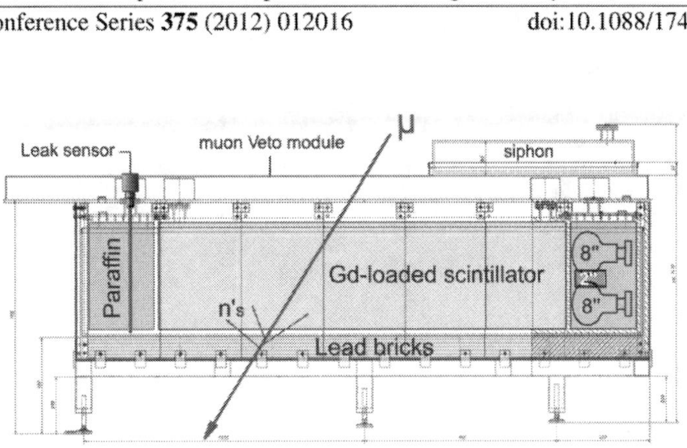

Figure 2. General scheme of the neutron counter (side view): 1 m^3 of Gd-loaded liquid scintillator viewed by 16 PMTs of 8-inch and 6 PMTs of 2-inch diameter. A layer of lead bricks below the liquid scintillator volume acts as an effective source of secondary neutrons from muon interactions. Directly on top of the liquid scintillator volume is a muon-veto module consisting of 5 cm thick plastic scintillator.

photomultiplier tubes (PMT) of 8-inch diameter. These PMTs are optimized to register the light produced after the neutron capture. In addition, the system is equipped with 6 smaller PMTs (2-inch diameter) to register muons crossing the neutron detector. The scintillator and PMTs are placed in one plexiglass container divided into three parts: the central one for the scintillator itself and two side ones filled with paraffin in which the PMTs are immersed. In order to enhance the neutron production, a 10 cm thick layer of lead bricks is put underneath the detector. On top of the counter, a plastic scintillator module of 3.15m length (same type as the muon veto of EDELWEISS) is installed. The complete system is positioned directly to the western wall of the EDELWEISS muon veto.

Figure 3(left) shows the rate of muon-sequences per day as a function of the number of secondary events. Data are based on 608 live days from April 2009 to August 2011, where 3507 muon candidates were detected out of which 123 had at least one secondary within 60 μs after passing through the scintillating liquid. Also shown is the expectation from MC simulations scaled to the rate of muons with one secondary hit. The time difference between prompt

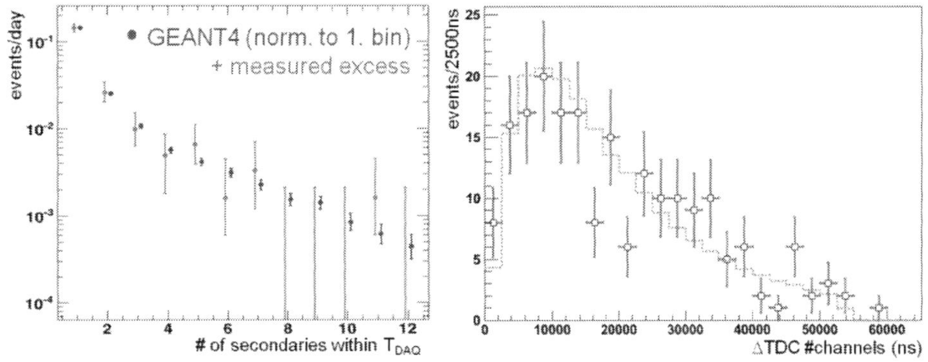

Figure 3. Left: Rate of muon-induced cascades as function of the number of secondaries after a muon candidate in comparison with an expectation from simulation (scaled); Right: time difference between the primary muon and secondary hits. The dotted histogram denotes expected time differences for neutron captures from calibration measurements with an AmBe source.

muon and secondary as depicted in Figure 3(right) follows closely the expectation of neutron captures as deduced in routine calibrations with an AmBe source where the prompt gamma with $E_\gamma = 4.4\,\mathrm{MeV}$ replaces the initial muon. Figure 4 shows the energy spectra of the prompt

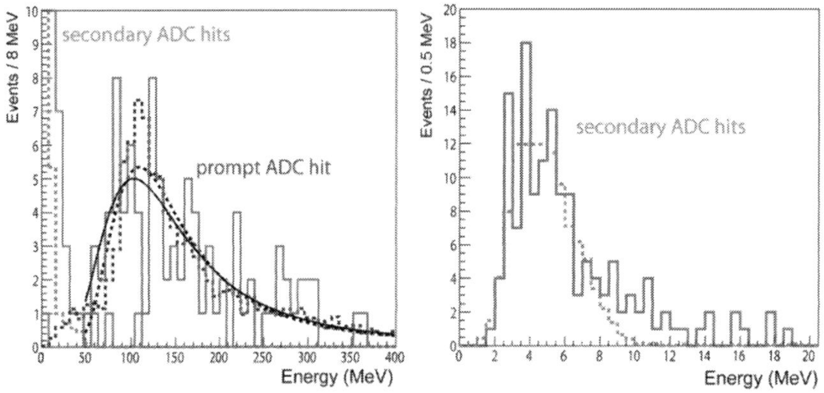

Figure 4. Energy of the muon-induced sequences: left: both prompt (blue) and delayed events (red); right: expanded spectrum of delayed events. Also shown as dashed histograms are results from MC simulations and Landau fits to the prompt energy spectra.

muon candidates following a Landau distribution (left panel; blue) and the delayed signature (left panel in red, expanded in the right panel) as expected from a neutron capture on Gd. Extensive MC simulations including a fully 3D description of the geometry of the setup as well as a muon generator for the LSM underground lab accompany the presented measurements to understand the neutron yield from energetic muons. As a preliminary result of the comparison of the measured rate with the GEANT4 packet version 9.2 result, GEANT seems to underestimate the neutron production by a factor of up to 2 while reproducing the multiplicity of secondary neutrons as well as the topology of cascade events (Fig. 3). Further investigations are ongoing in terms of simulations as well as continuing data taking.

4. Acknowledgements

The studies of the muon-induced neutron background were in part funded by the German Research Foundation (DFG) through the Transregional Collaborative Research Center SFB-TR27. Simulations of muon-induced signatures are funded in part by the German ministry of science and education (BMBF) within the *Verbundforschung Astroteilchenphysik* grant 05A11VK2. The dedicated neutron measurements (ambient and muon-induced) were partly supported by the EU contract RII3-CT-2004-506222 and the Russian Foundation for Basic Research (grant No. 07-02-00355-a).

References

[1] E. Armengaud et al, Phys. Lett. B**702**,5 (2011) 329.
[2] E. Armengaud for the Edelweiss Collaboration, *these proceedings*.
[3] A. Broniatowski et al., Phys. Lett. B**681** (2009) 305, (*Preprint* arXiv:0905.0753v1).
[4] C. Nones for the Edelweiss Collaboration, *these proceedings*.
[5] P. Loaiza for the Edelweiss Collaboration, *these proceedings*.
[6] H. Kraus et al., (*Preprint* arXiv:0903.1539); G. Gerbier for the Eureca Collaboration, *these proceedings*.
[7] G.S. Vidyakin et al., Instruments and experimental techniques, **32**(4), 809 (1989).
[8] S. Fiorucci et al., Astroparticle Physics **28** (2007) 143.
[9] S. Rozov et al., (*Preprint* arXiv:1001.4383v1).
[10] V. Chazal et al., Astroparticle Physics **9**, 163 (1998).
[11] V. Kozlov et al., Astroparticle Physics **34** (2010) 97, (*Preprint* arXiv:1006.3098).

Journal of Physics: Conference Series **375** (2012) 012017 doi:10.1088/1742-6596/375/1/012017

Prospects for a Phase III SIMPLE Measurement

TA Girard[1,2,§], J Puibasset[3], M Felizardo[4,1,5], A C Fernandes[5,1], A R Ramos[5,1], J G Marques[5,2,1]

(for the SIMPLE Collaboration)

[1] Centro de Física Nuclear, Universidade de Lisboa, Av. Prof. Gama Pinto 2, 1649-003 Lisbon Portugal

[2] Departamento da Físíca, Universidade de Lisboa, 1749-016 Lisboa, Portugal

[3] Centre de Recherche sur la Matiére Divisée CNRS et Université d'Orleans, 45071 Orleans, 02 France

[4] Departamento de Física, Universidade Nova de Lisboa, 2829-516 Caparica, Portugal

[5] Instituto Tecnológico e Nuclear, Estrada Nacional 10, 2686-953 Sacavém, Portugal

[§] email: criodets@cii.fc.ul.pt

Abstract. Current plans of the SIMPLE project include a Phase III measurement based on a new device under development the past two years which contains a large superheated liquid "droplet" within a gel-sheathed vessel (effectively a bubble chamber). Apart from a change in refrigerant and instrumentation, the experiment is to be re-shielded to provide an additional factor 10^4 in neutron suppression. We briefly describe the chamber and its instrumentation, the projected run configuration, and expected results.

1. Introduction

The recently-reported Phase II SIMPLE measurements [1] yielded the most restrictive exclusion contour in the spin-dependent (SD) sector of WIMP-proton interactions from a direct search experiment to date, overlapping for the first time results previously obtained only indirectly. In the spin-independent (SI) sector, a contour challenging the recent CoGeNT result [2] was also obtained.

The measurements were executed in two Stages of 15 superheated droplet detectors (SDDs) each, each stage comprising ~ 0.2 kg active mass, over periods of ~ 100 days without recompression. The surrounding neutron shielding, following improvements between the two Stages, provided an on-detector background rate estimate of 0.253 evt/kgd, arising mainly from the U/Th content of the water-shield itself.

In order to obtain an improved sensitivity over the Phase II measurements in a fashion competitive with other experiments in the field, any next-phase measurement requires larger exposure measurements over shorter observation times with significantly improved neutron shielding. With a continued use of SDD's, this is beyond the capacity of the activity: detector fabrication alone requires two days per detector; the analysis of the matrix signals, more.

2. The Phase III plan

Phase III replaces the SDDs with a prototype bubble chamber (BC) variant of ~ 25x the current matrix SDD active mass in a volume of ~ 8 SDDs, combined with additional neutron shielding using purified water.

2.1. Change of detector

Unlike a SDD, a BC involves no distribution of superheated liquid in a gel matrix, so that the active volume ~ device volume. A 300 g BC variant [3], prototyped by SIMPLE in 2010, differs from that of COUPP [4] and more conventional devices in the use of a purified food-gel based-sheathing of the containment, providing the same smooth gel-freon interface as with the SDDs (to reduce the probability of spontaneous nucleation: recompression times of up to ~ 12 h have been observed [3]) and maintaining the containment radio-purity (α-induced events < 0.5 evt/kgd), while preserving the reduced cost of detector fabrication.

=>

(a) (b)

Fig. 1: (a) standard Phase II SDD, containing 10-20 g C_2ClF_5; (b) the 300g BC prototype of the about the same volume undergoing testing.

At present, the "home water filter" prototype is recompressed using a pressure-activated system; a new piston-based recompression prototype to separate the recompression gas (N_2) from the gel and superheated liquid has been developed. The prototype scale-up to a 5 kg liquid containment, constrained by commercial plastics and their ability to sustain a 10 bar recompression, is shown in Fig. 2, and differs from Fig. 1 (b) in having the acoustic instrumentation and recompression mounted at opposing ends of the chamber.

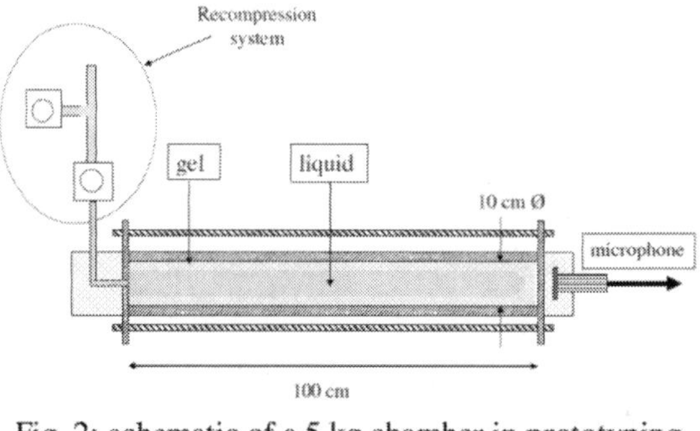

Fig. 2: schematic of a 5 kg chamber in prototyping.

Owing to dwindling stocks of C_2ClF_5 and its general unavailability in Europe, Phase III is anticipated to use C_3F_8 as the active superheated liquid. This implies operation at 10ºC and 2 bar to preserve the 8 keV or lower recoil energy threshold of Phase II; other refrigerants (C_2ClF_5, C_2ClF_2) can also be used with little change, whereas CF_3I and C_4F_{10} require some attention to the temperature control (which is to be effected by the water bath immersion as in Phase II).

2.2. Instrumentation

The loss of the viscoelastic gel matrix as the response medium precludes the use of the previous low frequency instrumentation (the signal deriving from the gel response to a bubble nucleation event). As recently demonstrated by COUPP [4], the primary particle-induced event signal obtains from the differing number of proto-bubbles formed in the superheated liquid by neutrons and α's. The proto-bubble signal however lies in the high frequency regime of ~130 kHz, requiring a replacement of the current sensors.

At present, the prototype BC is instrumented by an externally-polarized condenser (CM16/CMPA40-5V) microphone with a flat response over 10–250 kHz. Experiments are currently in progress to establish neutron-α discrimination criteria comparable to Phase II.

2.3. Change of shielding

Phase II was operated in the 60 m^3 GESA cavern at the 505m depth of the LSBB; its shielding configuration as described in [1] is shown in Fig. 3(a). The final primary source of on-detector neutrons was (α,n) and spontaneous fission neutrons from the U/Th contamination of the shield water itself.

Ultimately, a Phase IV experiment is to be relocated to the 1250 m^3, 505 m deep "capsule" of the LSBB, which although similar in structure to GESA (2 m concrete lining, 1 cm internal steel sheathing to form a Faraday cage) requires extensive revisions of its current internal configuration, which are not anticipated before 2013: accordingly, Phase III is intended to be executed in GESA.

To further decrease the on-detector neutron flux in GESA, we envision an increase of the current water shield thickness to 1.3 m, which is constrained by the cavern definition as seen in Fig. 3(b). The water was however obtained from locally-available, untreated tap water with U/Th concentrations of xxxx; treated tapwater already yields a factor 10 reduction in the radio-contaminants. The Phase III water is to be replaced by purified reactor containment water, which involves a resin-based tapwater treatment process which reduces the heavy metals content by 10^2-10^3. The design involves the immediate surrounding of the detectors, to include the thermal waterbath itself, with the purified water, with the previous water shield moved to the GESA periphery as an external baffle.

(a) (b)

Fig. 3: (a) GESA and its neutron shield (color) in Phase II configuration; (b) projected Phase III configuration; the indicated grey areas are either water or concrete, as included in the simulations.

Reducing the on-detector neutron background by ~ 10^4 provides an overall 10^3 improvement in experiment sensitivity, and suggests a neutron background rate well below zero for exposures of ~10^3 kgd.

3. Projected Phase III Sensitivities

Figure 4 shows the estimated levels of sensitivity in the two sectors of the dark matter search for exposures of 100 and 500 kgd, using two 5 kg prototypes and assuming the same particle-induced event discrimination as in Phase II and zero candidate events, in comparison with the current Phase II results [1].

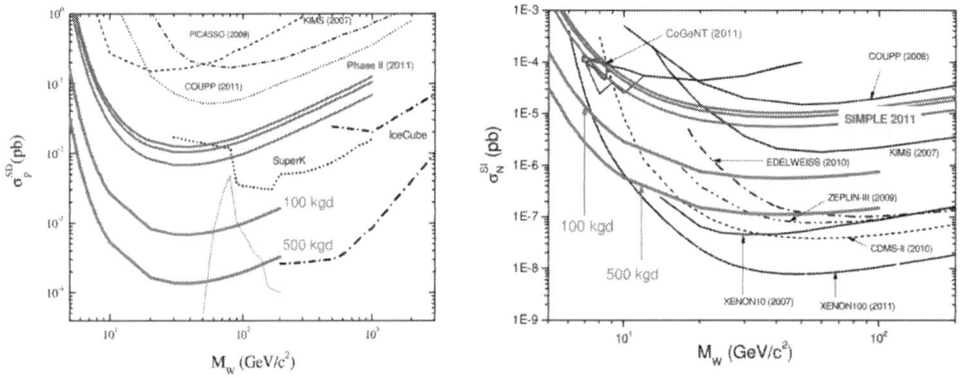

(a) (b)

Fig. 4: (a) projected sensitivity levels in the SD sector; (b) same in the SI sector for the same exposures.

In the SD sector, a two week exposure, assuming a 20% loss of measurement time due to recompressions, is expected to provide more than a factor 10 improvement in the current exclusion. For the SI sector, the same result provides a similar improvement, and should confirm/deny SIMPLE's current tension with CoGeNT. While this level remains above those established by XENON, CDMS, and CRESST-II, extension of the exposure to two months will address the situation.

4. Summary

The Phase III measurement, as described above, if realized in scope and time scale, is clearly competitive with other WIMP searches in the field, and would both confirm/deny the light WIMP possibility as well as serve to itself discover a WIMP candidate. The involved cost of the developments discussed above remains at the level of a few hundred k$. The Phase III plan, in generality, has recently been funded, with a completion date of December 2012.

Acknowledgments

This work was funded, in part, by the Nuclear Physics Center of the University of Lisbon, and by grants PDTC/FIS/83424/2006 and PTDC/FIS/115733/2009 of the Portuguese Foundation for Science and Technology (FCT). The activity of M. Felizardo was supported by grant SFRH/BD/46545/2008 of FCT.

References

[1] Felizardo M, et. al: 2011 arxiv:1106.3014**v1**.
[2] Aalseth C, et. al.: 2011 *Phys. Rev. Lett.* **107**, 141301.
[3] Puibasset, J: PhD thesis, Université de Paris 7 (2000, unpublished)
[4] Behnke E, et. al.: 2011 *Phys. Rev. Lett.* **106**, 021303.

Measurement of the differential neutron flux inside a lead shielding in a cryogenic experiment

N Coron[1], C Cuesta[2], E García[2], C Ginestra[2], J Gironnet[1], P de Marcillac[1], M Martínez[1], Y Ortigoza[2], C Pobes[2], J Puimedón[2], T Redon[1], T Rolón[2], M L Sarsa[2], L Torres[1] and J A Villar[2]

[1] Institut d'Astrophysique Spatiale. Bâtiment 121, Université Paris-Sud 11 and CNRS (UMR 8617), 91405 Orsay Cedex, France
[2] Grupo de Física Nuclear y Astropartículas. Universidad de Zaragoza. C/ Pedro Cerbuna 12, 50009 Zaragoza, Spain

E-mail: puimedon@unizar.es

Abstract. The ROSEBUD collaboration measured the differential flux of fast neutrons inside a shielding of lead irradiated with a source of ^{252}Cf using two scintillating bolometers of LiF and Al_2O_3 at 20 mK. We compare an unfolding method using a three parameter model for the fast component of the neutron flux with a multigroup method in which the energy interval of interest is divided into groups. Some issues regarding the neutron monitoring with LiF or Al_2O_3 in a cryogenic experiment searching for dark matter WIMPs are discussed.

1. Introduction

The observation of some events identified as nuclear recoils by experiments using hybrid bolometers for dark matter searches [1, 2, 3] are reported with a careful study of the background sources, in particular the neutron interactions in the involved experimental setups. The hybrid bolometers can separate the nuclear recoils from the electron recoils; but a substantial fraction of the former comes from neutron scattering with the nuclei of the detectors and the remaining, if any, could come from the scattering of dark matter particles. Then, the right estimate of the recoils induced by neutrons and the crosschecking of experimental data with Monte Carlo simulations are issues of paramount importance. Important efforts have been devoted to the measurements of the thermal [4] and muon-induced [5] neutrons in the laboratory, and the beneficial effect of a neutron shielding in a dark matter search was tested underground [6]. In this work we measured with two scintillating bolometers the differential flux of fast neutrons inside a setup similar to a small-scale search for dark matter [7], showing the feasibility of monitoring the fast neutron flux inside the shielding of future dark matter searches with hybrid bolometers. We also stress that besides this monitoring, the positive identification of dark matter would require a signature like the annual modulation of the signal [8, 9, 10, 11] and, furthermore, the events should follow suitable temporal and spatial patterns indicating that mechanical effects as fractures in the bolometers [12] are unlikely.

2. Experimental data and results

We irradiated, in the old Laboratorio Subterráneo de Canfranc, scintillating bolometers of LiF (33 g) and Al_2O_3 (50 g) inside a lead shielding of 25 cm thickness with a ^{252}Cf source outside the shielding.

The LiF, through the $^{6}Li(n,\alpha)^{3}H$ capture (Q=4.78 MeV), is a very interesting target for thermal and fast neutron monitoring [13], because the reaction products give, in the light–heat plot, a signal in a region where the recoils from hypothetical WIMPs are not expected [14, 15]. See in figure 1 the electron recoil events around the straight line passing through (0,0) mV and (1000,500) mV, the nuclear recoil events clustered mainly below (500,100) mV, the captures of thermal neutrons around (2000,400) mV and the resonant captures of fast neutrons around (2100,400) mV. The linearity of the heat signal with respect to the recoil energy is shown in [16]. The Al_2O_3 detects neutrons only by elastic scattering and it does not distinguish, in principle, neutrons from WIMPs. Figure 2 shows the heat–light plot of Al_2O_3, where the nuclear recoil events give lower light output than the electron recoil ones.

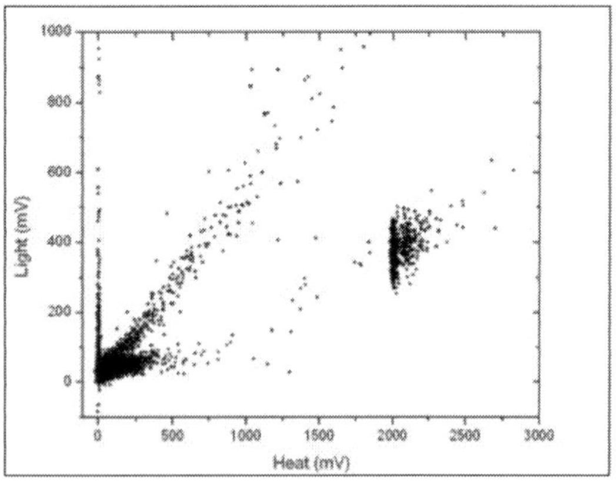

Figure 1. Heat–light plot of LiF irradiated with ^{252}Cf.

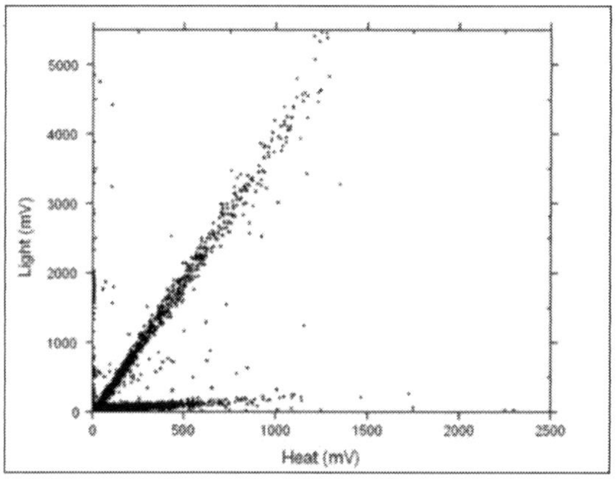

Figure 2. Heat–light plot of Al_2O_3 irradiated with ^{252}Cf.

The heat signals of Al_2O_3 were calibrated with a low activity source of ^{57}Co, assuming that the relative efficiency factor between nuclear and electron recoils is one [17], giving 37±9 keV at the 50 mV discrimination threshold. The heat signals of LiF in the region of the nuclear recoils were not calibrated similarly because the ^{57}Co peaks were not observed. The calibration with the observed peaks at 2000 mV and 2100 mV [18] gave 160±133 keV at the discrimination threshold of 250 mV. Therefore, the nuclear recoil events of LiF bolometer were not used to estimate the neutron flux.

In a previous analysis [18], assuming the same isotropic neutron flux around both crystals (they were ~20 cm apart), we modeled the neutron energy spectrum with the parameters Φ_0, α, T: $\Phi(E)dE = (\Phi_0/\Gamma(\alpha+1))\,(E/T)^{\alpha}\exp(-E/T)\,dE/T$ and using MCNP version 4C [19], we found the solution $\alpha = -0.9$, T=1.48 MeV and a flux of fast neutrons $\Phi(E > 0.1\text{ MeV}) = 0.20\pm0.02\text{ n s}^{-1}\text{ cm}^{-1}$.

In this work we have applied a more elaborated model based on a multigroup method in which the energy interval of interest is divided into groups. We have assumed that the energy spectrum is a piecewise constant function

$$\Phi(E)dE = \Sigma_k\,\Phi_k\,dE$$

where $\Phi_k \neq 0$ if $E_{k-1} \leq E \leq E_k$ and $\Phi_k = 0$ otherwise. We have chosen six energy groups: (0.05, 0.1), (0.1, 0.2), (0.2, 0.5), (0.5, 1.0), (1.0, 2.0) and (2.0, 5.0) MeV and we have estimated with MCNP–PoliMi [20] the response function of every energy group in both bolometers. After, we have fitted the heat spectra of the events produced by the neutron interactions to a linear superposition of the response functions and we have obtained for every scintillating bolometer the results plotted in figure 3. The experimental data do not give information for energies below 0.1 MeV, the LiF bolometer gives information up to 1.0 MeV and the Al_2O_3 bolometer up to 5 MeV. Figure 4 shows that the combined solution ($LiF+Al_2O_3$), assuming that the neutron flux is the same on both bolometers, is equivalent to the three parameters model.

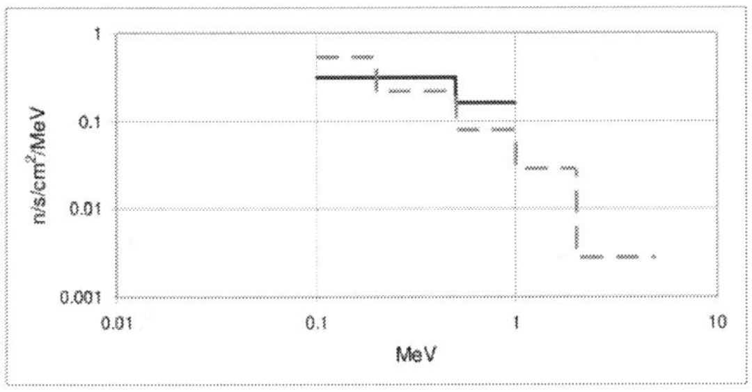

Figure 3. Estimates of the neutron flux using the model with energy groups to LiF data (solid line) and to Al_2O_3 data (dashed line)

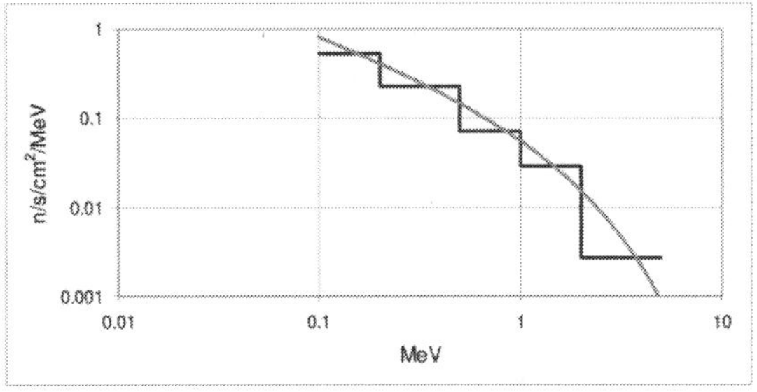

Figure 4. Comparison of the combined solution ($LiF+Al_2O_3$) obtained with six energy groups (histogram) and the obtained with the three parameters model (continuous line).

In order to crosscheck our results, we have estimated with MCNP-PoliMi the absorption spectrum in LiF (figure 5) and the nuclear recoil spectrum in Al_2O_3 (figure 6) produced by a fast neutron flux inside the shielding using the solution of the three parameters model (α=− 0.9 and T=1.48 MeV), with the total flux as a free parameter for each bolometer. The experimental and simulated spectra are in good agreement in both bolometers; the fast neutron fluxes (E > 0.1 MeV) are 0.218±0.016 and 0.204±0.025 n s^{-1} cm^{-2} for LiF and Al_2O_3, respectively, indicating that the model with three parameters describes quite well the neutron flux. Similar results are obtained with the model with six energy groups.

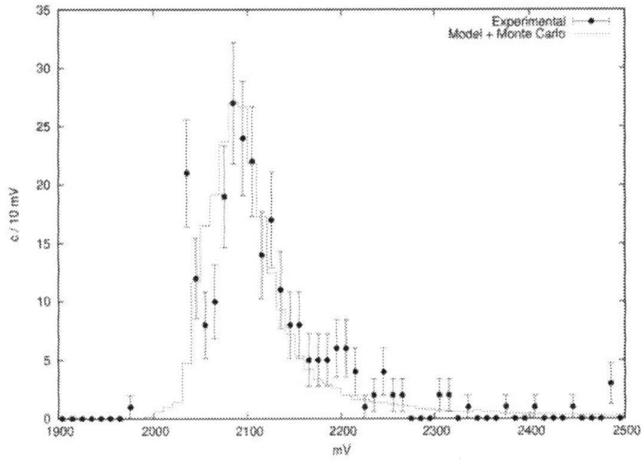

Figure 5. Experimental and simulated spectra of absorbed fast neutrons in LiF. The peak corresponding to the capture of thermal neutrons (~2000 mV) is not shown.

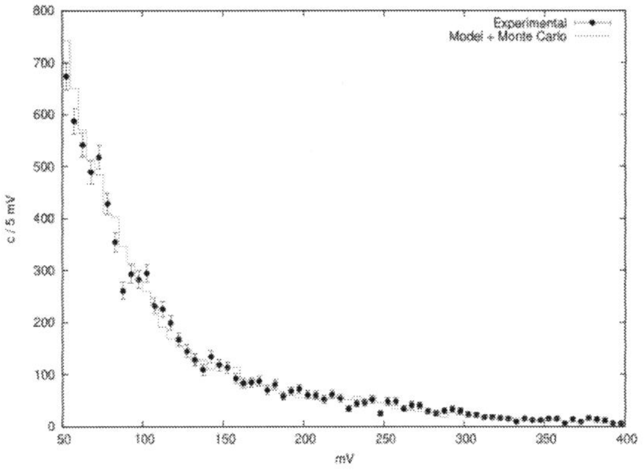

Figure 6. Experimental and simulated spectra of nuclear recoils in Al_2O_3. The ^{57}Co calibration gave 37±9 keV at the discrimination threshold at 50 mV.

3. Conclusions

We have measured inside a shielding of lead the differential flux of fast neutrons (E > 0.1 MeV) coming from a ^{252}Cf source placed outside the shielding. The source produced a rate of 1.7×10^7 captures y^{-1} kg^{-1} of LiF around the resonance at 0.24 MeV, and the estimated flux of fast neutrons (E > 0.1 MeV) on the bolometers was 0.2 n s^{-1} cm^{-2}. A simple scaling law indicates that a flux of 2×10^{-9} n s^{-1} cm^{-2} with similar energy dependence can be monitored by 100 kg of LiF or by

10 kg of enriched ^{6}LiF, giving ~20 captures y^{-1}. The analysis of the response of LiF to fission neutrons from internal sources is in progress.

The neutron monitoring with LiF of a dark matter search could be better if the nuclear recoils of LiF were well calibrated and its discrimination threshold was low enough; in this case, the compatibility of capture and elastic scattering of neutrons could be tested within the same detector, not with another near detector.

Acknowledgements

This work has been supported by the Spanish Ministerio de Ciencia e Innovación (MICINN) (FPA2008-03228) and the French CNRS/INSU (MANOLIA and BOLERO projects). The partial supports of the Gobierno de Aragón and of the Consolider-Ingenio 2010 Programme under grant Multidark CSD2009-00064 are also acknowledged. M. Martínez is supported by a postdoctoral grant from the P2I program funded by Université de Paris Sud, Y. Ortigoza is supported by EUI2009-03957 (MICINN) project contract, L. Torres is supported by a MICINN postdoctoral grant (Programa Nacional de Movilidad de Recursos Humanos 2008-2011) and T. Rolón thanks the support of UZ/BSCH/Fundación Carolina.

References

[1] The CDMS II Collaboration 2010 *Science* **327** 1619–21
[2] Angloher G *et al* 2011 *Preprint* astro-ph/1109.0702v1
[3] Armengaud E *et al* 2011 *Phys. Lett.* B **702** 329–35
[4] Rozov S *et al* 2010 *Preprint* astro-ph/1101.4383v1
[5] Kozlov V Yu *et al* 2010 *Astroparticle Physics* **34** 97–105
[6] Carmona J M *et al* 2004 *Astroparticle Physics* **21** 523–33
[7] Calleja A *et al* 2008 *J. Low Temp. Phys.* 151 848-53
[8] Drukier A K, Freese K and Spergel D N 1986 *Phys. Rev.* D **33** 3495–508
[9] Sarsa M L *et al* 1997 *Phys. Rev.* D **56** 1856–62
[10] Bernabei R *et al* 2010 *Eur. Phys. J.* C **67** 39–49
[11] Aalseth C *et al* 2011 *Phys. Rev. Lett.* **107** 141301
[12] Åström J *et al* 2006 *Phys. Lett.* A **356** 262–6
[13] Knoll G F 2000 *Radiation detection and measurement* 3rd ed. (New York: John Wiley & sons) pp 507, 546
[14] de Marcillac P *et al* 1993 *Nucl. Instrum. Methods* A **337** 95-100
[15] Coron N *et al* 2003 *Nucl. Instrum. Methods* A **520** 159-62
[16] Gironnet J 2010 *Spectrométrie de neutrons rapides par bolomètres à cible lithium pour la réduction du fond des expériences de détection directe de matière noire.* http://tel.archives-ouvertes.fr/tel-00633437/fr/
[17] Coron N *et al* 2008 *Phys. Lett.* B **659** 113-8
[18] Coron N *et al* 2010 *Journal of Physics: Conference Series* **203** 012139
Coron N *et al* 2011 *Proceedings of Science* PoS(IDM2010)054
[19] Briesmeister J F (ed) 2000 MCNP - *A general Monte Carlo N-Particle transport code, version 4C*, LA-13709-M
[20] Pozzi S A, Padovani E and Marseguerra M 2003 *Nucl. Instrum. Methods* A **513** 550-8

Study of nuclear recoils in liquid argon with monoenergetic neutrons

C. Regenfus, Y. Allkofer, C. Amsler, W. Creus, A. Ferella, J. Rochet, M. Walter

Physik-Institut der Universität Zürich, CH–8057 Zürich, Switzerland

E-mail: `regenfus@cern.ch`

Abstract. In the framework of developments for liquid argon dark matter detectors we assembled a laboratory setup to scatter neutrons on a small liquid argon target. The neutrons are produced mono-energetically (E_{kin}=2.45 MeV) by nuclear fusion in a deuterium plasma and are collimated onto a 3" liquid argon cell operating in single-phase mode (zero electric field). Organic liquid scintillators are used to tag scattered neutrons and to provide a time-of-flight measurement. The setup is designed to study light pulse shapes and scintillation yields from nuclear and electronic recoils as well as from alpha particles at working points relevant for dark matter searches. Liquid argon offers the possibility to scrutinise scintillation yields in noble liquids with respect to the population strength of the two fundamental excimer states. Here we present experimental methods and first results from recent data towards such studies.

1. Introduction

Over the last years liquid xenon TPCs [1, 2] established themselves among the leading technologies for WIMP searches. Furthermore considerable effort is being made by various groups to bring the liquid argon (LAr) sector to a competitive level in this respect [3]. Liquid argon has the potential to be a large and sensitive multi purpose detector due to its low ionisation potential and large abundance on earth. Interest in the low energy frontier of this technology increased by recent progress in the production of ^{39}Ar depleted argon from dwell gases. Here we describe research activities which are part of the design study for a next generation dark matter facility, DARWIN [4], presently proposed as a combined liquid xenon - liquid argon installation.

WIMPs, hypothesised to be distributed as a thermalised halo in our galaxy, should produce nuclear recoils in any target on earth, which can be detected and isolated in noble liquids through their characteristic excitation and ionisation patterns. However the complex microscopic processes which lead eventually to the scintillation and charge signals are not very well understood at low energies and are currently the subject of large experimental controversy. The experimental determination of light and charge yields at keV energies and possibly below is therefore of highest interest for this research since it defines the energy scale and sensitivity of the experiment. The uncertainty in the signal calibration of nuclear recoils still yields the largest contribution to the systematical error for liquid xenon WIMP searches [1, 5] despite the large efforts of various groups [5, 6] to determine this quantity. In the liquid argon sector the situation is even worse, only scarce information about the scintillation efficiency can be found in literature [7, 8] and no comprehensive measurement of the charge yield exists at all.

The energy dependent light yield Y_{nr} of nuclear recoils is commonly described by the unitless quantity $\mathcal{L}_{eff}$, the relative scintillation efficiency, determined in relation to the yield of recoiling electrons (Y_{er}) of same kinetic energy. By convention $\mathcal{L}_{eff}$ is measured at zero electric field. The value of $\mathcal{L}_{\text{eff}}$ is expected to decrease towards low energies, mainly due to the increasing fraction of nuclear stopping power, transferring energy to heat, described by Lindhard's theory [9, 10]. Additional (luminescence) quenching, which is less well understood, arises at high ionisation densities by interactions among the excited and ionised states created in the process of energy loss to electrons (electronic stopping). The light yield of recoiling electrons is not affected by nuclear effects and shows a linear response for energies above some tens of keV [11]. This allows for the convenient comparison of the energy scales of both recoil types, where E_{ee} represents the (equivalent) energy of an electron to produce the same amount of light as a recoiling nucleus with Energy E_{r}. Electron yields are traditionally determined at the 122 keV line of ^{57}Co. Electric fields reduce the recombination of free charge carriers and decrease scintillation light [12]. In the estimation of light yields this effect is taken into account by two energy and field dependent correction functions S_{n} and S_{e}, for nuclear and electronic recoils, respectively, which are unknown for liquid argon. In liquid xenon [13, 14] values for S_{n} are close to unity while S_{e} drops to below 0.5 at fields >1 kV/cm. This is due to screening effects at the large ionisation densities of nuclear recoils. Thus the relative scintillation efficiency can be calculated by $\mathcal{L}_{\text{eff}} = Y_{\text{nr}}/Y_{\text{er}} \cdot S_{\text{e}}/S_{\text{n}} = E_{\text{ee}}/E_{\text{r}} \cdot S_{\text{e}}/S_{\text{n}}$. In this work the values for S_{e} and S_{n} are set to unity.

Liquid argon allows for a separate study of fast and slow components in the scintillation signal due to the large difference in their lifetimes. The analysis of pulse shapes [15] is an important tool for background discrimination in LAr dark matter detectors. Here we apply pulse shape analysis to light yield studies in regions of both, low and high ionisation densities in a noble liquid. In the following we present a brief overview of the experimental method and show first results obtained from data recorded under neutron, alpha and gamma irradiation in liquid argon at zero electric field.

2. Experimental Setup

We induce nuclear recoils in the liquid argon cell by fast neutrons (2.45 MeV) from a deuterium-fusion-generator made by the company NSD-Fusion, Germany. It delivers up to $2 \cdot 10^6$ isotropically emitted mono-energetic neutrons from the two-body reaction $\text{dd} \rightarrow {}^3\text{He} + \text{n}$ ($\text{dd} \rightarrow \text{t} + \text{p}$ is equally probable). The overall setup was developed in collaboration with the producer. The fusion rate in the deuterium plasma is controlled by an electrical DC field generated by an adjustable constant current (1-15 mA) HV supply. The voltage to keep the plasma current stable (30-120 kV) is regulated indirectly by the pressure of the deuterium gas which is released from heated getter disks storing deuterium on their surfaces. Higher voltage (lower gas pressure) produces higher fusion rates but also a harder bremsstrahlungs background spectrum. The environment is shielded from neutron and X-ray radiation by means of a 1600 kg polyester cylinder with 2 mm Pb cladding (fig. 1) keeping the radiation dose far below the limit of 2.5μSv/h imposed by CERN's radiation protection requirements. A safety area with access control and radiation interlock complete the setup. A polyethylene

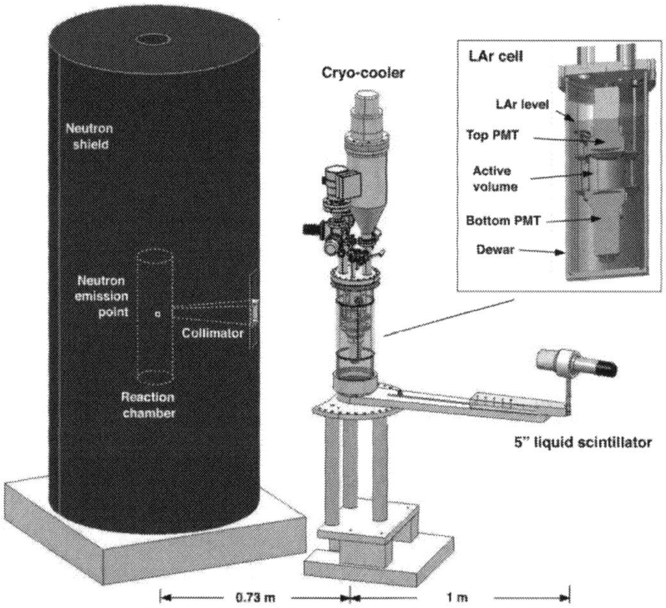

Figure 1. Setup of the scattering experiment with neutron generator, shield and zoom on the LAr cell.

collimator with square cross section allows for the emission of neutrons in a fixed solid angle of about 0.2% of 4π, covering fully the sensitive volume of the cryogenic LAr cell. For the measurements described here the cell was located at a distance of about 73 cm from the neutron emission point while the liquid scintillator counter (LSC) could be set at various scattering angles on a 1 m long arm rotatable around the centre of the cell. The nominal scattering angle is given from the two axes of neutron emission point to the centre of the LAr cell and further on to the centre of the liquid scintillator. Errors on angles or mechanical alignments were estimated to be below 0.5 degree. The internal structure of the cryogenic cell is shown in the inset of fig. 1. Two TPB coated 3" PMTs (Hamamatsu R6237-MOD) with bialkali photo cathodes and Pt underlay (QE≈15%) are arranged face to face at a distance of about 47 mm forming a cylindrical sensitive volume of roughly 0.2ℓ. This volume is defined by a thin Al cylinder holding the TPB coated reflector foil[1] to wave shift the VUV scintillation light. A small ^{210}Po alpha source of about 40 Bq activity is installed in the centre of the cell. The source was coated with some tens of μm plastic (Paraloid B-72) to spread the energies of emitted alphas over a broad range. The Al cylinder is polarised to the same voltage as the photo cathodes of the PMTs to keep the internal electrical field close to zero. A $60\,\ell$/s turbo pump is used to evacuate the chamber to typically 10^{-6} mbar prior to filling with argon gas class 60 (impurities ≤ 1.3 ppm). A membrane pump provides for continuous recirculation via two in parallel mounted OXISORB-W cartridges which reduce the O_2 and H_2O levels to <5 and <30 ppb respectively. The gas is condensed on top of the chamber at the cold head of a Sumitomo CH210 cryocooler system with about 80 W cooling power driven by a Sumitomo F-70CH helium compressor. A LabView based slow control system regulates the cold head temperature and records temperatures, pressures and liquid levels. The analogue signals from the two PMTs are each split into 2 inputs of a LeCroy WavePro 35Zi DSO, sampled with 5000 points at 1 GS/s and stored to the hard drive. The splitting of the signals permits a large dynamic range to fully reconstruct alpha particles while maintaining high signal over noise ratios for single photons. Cosmic signals however can saturate the FADC [3]. Event triggers of coincidences and delays between PMT signals and external detectors were produced by means of the programable trigger logic in the oscilloscope. Signals of neutrons in the LSC are processed in a dedicated analogue pulse shape discriminator (Mesytec MPD4 [18]) and fed into the external input of the oscilloscope.

3. Data reconstruction and light yield

Scintillation signals in liquid argon are determined by numerical integration of the digitised photo currents of the two PMTs, normalised to their mean single photon charges. Due to the long integration times in LAr an iterative method was developed to calculate precise pedestal values on an event by event base. The sum of both channels yields the raw integrated pulse height (*IPH*) in units of photo electrons (pe). This value is furthermore corrected for the finite integration time, losses by software thresholds (clusterfinder) and most importantly by losses due to impurities in the liquid argon (see next chapter). Light yield calibrations were performed periodically during data taking by means of a strong external ^{241}Am source producing a prominent 60 keV photo peak in the integrated pulse height spectrum of the data (fig. 2 left). The gains of the PMTs were determined

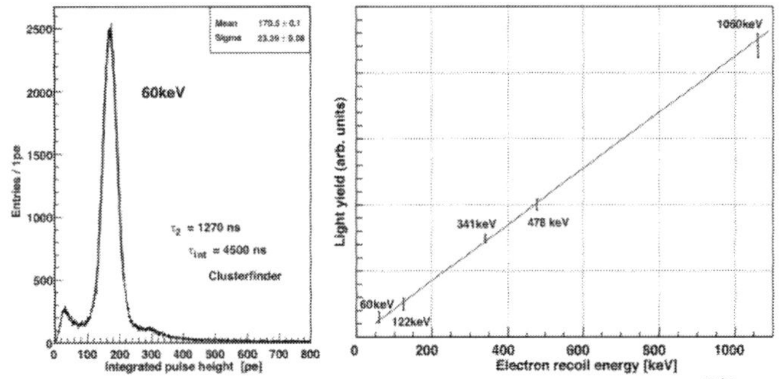

Figure 2. *Left: 60 keV line from the raw signals of the ^{241}Am data. Right: light yield of various photon sources.*

[1] 1 mg/cm^2 tetra-phenyl-butadiene (TPB) on Tetratex; PMT coatings 0.08 mg/cm^2 TPB. See [16, 17] for details

from LED pulses or single photon signals collected from event tails in the data. Single photon distributions are analysed for most probable values and corrected for asymmetry (+14.5%) to find the mean single photon charge. The gains during several weeks of data taking in summer 2011 were found to be stable within 2.7%. An average light yield of 3.75 ± 0.1 pe/keV has been determined over the whole run time. To check the linearity of the system we employed various external photon sources, ^{57}Co (122 keV photo peak), ^{22}Na (511 and 1275 keV Compton edges), as well as ^{137}Cs (662 keV Compton edge). Figure 2 (right) shows a linear fit to the raw measurements of these data which were taken under similar conditions.

4. Impurity effects – correction of light yields

In liquid argon VUV fluorescence from the so-called second continuum is the dominant mechanism for light emission under excitation. The light pulse shape is well described by the sum of two exponentials originating in the radiative decays of two fundamental excimer states. Atomic selection rules are the cause for a large difference in their lifetimes, τ_1 and τ_2, approximately 6 and 1600 ns, for singlet and triplet states, respectively. This feature is used for background discrimination in LAr dark matter detectors thanks to the ionisation density effect [15] modifying the pulse shape according to specific energy losses of particles. Suppression factors up to 3 orders of magnitude are reached [22] for recoil energies above 30keV$_r$ using a likelihood based discrimination method. The improvement of such algorithms is also part of our research but will be described elsewhere.

In the following we decompose individual or averaged signal traces by means of the sum of two exponential decays convoluted with a gaussian which describes general time spreads (2.9 ns). We denote the integrals of fast and slow scintillation components, representing the occurrences of singlet and triplet states, with A and B, respectively; $A+B$ being the total measured light yield Y. Furthermore we denote the fraction $A/(A+B)$ as the component ratio CR, representing the relative strength of the fast portion of the scintillation light. We want to stress that values for CR are similar but somewhat different from the common variable for the prompt light fraction f_p, determined from short and long signal integrations.

Due to their long lifetime triplet states undergo various collisions with neighbours before they eventually decay. Thereby interactions with impurities can cause the (non-radiative) destruction of these states and hence induce losses in the scintillation light (impurity quenching [20, 21]). This process occurs in competition to the natural decay of the states and results in an apparent reduction of the triplet lifetime τ_2 in the data [2]. On account of the normalisation of the exponential PDF the corrected light yield [20] is found from the relation

$$Y_{cor} = A + B_{cor} = A + B/\tau_2 \cdot \tau_2^{max} ,$$

where B_{cor} is the extrapolated yield of the triplet component and τ_2^{max} the undisturbed lifetime of triplet states in LAr (1600 ns [15, 19]). The fraction B/τ_2, corresponding to the triplet decay rate at t=0, is effectively invariant versus variations of purity and completely determined by the fit. In the same way the value for the corrected component ratio is given by $CR_{cor} = A/(A+B_{cor})$. Corrected light yields are conveniently calculated by $Y_{cor} = A/CR_{cor}$. As an example fig. 3 shows the two components of such fits to averaged signal traces of 60 keV calibration data versus the value found for τ_2 under

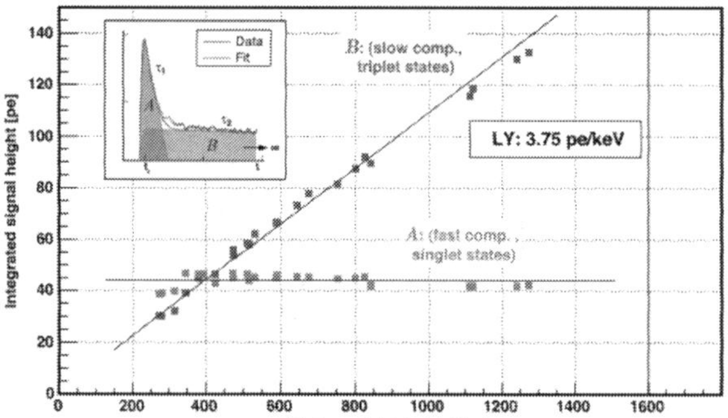

Figure 3. *Yield of the fast and slow scintillation components under different purity conditions.*

[2] In LAr the value for τ_2 can be used to estimate purity levels in respect to scintillation yields [19]. In this work purity levels are predetermined by residual H_2O on internal surfaces of the vacuum and gas system

different purity condition. Values for A remain unaffected (for purity conditions of $\tau_2 > 400\,\mathrm{ns}$) while B increases proportional to the measured τ_2. From these results we conclude that other quenching mechanisms, like VUV absorption by impurities, are not affecting light yields under the present geometrical conditions and purity levels. All yields as well as component ratios presented in the following are corrected for losses by impurities by the method presented here.

5. Time structure of the scintillation light

The well fitting exponential decay model assumes production and modification of excimer states being completed on time scales shorter than the measured decay constants. Moreover the experimentally observed invariance of lifetimes, i.e. their independence on ionisation densities, rules out [23] the participation of these states in the fast processes of luminescence quenching during production [3]. However, in the temporal transition region between fast and slowly decay modes, two component fits systematically underestimate the data. This situation can be improved by the introduction of a third component in the fit function. The relative contribution of such an extension scales with the value for CR and ranges between 3 and 7% of the total integral [24], lifetimes vary between 50 and 200 ns. The nature of this effect is presently not understood, possible explanations comprise PMT-after-glow and recombination light. In the context of this work we exclude the transition region from the fit, until the results of a systematic study on this effect, e.g. by measuring the electric field dependence of light pulse shapes.

In the following we compile measurements of component ratios determined from event-wise likelihood fits for recoiling Ar nuclei, alpha particles and electrons. The former were induced by 2.45 MeV neutrons from the generator, the second were emitted from the built-in ^{210}Po source, and the latter were induced by 511 keV photons from an external ^{22}Na source, respectively. As already mentioned the ^{210}Po source was coated with a thin layer of plastic to degrade and spread the energy of emitted alphas. Data for neutrons and alphas was purely triggered on signals in the liquid argon, while an additional coincidence with an external detector was required for the photon data to tag the emission of two 511 keV photons. This data was also used to determine the time calibration for the time-of-flight (TOF) measurement for scattered neutrons as well as the trigger roll off at small signal sizes. The latter is derived from the lower end of the Compton spectrum. While photon and alpha data are practically background free, nuclear

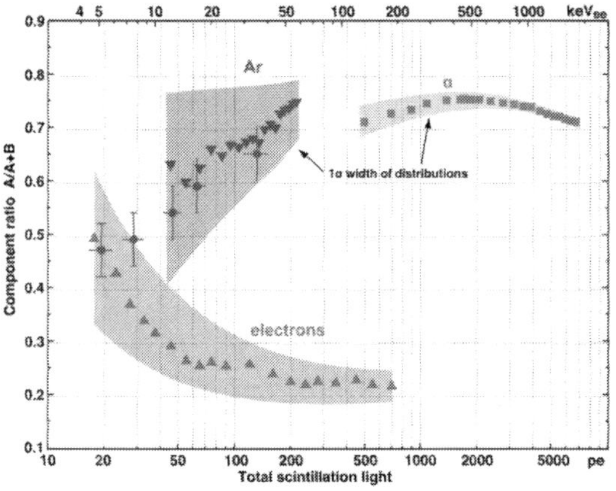

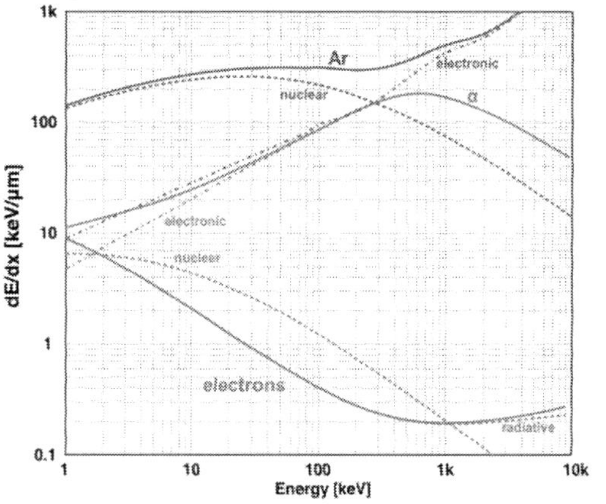

Figure 4. *Left: component ratio vs. signal size for argon, alpha and electron recoils. The grey zones illustrate the spread. Right: total stopping power (solid lines) and contributions (dashed lines).*

recoils suffer from a contamination with inelastic collisions from neutrons with argon nuclei producing photons. Due to the small geometric dimensions and the low kinetic energies however these events are estimated not to contribute more than 7% to this data. A Monte Carlo study

[3] Contrariwise quenching by impurities is a relatively slow process of destruction of (long lived) excimer states

is in progress. The software framework was tested for unbiased reconstruction of pulse shapes in respect to small signal heights. Therefor fixed-ratio test pulses were generated in the energy range of interest and analysed for flatness of *CR*. No significant deviation of the flat response could be found. Figure 4 to the left shows the component ratio versus the total amount of scintillation light produced by these particles. The two prominent bands from electrons (upward triangles) and nuclear recoils (downward triangles)merge for energies below about 10keV$_{ee}$. Grey zones behind the data points illustrate the 1σ widths of the distributions, derived from gaussian fits of vertical slices. Errors are dominated by systematics introduced by the analysis method and are estimated to be roughly 0.25 of the distribution widths. The values at low energies for argon recoils are also shown (circles) and derived from fits to mean traces (see chap. 7). The right plot summarises the values of total stopping power for the same particles as well as the nuclear and electronic contributions. This data was collected from the websites of SRIM, ASTAR and ESTAR [25], respectively.

The correlation among both plots is easily observed, larger ionisation densities generally lead to larger fraction of singlet states in the scintillation light. This trend is also true for the regions towards smaller kinetic energies as well as for the maximum in the curves for alpha particles. Interesting is also the apparent value close to 0.25 for lowest ionisation densities, corresponding to a uncorrelated production of triplet and singlet states. According to their statistical weights just a factor of 3:1 is expected for the relative ratio in their populations. If (luminescence) quenching effects are neglected, deviations from the minimal value of 0.25 imply either the enhanced production of singlet states on the exciton level or the transfer from triplet to singlet states shortly after their production in the area of high ionisation density. Quenching effects among excimer states are excluded for the upper mentioned arguments of lifetime invariance. The physical relation between the specific production of excimer states and the prevailing specific energy deposit can be unfolded from these data if the integrated nature of the measurement is taken into account. In the next chapter we apply this method to alpha particles by using a simple linear relation for the relation between component ratio and linear energy transfer.

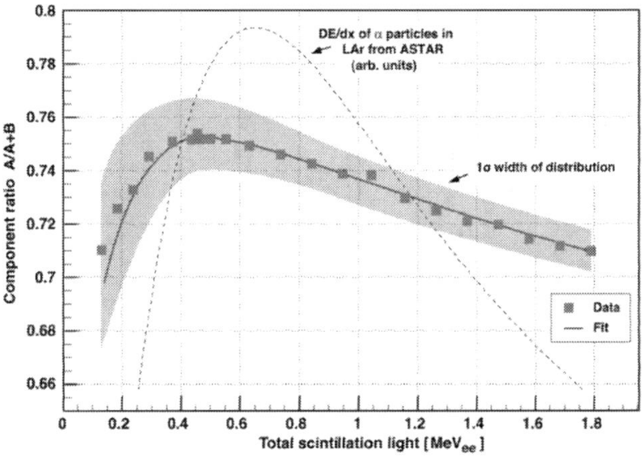

Figure 5. *Component ratio of alpha particles in LAr and fit vs. total scintillation light in units of* MeV$_{ee}$.

6. Scintillation yield of alpha particles in LAr

In this section we extract the value for the relative scintillation yield $\mathcal{L}_{\mathrm{eff}}^{\alpha}$ of alpha particles by comparison of the shapes of the component ratio to linear energy transfer (*LET*). In good approximation the latter is assumed to be identical to the electronic stopping power $\mathrm{d}E/\mathrm{d}x\,(E)$. For the fit function we use a 10th degree polynomial parametrisation of the ASTAR curve from fig. 4. The integrated nature is taken into account in a linear relation to the $\mathrm{d}E/\mathrm{d}x$ curve by the following relation,

$$CR(E) = \int_0^{x_{E=0}} \frac{\mathrm{d}CR(E)}{\mathrm{d}E} \cdot \mathrm{d}E/\mathrm{d}x\,(E)\,\mathrm{d}x \quad \text{with} \quad \frac{\mathrm{d}CR(E)}{\mathrm{d}E} = p_1 \cdot \mathrm{d}E/\mathrm{d}x\,(p_2 \cdot E) + p_3,$$

where E is energy, x the path of the alpha particle and p_i, $i = 1..3$ are the three free fit parameter defining the convolution of *LET* and *CR*. While p_3 describes an empiric vertical offset, p_2 is the parameter of interest introducing a scale for the energy dependent yield (quenching). The naive comparison of the maxima of both curves already gives a first estimate on the value for $\mathcal{L}_{\mathrm{eff}}^{\alpha}$, being roughly 0.75 . Fig.5 shows the measurements for CR obtained by event-wise likelihood fits of the signal shape from alpha particles. The 1σ spread is again illustrated by the grey band. In general the fit (solid line) describes the data well but starts to deviate in the area of low energies probably due to our simple linear model. Also shown is the curve of the electronic stopping power for alpha particles in LAr from ASTAR (dotted line), which actually deviates less than 2% from the total stopping over this energy range. A value of 0.74±0.04 is found for the parameter p_2 from the fit. This corresponds to an estimate for the mean relative scintillation efficiency for alpha particles in LAr in a range of 0.18-2.5 MeV_r. The result agrees within the errors with the value of q_e=0.71 given in [11] for 5 MeV alpha particles assuming a nuclear quenching of $q_n = 0.98$. A similar value was also obtained in [23]. Due to their relatively small mass alpha particles in this energy range are only little affected by effects of nuclear quenching. The light loss of roughly 25% is hence entirely due to luminescence quenching effects.

7. Light yield of nuclear recoils in LAr

In this section we present the data taken in the experimental arrangement of the neutron generator described in chap. 2 . A 5" liquid scintillator (LSC) was used for triggering positioned under the nominal angles of 30, 40, 50, 60 and 90 degrees corresponding to 16.4, 28.5, 43.4, 60.5 and 120 keV recoil energies, respectively. Due to the geometrical extensions of the detectors scattering angles are distributed around the nominal angles roughly box-like (FWHM≈10.8 deg). In most cases the neutron generator was operated at 80 kV and 10 mA corresponding to the emission of about $2{\cdot}10^5$ neutrons per second in 4π. The settings were chosen mainly to compromise about background induced by bremsstrahlung (accidental coincidences). Under these settings about 1 neutron per minute scatters off an argon nucleus in the active volume and is detected in the liquid scintillator 1 m downstream. The direct line of sight between the exit of the collimator and the LSC was obstructed with a 20 cm thick sheet of polyethylene. Background rates amount roughly to 5 per minute originating mainly from two sources; firstly cosmic muons saturating the LSC output and faking a neutron signal in the analog pulse shape discriminator and secondly accidental coincidences between diffusively scattered neutrons and bremsstrahlung.

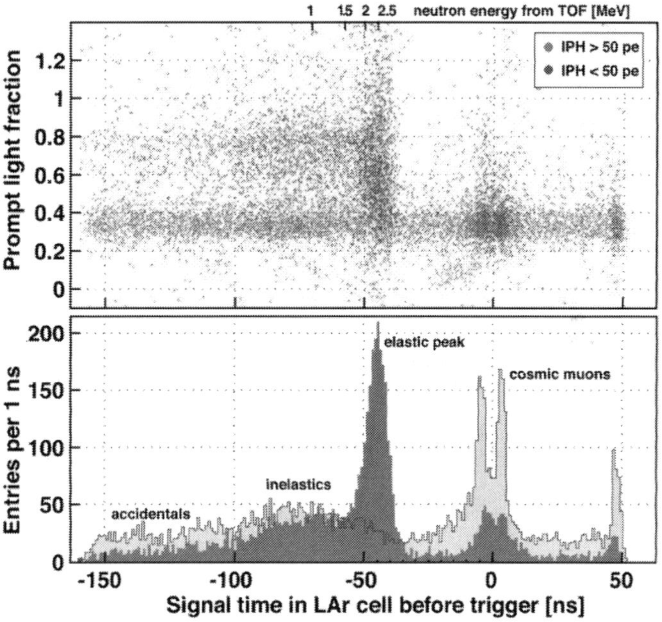

Figure 6. *f_p vs. signal time in LAr with projection on the x-axis, if a neutron triggered the liquid scintillator counter.*

A trigger is generated if in a window of -150...+50 ns around the arrival time of a neutron in the LSC an event was recorded by the liquid argon cell. Such an event is defined as both PMTs showing signals above 0.2 photoelectron pulse height in a 50 ns time coincidence. As an example fig. 6 shows the 19k events from the 30 deg scattering data. The upper plot displays the prompt light fraction of raw signals versus the time difference between the signal in the LAr cell and the arrival time of a neutron in the LSC (TOF). Events above 50 pe of IPH are printed in blue while events below are printed in brown. The histogram on the bottom shows the projection to the x-axis of the same events. The small signal distribution (brown) is dominated by elastically scattered neutrons around the nominal flight time at roughly -45 ns. Also visible is a broad distribution of inelastic scatters where neutrons loose a substantial fraction of their energies. A small uniform background originates presumably in accidental coincidences of bremsstrahlung photons in the outer LAr volume. Two principle cuts are applied to select events. Firstly a time cut of a 5 ns window to the right of the most probable value of the elastic scatter peak and secondly the (loose) limit on *IPH* (typ. 50 pe). The latter removes very efficiently accidental coincidences with photons, inelastic scatters with large energy deposit in the LAr as well as cosmic muons. In the case of the 30 deg data we end up with roughly 1k events, shown in fig. 7. From these distributions we determine the value for *A* and *CR*, the former from the distribution mean and the latter by a fit to the mean trace of selected events. A more precise study by comparing the measured distributions to the ones generated by MC is in progress. We estimate roughly 15% background in the selected data (incl. double scatters). From a preliminary analysis of the 5 scattering angles we determine light yields as well as component ratios with errors of typically 15%. The errors are estimated from the reconstruction of *A* and *CR* and calculated according to the standard propagation law. Figure 8 to the left shows our measurements in comparison to [8] and the one averaged value

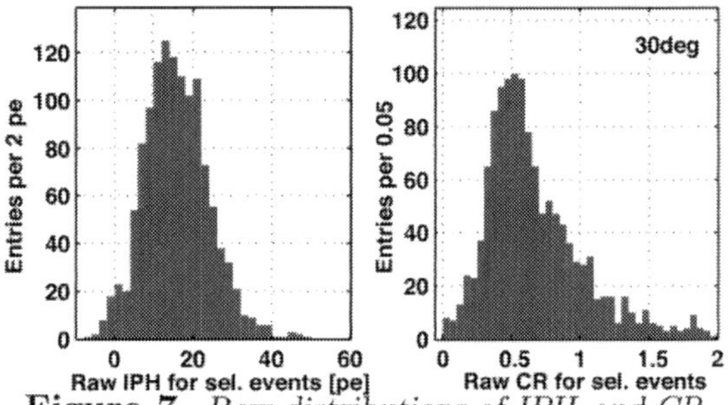

Figure 7. *Raw distributions of IPH and CR after the time and IPH cut.*

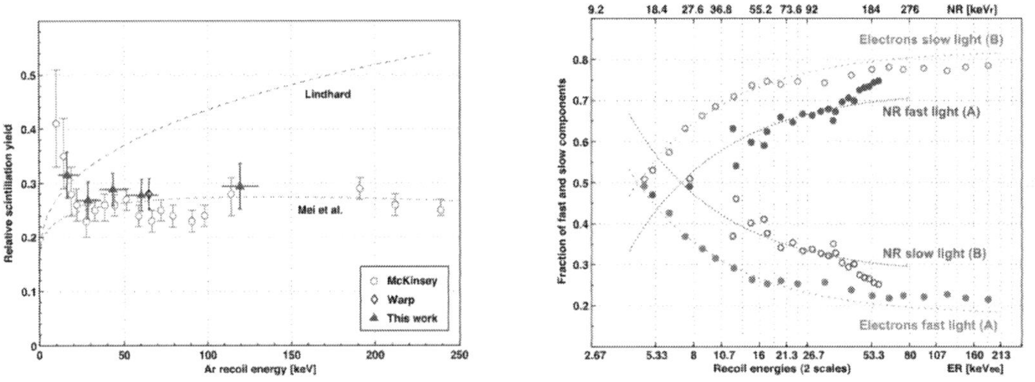

Figure 8. *Left: values for $\mathcal{L}_{\mathrm{eff}}$ with a comparison to theoretical curves and other published data. Right: relative scintillation components for nuclear recoils and electrons plotted against the two energy scales.*

from [7]. The measurements still allow for a flat interpretation of $\mathcal{L}_{\mathrm{eff}}$, leading to a mean value of $<\mathcal{L}_{\mathrm{eff}}> = 0.29 \pm 0.03$ for nuclear recoils at energies above 20keV$_\mathrm{r}$ or 6keV$_\mathrm{ee}$. At the present state of the analysis we can neither exclude nor confirm the increase of $\mathcal{L}_{\mathrm{eff}}$ at the low energy edge of the data from [8]. However we plan to upgrade the LAr cell with higher QE PMTs as well as to improve significantly on the argon cleaning system.

LAr features the possibility of a separate study of singlet and triplet contributions under quenching effects. This is shown by the relative values of *A* and *B* in fig. 8 (right), for nuclear recoils and electrons, respectively. The two energy scales are added to top and bottom. The dashed lines correspond to linear fits of the absolute values of *A* and *B* and are meant to guide the eye. A comparison with the theoretical description of luminescence quenching in LAr by a simple saturation law combined with the Lindhard model [11] favours the assumption of a constant value for $\mathcal{L}_{\mathrm{eff}}$. Hence these measurements are compatible with an interpretation of collisional spin change of triplet state excimers or a preferred production of singlet states excimers under higher ionisation densities during the selftrapping process of the excitons.

Summary and conclusions

We confirmed scintillation light quenching in liquid argon by impurities being of the same physical nature as observed in gaseous argon. In analogy to this we use measured lifetimes and component ratios to reconstruct purity independent results. Pulse shape analysis under different particle impacts, ionisation densities and correlated quenching effects can deliver new insight into quenching effects. We presented preliminary results from data taken in liquid argon under excitation with mono-energetic neutrons, alpha particles and photons at zero electric field. We confirm values for CR monotonically rising with values for LET. Moreover changes of CR seem to be decoupled from quenching processes and happen in regions with constant specific light yield. A component ratio close to the one expected from statistically populated singlet and triplet states (1/(1+3)=0.25) is observed for recoiling electrons in the region of lowest ionisation densities (uncorrelated production). For nuclear recoils we find the component ratio rising from values around 50% to 75% in an energy range from 20 to 200 keV. For alpha particles in the MeV range we determine a relative scintillation yield $\mathcal{L}_{\text{eff}}^{\alpha}$ of 0.74±0.04 (luminescence quenching) and observe a slightly smaller CR as for recoiling nuclei. A preliminary analysis of the relative scintillation yield $\mathcal{L}_{\text{eff}}$ for nuclear recoils at energies between 16 and 120 keV shows a flat response within present errors. A mean value $<\mathcal{L}_{\text{eff}}> = 0.29 \pm 0.03$ was found. No conclusive results for energies below that region can be drawn at the present state of the analysis. In the near future we plan to upgrade the cell with PMTs of larger quantum efficiencies and to improve the cleaning system. At a later stage we plan to add an internal electric field and to extract the ionisation charge (dual phase) to determine field and energy dependences in liquid argon of both, the light and the charge yields, at working points relevant for dark matter searches.

Acknowledgments

This work is supported by grants from the Swiss National Science Foundation and the Aspera funded DARWIN project.

References

[1] Aprile E *et al.*, 2011 arXiv:1104.2549v3 [astro-ph.CO]
[2] Aprile E *et al.*, 2011 arXiv:1107.2155v1 [astro-ph.IM]
[3] Regenfus C, 2010, DOI: http://dx.doi.org/10.3204/DESY-PROC-2010-03/regenfus_christian
[4] Baudis L, 2010 arXiv:1012.4764v1 [astro-ph.IM] and http://darwin.physik.uzh.ch/
[5] Plante G *et al.*, 2011 arXiv:1104.2587v1 [nucl-ex] (and references therein)
[6] Manalaysay A, 2010 arXiv:1007.3746v1 [astro-ph.IM] (and references therein)
[7] Brunetti R *et al.*, 2005 *New Astr. Rev.* **49** 265
[8] Gastler D *et al.*, 2011 arXiv:1004.0373v2 [physics.ins-det]
[9] Lindhard J and Scharff M, 1961, *Phys. Rev.* **124** 128
[10] Lindhard J, Scharff M and Schiøtt E, 1963, *Mat. Fys. Medd. Dan. Vid. Selsk.* **33/14** 1
[11] Mei D *et al.*, 2008 *Astropart. Phys.* **30** 12
[12] Kubota S *et al.*, 1978 *Phys. Rev. B* **17** 2762
[13] Aprile E *et al.*, 2006 *Phys. Rev. Lett.* **97** 081302
[14] Manzur A *et al.*, 2010 *Phys. Rev. C* **81** 025808
[15] Hitachi A *et al.*, 1983 *Phys. Rev. B* **27/9** 5279
[16] Boccone V *et al.*, 2009 *JINST* **4** P06001
[17] Walter M, Diploma-Thesis Universität Zürich, 2011,
[18] see: http://www.mesytec.com/datasheets/MPD-4.pdf
[19] Amsler C *et al.*, 2010 *JINST* **5** P11003
[20] Amsler C *et al.*, 2008 *JINST* **3** P02001
[21] Acciarri R *et al.*, 2009 *NIM A* **607** 169
[22] Lippincott W *et al.*, 2008 *Phys. Rev. C* **78** 035801 and 2010 *Phys. Rev. C* **81** 039901(E)
[23] Hitachi A *et al.*, 1992 *Phys. Rev. B* **46** 11463
[24] Boccone V, Dissertation Universität Zürich, 2010 *CERN-THESIS*-2010-090
[25] http://www.srim.org, http://physics.nist.gov/PhysRefData/Star/Text/ASTAR.html , ... /ESTAR.html

12th International Conference on Topics in Astroparticle and Underground Physics (TAUP 2011) IOP Publishing
Journal of Physics: Conference Series **375** (2012) 012020 doi:10.1088/1742-6596/375/1/012020

Neutron background simulation for the CRESST–II experiment

S Scholl[1] **and J Jochum**

Kepler Center for Astro and Particle Physics, Eberhard–Karls–Universität Tübingen, Auf der Morgenstelle 14, 72074 Tübingen, Germany

E-mail: scholl@pit.physik.uni-tuebingen.de

Abstract. The CRESST collaboration reported an excess signal in their latest physics run. After removal of expected electromagnetic background, low energy αs and degraded lead recoils from the signal region, a mean value of 35.4 excess events remain. In this work, we investigate if neutron induced nuclear recoils can explain the reported excess. Three results are found: The total event rate in the simulation of $2.04^{+0.23}_{-0.15} \times 10^{-3}\ {}^{\mathrm{cts}}/_{\mathrm{kgd}}$ is much lower than the one inferred from the remaining excess events of $4.85 \times 10^{-2}\ {}^{\mathrm{cts}}/_{\mathrm{kgd}}$ in the experimental data. Additionally, the dominance of oxygen recoils ($\sim$ 90 %) predicted by the simulation is in disagreement with the favored experimental lightyield distribution of the excess signal. Furthermore, the experimentally observed fraction of higher multiplicities is only half of the $\sim$ 20 % obtained by the simulation. Especially the experimental absence of double detector hits in the presence of higher multiplicities is in clear contradiction to the results of the simulation. In combination, these three discrepancies between the experimental and the simulated background allows the conclusion that the observed excess signal cannot be explained by neutron induced nuclear recoils alone.

1. Introduction

The latest result of the CRESST–II experiment reports an excess signal in the signal region of the experiment [1]. In order to determine if this excess signal is the result of neutron induced background, simulations of the various neutron sources have been conducted. The results of the simulation are checked against the experimental results considering the total expected background rate, the distribution of the multiplicity, i.e. the number of detector modules hit by particles simultaneously and the distribution of the lightyield (LY) of the detector hits.

2. Setup of the simulation

The simulation are conducted in GEANT4.9.4.p02 [2; 3]. Neutron–nuclear interaction is implemented via the data driven G4NeutronHP class using the G4NDL3.14 data base providing the cross–section data. The inelastic neutron scattering process is modified to correct the reaction kinematics and to ensure that the recoiling nucleus is always generated as reaction product. A description of the setup including a sketch can be found in Ref. [1], it has been accurately included into the simulation, especially the carousel holding the 18 detector modules. Holding the coldbox with the carousel, the cryostat is located along the central axis. The outmost shield

[1] Present address: Max–Planck–Institut für Physik, Föhringer Ring 6, 80809 München, Germany

Table 1. Simulated neutron sources, giving the required neutron flux per year, the starting volume and the production mechanism.

Source	Primary n required per a	Starting Volume	Production mechanism	Contamination in ppb
Ambient	1.27×10^7	outside	s.f. + (α, n) in rock	
Lead	1790	Pb shield	s.f.	U, Th: < 0.2
Copper	76.09	Cu shield	s.f. + (α, n) on Cu	U, Th: 0.02
PE	8570	PE shield	s.f. + (α, n) on PE	U: 0.5, Th: 1.3
Cosmogenic	10087	outside	cosmogenic	

is made of 42.25 cm of polyethylene (PE) to moderate incoming neutrons. Inside this shielding a muon veto of 10 cm thickness is found enclosing the radon box which is flooded by nitrogen to avoid Rn contamination. The Rn box contains an outer electromagnetic shield made of 20 cm of Pb enclosing an inner shield of 14 cm Cu which houses the coldbox holding the detector carousel. Tab. 1 presents the simulated neutron sources, the flux and the spectrum for the ambient neutrons have been taken from the listing for dry concrete of the LNGS hall given in Ref. [4] and are in accordance with the measurement presented in Ref. [6]. The contamination of the PE and the lead has been measured at the chemical laboratory of the LNGS, resulting in an upper limit of 0.2 ppb of U and Th for lead and a contamination of 0.5 ppb U and 1.3 ppb Th in the PE. A contamination of 0.02 ppb has been assumed for the Cu shielding, this is a contamination level comparable to the measured values of electroformed Cu given in Ref. [5]. A simulation of the primary muons is not presented in this work, instead we present a simulation starting from the calculated spectrum of the muon induced neutrons in the LNGS laboratory. The flux and spectrum of the muon induced neutrons is taken from Ref. [7].

The constraints used for analysis of the experimental data are also applied to the simulated data. 8 out of 18 installed modules are used for the analysis of the signal region and 9 modules are used as veto modules only. A standard detection threshold of 5 keV is applied to each detector module, the signal region extends from an individual separation threshold as given in Tab. 2 to 40 keV. The nuclear lightyields are the same as used in the experimental analysis ($LY_O = 0.102$, $LY_{Ca} = 0.0638$ and $LY_W = 0.0391$). Only single hits in the signal region of one detector module are accepted.. The muon veto rejects events if the sum of all energy depositions in the muon veto exceed 15 MeV, removing 35 % of the events induced by cosmogenic neutrons. A successful test of the simulation has been done by comparing the spectral shape of the recoil spectrum and the detector hit multiplicities with calibration run data.

Table 2. The separation thresholds as applied to the detectors. These values have been taken from analysis of the experimental data given in Ref. [1].

Channel	Separation threshold	Channel	Separation threshold	Channel	Separation threshold	Channel	Separation threshold
05	12.3 keV	29	12.1 keV	43	15.5 keV	47	19.0 keV
20	12.9 keV	33	15.0 keV	45	16.2 keV	51	10.2 keV

Table 3. Background rates obtained for the neutron sources presented in Sec. 2. *Oxygen* events have $0.084 < LY_O < 0.162$, *calcium* events are defined by $0.052 < LY_{Ca} < 0.084$ and *tungsten* events show LY_W of less than 0.052. The error interval denotes the 90 % C.L.

Source	Background rate in $^{cts}/_{kgd}$	Oxygen in $^{cts}/_{kgd}$	Calcium in $^{cts}/_{kgd}$	Tungsten in $^{cts}/_{kgd}$
Cosmogenic	$1.25^{+0.06}_{-0.06} \times 10^{-3}$	$1.14^{+0.06}_{-0.06} \times 10^{-3}$	$9.22^{+1.78}_{-1.55} \times 10^{-5}$	$2.07^{+0.94}_{-0.70} \times 10^{-5}$
Lead shielding	$4.28^{+0.11}_{-0.10} \times 10^{-4}$	$3.93^{+0.11}_{-0.10} \times 10^{-4}$	$3.05^{+0.30}_{-0.28} \times 10^{-5}$	$4.94^{+1.33}_{-1.10} \times 10^{-6}$
Ambient	$2.50^{+2.20}_{-1.33} \times 10^{-4}$	$2.50^{+2.20}_{-1.33} \times 10^{-4}$	$0.00^{+8.23} \times 10^{-5}$	$0.00^{+8.23} \times 10^{-5}$
Cu shielding	$8.56^{+0.37}_{-0.35} \times 10^{-5}$	$7.58^{+0.35}_{-0.33} \times 10^{-5}$	$8.17^{+1.24}_{-1.15} \times 10^{-6}$	$1.71^{+0.60}_{-0.47} \times 10^{-6}$
PE shielding	$2.68^{+0.28}_{-0.26} \times 10^{-5}$	$2.51^{+0.27}_{-0.25} \times 10^{-5}$	$1.55^{+0.80}_{-0.58} \times 10^{-6}$	$1.93^{+4.15}_{-1.58} \times 10^{-7}$
Total	$2.04^{+0.23}_{-0.15} \times 10^{-3}$	$1.88^{+0.23}_{-0.14} \times 10^{-3}$	$1.32^{+0.84}_{-0.16} \times 10^{-4}$	$2.76^{+8.28}_{-0.71} \times 10^{-5}$

3. Comparison between experiment and simulation

The 67 total events in the signal region are reduced to $36.9^{+10.6}_{-10.9}$ events for the M1 solution and $33.9^{+10.1}_{-8.8}$ for the M2 solution by the likelihood fit as γ leakage, degraded αs and Pb recoils are subtracted [1]. For the mean value of 35.4 excess events and an exposure of 730 kgd, the background is $4.85 \times 10^{-2}\ ^{cts}/_{kgd}$. On top of those 35.4 single events, no double hit events but two triple hit events and one five–fold coincidence have been observed. The mean lightyield (LY) of the observed excess events for the M1 solution is 0.05 and 0.09 for the M2 solution.

The observed rate of the excess signal of $4.85 \times 10^{-2}\ ^{cts}/_{kgd}$ exceeds the flux of the neutron sources considered in Tab. 3 by a factor of 24.

Tab. 4 presents the distribution of events in the nuclear recoil bands in the signal region, which is incompatible with the observed mean LY ~ 0.05 of the M1 solution. The distribution of the M2 solution with a mean LY of ~ 0.09 cannot be distinguished from the predictions of the simulation, however.

Using the distribution for the weighted sources for 35.4 single,7.5 double, 1.78 triple events and 0.10 five–fold coincident events are expected. Tab. 5 shows the resulting multiplicity distriutions from the simulation, the absence of any double detector hit events is incompatible with the combined simulation result at a C.L. of 99.95 %.

Table 4. Lightyield distribution for perfect detector resolution given for signal region. The LY bands and the weights are as given in Tab. 3, the error interval contains the 90 % C.L.

Source	Mean lightyield	Oxygen in %	Calcium in %	Tungsten in %
Cosmogenic	0.094	$90.96^{+0.40}_{-0.42}$	$7.37^{+1.30}_{-1.16}$	$1.66^{+0.73}_{-0.55}$
Pb shielding	0.094	$91.72^{+0.20}_{-0.20}$	$7.13^{+0.66}_{-0.62}$	$1.15^{+0.31}_{-0.25}$
Ambient	0.097	$100_{-24.73}$	$0.00^{+24.73}$	$0.00^{+24.73}$
Cu shielding	0.093	$88.47^{+0.45}_{-0.47}$	$9.54^{+1.24}_{-1.14}$	$2.00^{+0.68}_{-0.54}$
PE shielding	0.096	$93.52^{+0.59}_{-0.67}$	$5.76^{+2.73}_{-2.06}$	$0.72^{+1.51}_{-0.59}$
Weighted sources	0.094	92.16	6.49	1.35

Table 5. Multiplicity distribution of events in the signal region as detailed in Sec. 2. Events induced by cosmogenic neutrons reach much higher multiplicities than shown. The error interval denotes the 90 % C.L.

Source	Multiplicity in % of all events				
	1	2	3	4	5
Cosmogenic	$72.43^{+0.94}_{-0.97}$	$20.54^{+1.50}_{-1.45}$	$5.52^{+0.98}_{-0.87}$	$1.02^{+0.50}_{-0.37}$	$0.36^{+0.34}_{-0.20}$
Lead shielding	$81.89^{+0.37}_{-0.38}$	$15.80^{+0.78}_{-0.76}$	$2.00^{+0.35}_{-0.31}$	$0.29^{+0.15}_{-0.11}$	$0.02^{+0.07}_{-0.02}$
Ambient	$100_{-24.73}$	$0.00^{+24.73}$	$0.00^{+24.73}$	$0.00^{+24.73}$	$0.00^{+24.73}$
Copper shielding	$82.29^{+0.60}_{-0.62}$	$15.00^{+1.30}_{-1.24}$	$2.33^{+0.60}_{-0.54}$	$0.32^{+0.31}_{-0.18}$	$0.05^{+0.20}_{-0.05}$
PE shielding	$82.99^{+1.36}_{-1.48}$	$14.02^{+3.20}_{-2.83}$	$2.69^{+1.91}_{-1.27}$	$0.30^{+1.10}_{-0.28}$	$0.00^{+0.68}$
Weighted sources	78.35	16.71	3.94	0.70	0.22

4. Conclusion

The observed excess signal of the latest CRESST–II run 32 cannot be explained by neutron induced background since the simulated background falls short by a factor of 24. This discrepancy alone is not a decisive argument, as a point like neutron source close to the detectors could evade detection while rising the background to the observed level. The simple comparison of the mean lightyield is in disagreement with the M1 solution presented in Ref. [1] while being compatible with the M2 solution. The most striking argument against a neutron induced excess signal is the observed absence of double detector hits while observing higher multiplicities which cannot be explained by any conceivable neutron source. In combination, these three results exclude neutrons as the origin of the observed excess signal.

Acknowledgments

This work was supported by funds of the SFB Transregio 27 (Neutrinos and Beyond) of the Deutsche Forschungsgemeinschaft DFG , the Europäisches Graduiertenkolleg Basel, Graz und Tübingen (Hadronen im Vakuum, in Kernen und in Sternen) sponsored by DFG, Fonds zur Förderung der wissenschaftlichen Forschung and Schweizer Nationalfonds and the Network on Direct Dark Matter Detection N3 of the EU FP6 project ILIAS.

References

[1] Angloher G *et al.* 2012 *Eur Phys J C* **72** 1 –22

[2] Agostinelli S *et al.* 2003 *Nucl Instrum Meth A* **506** 250–303

[3] Allison J *et al.* 2006 *IEEE T Nucl Sci* **53** 270–278

[4] Wulandari H, Jochum J, Rau W and von Feilitzsch F 2004 *Astropart Phys* **22** 313–322

[5] ILIAS Database on radiopurity of materials
URL `http://radiopurity.in2p3.fr/search.php?Material=copper`

[6] Belli P *et al.* 1989 *Il Nuovo Cimento A* **101** 959–966

[7] H. Wulandari *et al.* 2004 IAU Symposium **220**, Dark Matter in Galaxies, Eds: S. D. Ryder, D. J. Pisano, M. A. Walker, and K. C. Freeman. San Francisco: Astronomical Society of the Pacific 491– 495

Journal of Physics: Conference Series **375** (2012) 012021 doi:10.1088/1742-6596/375/1/012021

Background discrimination in new iZIP detectors at SuperCDMS

Silvia Scorza for the SuperCDMS Collaboration

Department of Physics, Southern Methodist University, Dallas, TX 75275-0175

E-mail: sscorza@mail.smu.edu

Abstract. SuperCDMS is a direct dark matter search experiment in the Soudan Underground Laboratory (Minnesota, USA). Cryogenic germanium detectors are used to identify rare nuclear recoils induced by elastic scattering of WIMPs from our Galactic halo. The detectors measure phonon and ionization signals simultaneously, allowing an event-by-event discrimination between the electronic recoils, tracers of electromagnetic background, and nuclear recoils originated by neutrons and WIMPs. To further increase the background discrimination power, novel detectors with a special interleaved electrode scheme have been tested and installed. I will present the background rejection performances achieved, giving the resulting improvement in the sensitivity of the experiment for spin-independent WIMPs interactions.

1. Introduction

First indications for the existence of dark matter were already found in the 1930s [1]. By now there is strong evidence [2] to believe that a large fraction (more than 80%) of all matter in the Universe is dark (interacts very weakly with electromagnetic radiation, if at all - no photon coupling) and that this dark matter is predominantly non-baryonic. Weakly Interacting Massive Particles (WIMPs) are one of the leading candidates for dark matter. WIMPs are stable particles which arise in several extensions of the Standard Model of electroweak interactions [3]. Typically they are presumed to have masses between few tens and few hundreds of GeV/c^2 and a scattering cross section with a nucleon below 10^{-6} pb.

The SuperCDMS experiment (Cryogenic Dark Matter Search), situated in the Underground Laboratory in the Soudan Mine (Minnesota), is dedicated to the direct detection of WIMPs. The direct detection principle consists in the measurement of the energy released by nuclear recoils produced in an ordinary matter target by the elastic collision of a WIMP from the Galactic halo. The main challenge is the expected extremely low event rate (<1 evt/kg/year) due to the very small interaction cross section of WIMPs with ordinary matter. Another constraint is the relatively small deposited energy (<100 keV). In order to measure low energy recoils, SuperCDMS employs cryogenic detectors (high purity Ge crystal) with simultaneous measurements of phonon and ionization signal. The ionization signal, corresponding to the collection on electrodes of electron-hole pairs created by the energy loss process, depends on the particle type whereas the phonon signal reflects the total energy deposit.

SuperCDMS aims to reach a sensitivity in the WIMP-nucleus interaction detection better than 0.003 counts/kg·d for recoil energy above 10 keV. To reach this goal, background rejection and discrimination are necessary. SuperCDMS backgrounds includes gamma particles, beta

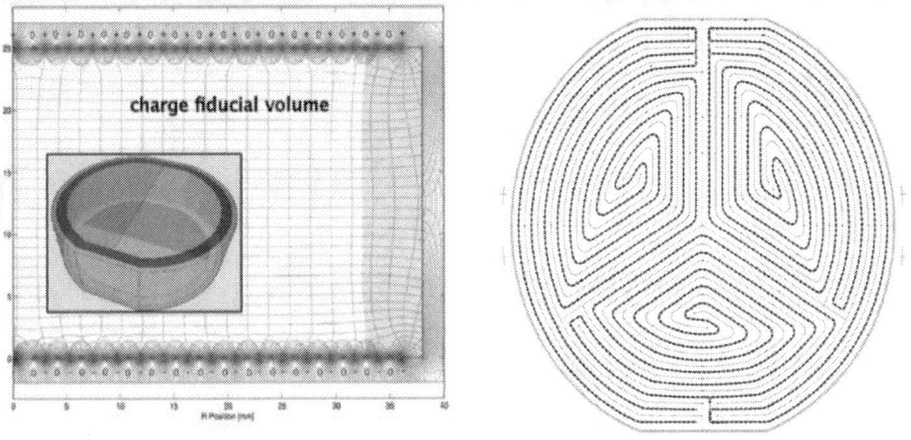

Figure 1. Left panel: Finite Element Model (FEM) calculation (azimuth symmetry axis at R=0) of electric field lines and equipotential surfaces for the case of +2 V (-2 V) applied to top (bottom) electrodes of Ge 1-inch thick iZIP The charge fiducial volume (non-shadowed area) is defined by the inner charge electrodes. Inset: A schematic of new iZIP phonon channels geometry which has two inner phonon channels and one outer phonon channel on each surface. Right panel: The top side layout interleaved charge and phonon mask design where the outer two charge (phonon) rings form the charge (phonon) guard ring.

particles and neutrons from cosmic rays and natural radioactivity. Thanks to simultaneous measurements of charge and phonon signals iZIP detectors can discriminate gammas from WIMPs at high efficiency, because WIMPs interact with nuclei and gammas with electrons.

The main limiting background of the experiment comes from interactions occurring just underneath the collecting electrodes: essentially low energy β-rays due to ^{210}Pb contamination of the detector surface and/or in the vicinity of the detectors. The incomplete charge collection of these events can mimic nuclear recoils.

2. iZIP detector design

The iZIP detector consists of a 3-inch diameter x 1-inch thick Ge substrate (7.6 x 10^{10} cm^{-3} purity). The electron-hole pairs collection is provided by four charge channels (inner and outer on both sides). The phonon measurement is provided by eight photolithographically channels. There are four channels on each side of the detector: three inner phonon channels and one outer phonon ring channel for a better radius determination. The bottom channels are rotated 60 degrees from the top channels. The high impedance charge signal electrodes are interleaved with low impedance phonon sensors on both sides (right panel of Fig. 1)[4]. The electrode formed by the phonon sensors is at ground and not instrumented for ionization.

With the iZIP detector technology, surface events are tagged by the presence of charge on only one side charge electrode of the detector. For events occurring in the bulk of the crystal, the iZIP measures the number of electrons that travel across the detector (left panel of Fig. 1) to one charge electrode while at the same time measuring the number of holes traveling in the other direction to the electrode on the opposite side of the crystal. The two charge signals are therefore symmetric.

3. Background Power Rejection

The results here shown have been obtained exposing an iZIP detector to an internal ^{109}Cd ($\sim$100 Hz) source, and external ^{252}Cf neutron and ^{133}Ba gamma sources at a surface detector test facility.

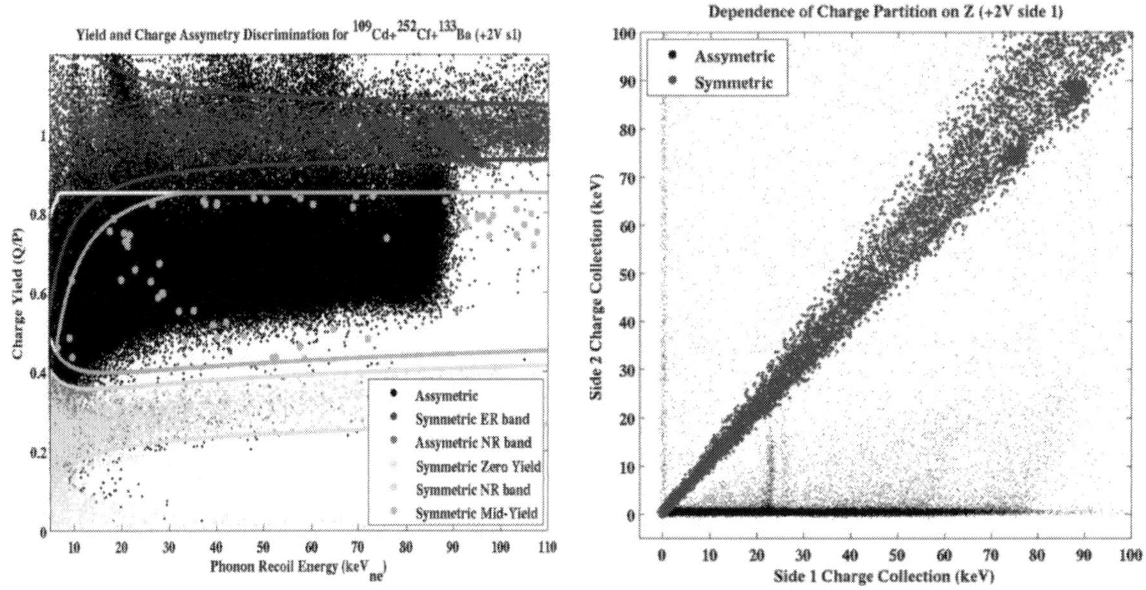

Figure 2. Left panel: Projection in the (Y, E) plane of the events recorded during neutron and gamma calibrations. The thick lines represent the 90% nuclear (in green) and electronic (in blue) recoil zones. Light orange lines represent the mid-yield region where inelastic neutron scatters have been recorded. Black events are excluded by the symmetric cut. They represent β-rays due to ^{109}Cd source. Right panel: Charge signals on the source side electrode (x axis) versus opposite side electrode (y axis). Source side electrode at +2V (collects electrons) is shown. Events which pass the symmetric charge cut are in blue.

Left panel of Fig. 2 shows the ionization yield (ratio of ionization signal to phonon signal normalized to electron recoil events from the gamma source) as a function energy of events recorded for calibration performed with neutron and gamma sources. Surface samples are contaminated with cosmogenic neutron in the nuclear recoil band ($\sim$7 evt/h). The electron signal from a Ba calibration run can be found in the right panel of Fig. 2. It shows the ability to reject surface events from bulk ones just performing a cut based on charge signal: "symmetric cut" means that the charge signal is equally shared on both sides of the detector. Three separate populations: bulk electron recoil, bulk nuclear recoil (taking advantage of different yield value between electronic and nuclear recoil) and surface electron recoil were defined using the the charge symmetric cut.

Direct discrimination studies on surface events which have leaked into the nuclear recoil are disturbed by the high neutron background due to cosmogenic. However, we found that low yield surface events could be rejected with a misidentification rate of $< 1\times10^{-4}$ (90%C.L.) by the symmetric charge cut.

As with the standard CDMS detectors described in detail in [5], the phonon measurement provides z-position information for each event via the timing difference between rising edges of the phonon pulses. Phonon timing rejection capabilities of the iZIP detector have been tested

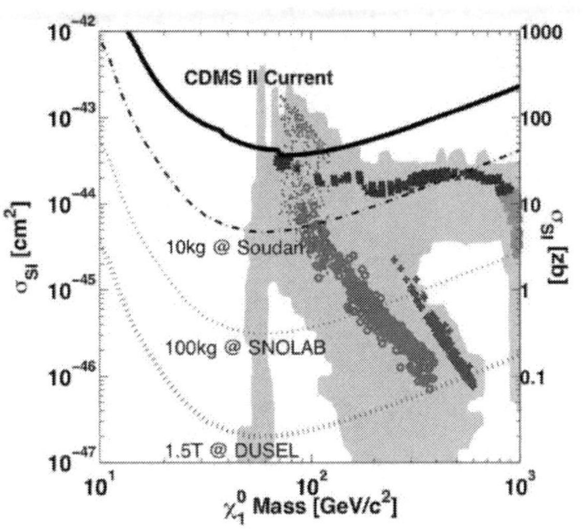

Figure 3. Limits on the cross-section for spin-independent scattering of WIMPs on nucleons as a function of WIMP mass from CDMS published data and SuperCDMS perspectives.

and demonstrated to be preserved from the standard ZIP detector [6].

Finally, by combining charge symmetry yield and phonon timing rejection, the iZIP design easily meets the discrimination requirements for the next stage of SuperCDMS experiment (expected sensitivity shown in Fig. 3).

4. Conclusion and Perspectives

The result sshown at surface test facility demonstrate the very high rejection capabilities of these robust detectors due to the improved electrode scheme for charge signal and the new layout phonon sensors. Provided the surface event rejection is efficient enough (by both charge and phonon signal cuts), the combination of fiducial mass to be installed and neutron screening/rejection should enable to reach sensitivities at the level ~few 10^{44}cm^2 in the next year (see Fig.3).

References

[1] Zwicky F 1933 *Helv. Phys. Acta* **6** 110–127
[2] Bertone G, Hooper D and Silk J 2005 *Phys. Rept.* **405** 279–390 (*Preprint* hep-ph/0404175)
[3] Hellis J, Hagelin J, Nanopoulos D, Olive K and Srednicki M 1984 *Nucl. Phys.* **B238** 453
[4] Brink P *et al.* 2006 *Nucl. Instr. and Meth. A* **559** 414–416
[5] Akerib D *et al.* 2005 *Phys. Rev.* **D72**
[6] Pyle M *et al.* 2009 *LTD13 Conference Proceedings* **1185**

Journal of Physics: Conference Series **375** (2012) 012022 doi:10.1088/1742-6596/375/1/012022

The XMASS 800kg detector

Jing Liu, for the XMASS collaboration
Kamioka Satellite, Institute for the Physics and Mathematics of the Universe, University of Tokyo, 456 Higashi Mozumi, Kamioka-cho, Hida-shi,Gifu 506-1205 Japan
E-mail: jing.liu@ipmu.jp

Abstract.
The XMASS 800kg detector, aiming primarily at the dark matter search, is currently under commissioning at the Kamioka Underground Observatory, Japan. The construction and performance of the detector are briefly reviewed.

1. Introduction

The XMASS experiment [1] is proposed to detect low energy solar neutrinos, neutrinoless double beta decay and dark matter particles using liquid xenon (LXe) scintillator. The scintillation lights generated by nuclear or electronic recoils in LXe are detected by surrounding photomultiplier tubes (PMTs). No electric field is applied to extract ionization signals from LXe. This makes it possible to realize a large scale experiment sensitive enough to the listed rare physic processes. Current XMASS detector contains 856 kg LXe in its active volume, hence named as *XMASS 800kg detector*. It is dedicated to dark matter search due to its limited size.

The experiment is located at the Kamioka Underground Observatory, Japan. The construction of experimental hall, 15 m wide, 15 m high and 21 m long, was completed in August 2008. The muon flux in the hall is down to $6 \times 10^{-8}/\mathrm{cm}^2/\mathrm{s}/\mathrm{sr}$, thanks to a mean overburden of 2700 m.w.e. A cylindric water tank, 10 m in height, 10 m in diameter, was constructed in March 2009. It shields the inner detector from external neutron and γ radiation. Its inner wall is equipped with 72 20-inch PMTs to detect Čerenkov light from cosmic ray muons. The water is constantly circulated and purified in order to keep the Rn concentration lower than 1 $\mathrm{mBq/m^3}$. The construction of inner detector was completed in October 2010. The performance of the detector and analysis methods are evaluated using commissioning data.

2. Inner detector

The inner detector is located in the center of the water tank. It contains 1080 kg of LXe in total, out of which, 856 kg is surrounded by 642 PMTs. The PMTs are supported by a copper frame shaped as a pentakis-dodecahedron with a diameter of about 1 meter, as shown in Fig.1. The photo-cathodes of PMTs cover 62% of the inner surface of the detector.

3. Calibration system

Low energy γ sources can be inserted into the inner detector using a calibration system as shown in Fig. 2. A gate valve is used to isolate the inner detector when a source is mounted to the bottom end of a OFHC copper rod. The top end of the rod is connected to a step motor by

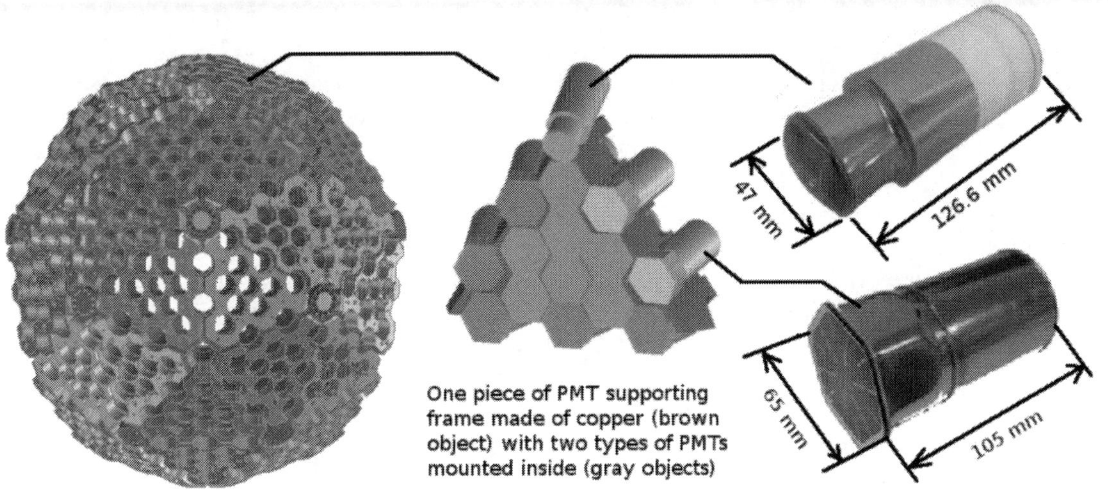

Figure 1. Structure of XMASS 800kg inner detector.

thin stainless steel wire. The z position of the source is controlled by the motor at < 1 mm accuracy. The gate valve is open after the top part of the system is vacuumed. The top PMT can be lifted up and rotated away, allowing the source to be lowered into LXe.

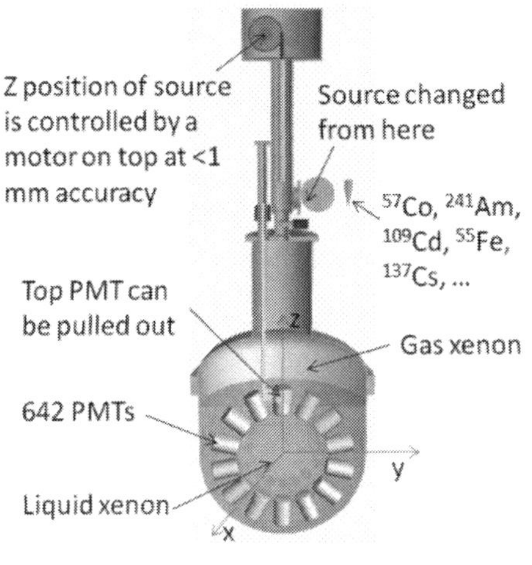

Figure 2. The low energy γ calibration system. The top PMT can be lifted up and rotated away, allowing a low energy γ source to be lowered into the active volume of the inner detector. The source can be fixed at different z positions. The precision is within 1 mm.

There is also an external γ calibration system. It consists of a "U" shaped soft hose mounted outside of the inner detector. A γ source can be moved around the detector through the hose. This makes it possible to study the detector response to external radiation. An external neutron calibration system is proposed. It will consist of a thin vertical pipe in the water tank, working as a collimator, and a plastic scintillator, working as a coincident detector.

4. Vertex reconstruction

The position and energy of an interaction point can be reconstructed using the number of photoelectrons (nPE) observed by each PMT. Assuming a position of the vertex, the expected nPE distribution in each PMT can be obtained by Geant4 [3] simulation. This distribution is normalized so that it can be used as a probability density function. The probability, p_i, of PMT i recording nPE can be calculated afterwards. The likelihood of the vertex in the assumed position $\vec{x}$ is the product of all p_i, $\mathcal{L}(\vec{x}) = \prod_{i=0}^{641} p_i(n)$. The most likely position is obtained by maximizing $\mathcal{L}$.

The performance of the vertex reconstruction program is evaluated using a ^{57}Co source fixed at $z = -40, -30, ..., 40$ cm in the active volume of the inner detector. The results are shown in Fig. 3. The left plot shows the distribution of reconstructed vertices on the z-y plain. The top right plot shows the same distribution along z axis. The achieved position resolution in terms of RMS is 1.4 cm at $z = 0$ cm and 1.0 cm at $z = 20$ cm for the 122 keV γ-rays from ^{57}Co. The bottom right plot shows the energy spectrum of the ^{57}Co source at the center of the detector (z = 0 cm). The light yield calculated from the 122 keV peak is 15.1 ± 1.2 PE/keV. The energy resolution in terms of RMS is 4% at 122 keV. Both position and energy distributions can be well produced by Geant4 simulation as shown by the blue histograms overlaid on the right plots.

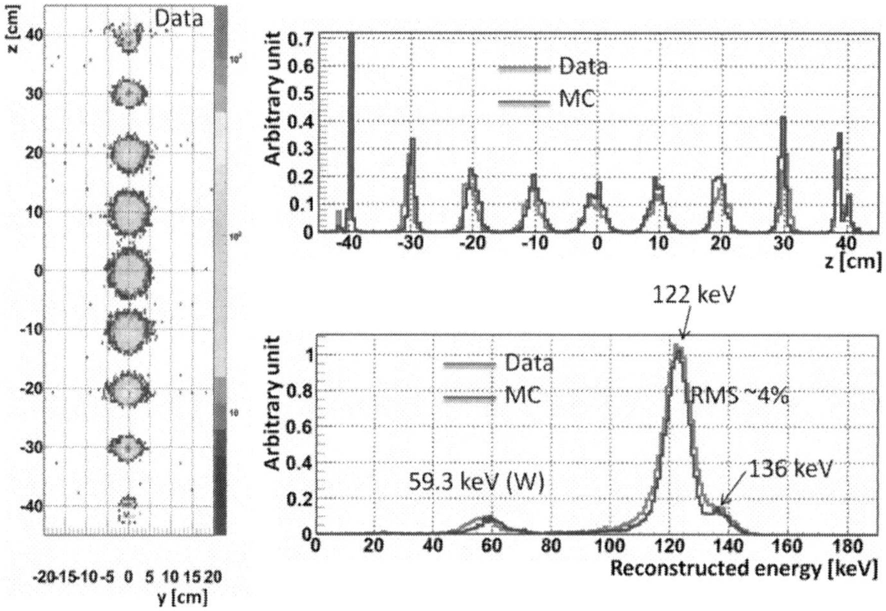

Figure 3. Results of the vertex reconstruction. The left and top right plots show the position distribution of reconstructed vertices of events collected with a ^{57}Co source at different z positions. The bottom right plot shows the energy distribution of the same source in the center of the detector.

5. Evaluation of background

The external background is expected to be dominated by γ-rays from PMTs. The average radioactivity in each PMT is measured to be 0.7 mBq for uranium chain, 1.5 mBq for thorium chain, 2.9 mBq for ^{60}Co and less than 5.1 mBq for ^{40}K. Simulation shows that the background events induced by the PMTs are less than 10^{-4} count/day/kg/keV in the inner most volume with a radius of 20 cm.

Possible internal background includes ^{222}Rn, ^{220}Rn and ^{85}Kr. Krypton can be distilled out of LXe due to its lower boiling point. This was done using a distillation tower purification system developed locally. [2] The remaining krypton contamination is measured to be less than 2.7 ppt at 90% confidence level with a gas chromatography followed by an atmospheric pressure ionization mass spectrometer. In order to remove radon, gaseous xenon was purified by SAES getters with a flow rate of 30 L/min before being filled into the inner detector. Xenon circulation systems in both gaseous and liquid phases for removing radon by filters are under preparation. The contamination of ^{222}Rn can be identified by the time coincidence between two consecutive decays of its daughter nuclei: 1. ^{214}Bi β-decays into ^{214}Po, and 2. ^{214}Po α-decays into ^{210}Pb with a half life of 164 μs. The measured value is 8.2 ± 0.5 mBq in total. Similarly, the contamination of ^{220}Rn can be identified by the time coincidence between ^{220}Rn α-decay into ^{216}Po and ^{216}Po α-decay into ^{212}Pb with a half life of 140 ms. The measured value in total is less than 0.28 mBq at 90% confidence level.

6. Sensitivity

The right plot in Fig. 4 shows the expected energy spectrum assuming 1 year of exposure, flat background at a level of 10^{-4} count/day/kg/keV, WIMP-nucleon spin independent cross section of 10^{-44} cm^2, 50 GeV WIMP and the relative scintillation efficiency of 0.2. The sensitivity of WIMP-nucleon cross section as a function of WIMP mass is shown in the left plot.

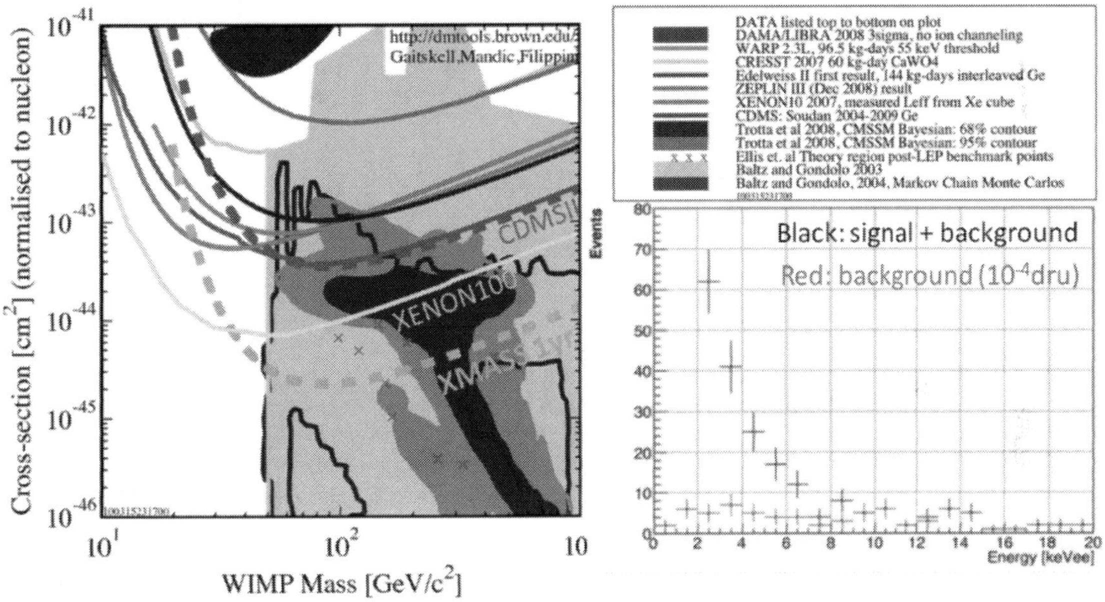

Figure 4. Left: Expected sensitivity of WIMP-nucleon spin independent cross section as a function of WIMP mass. Right: Expected energy spectrum after 1 year of measurement.

7. Conclusion

The construction of XMASS 800 kg detector was completed in October 2010 and commissioning runs are ongoing. The measured light yield is 15.1 ± 1.2 photoelectron/keV. The expected external and internal background is confirmed to be under control. The position resolution of the vertex reconstruction program is about 1.0 cm for the 122 keV γ-rays from ^{57}Co.

References

[1] Y. Suzuki *et al.*, hep-ph/0008296.
[2] K. Abe *et al.*, Astropart. Phys. **31** (2009) 290.
[3] S. Agostinelli *et al.*, Nucl. Instr. and Meth. A **506** (2003) 250.

Dark matter search with PICASSO

V.Zacek[1], S. Archambault[1], E. Behnke[2], J. Behnke[2], M. Das[3], A. Davour[4], F. Debris[1], N. Dhungana[5], J. Farine[5], S. Gagnebin[6], H. Hinnefeld[2], C.M. Jackson[1], A. Kamaha[4], C. Krauss[6], M. Lafrenière[1], M. Laurin[1], I. Lawson[7], C. Lévy[4], L. Lessard[1], I. Levine[2], D. Marlisov[6], J.-P.Martin[1], S. Kumaratunga[1], R. MacDonald[6], P. Mitra[6], P. Nadeau[5], A. Noble[4], M.-C. Piro[1], A. Plante[1], R. Podviyaniuk[5], S. Pospisil[8], S. Seth[3], O. Scallon[1], N. Starinski[1], I. Stekl[8], N. Vander Werf[2], U. Wichoski[5], T. Xie[4]

[1]Université de Montréal, Montréal, Canada, [2]Indiana University South Bend, South Bend, USA, [3]Saha Institute of Nuclear Physics, Kolkata, India, [4]Queens University, Kingston, Canada, [5]Laurentian University, Sudbury, Canada, [6]University of Alberta, Edmonton, Canada, [7]SNOLAB, Lively, Canada, [8]Czech Technical University in Prague, Prague, Czech Republic

E-mail: zacekv@lps.umontreal.ca

Abstract. PICASSO at SNOLAB searches primarily for spin-dependent WIMP interactions on ^{19}F using the superheated droplet technique. This technique is based on the bubble chamber principle, where phase transitions in superheated liquid droplets can be triggered by WIMP induced nuclear recoils. The physics of the detection process allows a highly efficient suppression of backgrounds from cosmic muons, γ- and β-rays. We will discuss qualitatively recent progress in PICASSO and its sensitivity reach for spin-dependent and spin-independent WIMP searches.

1.Introduction

PICASSO (Project In Canada to Search for Super symmetric Objects) is an experiment searching for cold dark matter through the direct detection of WIMPs primarily via their *spin-dependent* interactions with nuclei. It uses a superheated liquid fluorocarbon, C_4F_{10}, as the target material and searches for WIMP interactions on ^{19}F using a variant of the classic bubble chamber technique. ^{19}F is, next to free protons and ^3He, the most favorable nucleus for direct detection of spin-dependent interactions. The nuclear form factor of ^{19}F enhances the signal by nearly an order of magnitude compared to other frequently used targets (Na, Si, Al, Cl). Also, the expectation values for the nucleon spin can be predicted with good precision in ^{19}F. Moreover, the use of the light target nucleus ^{19}F together with a low detection threshold of $\approx$ 1 keV for recoil nuclei, renders PICASSO particularly sensitive to low mass WIMPs(below 20 GeV/c^2). This feature gives PICASSO some sensitivity in the region of the spin-independent sector where recently the CRESST, DAMA/LIBRA and CoGeNT collaborations reported an excess of events and annual count rate modulations, respectively [1,2,3].

2. The superheated droplet technique

The detector medium in PICASSO is a polymerized water saturated emulsion of C_4F_{10} droplets about 200 μm in diameter. Since C_4F_{10} has a boiling temperature T_b = -1.7 ^0C at a pressure of 1.013 bar these droplets are in a moderately metastable superheated state at ambient pressure and temperature. A heat spike created by the energy deposition of a charged particle traversing a liquid droplet triggers a phase transition if the energy deposition occurs within a certain critical length (of the order of hundreds of nm) and the deposited energy exceeds a certain critical energy (in the keV range). Details of the detector principle are explained in [4]. The phase transition is of explosive nature, such that each bubble nucleation is accompanied by an acoustic signal in the audible and ultrasonic region, which is recorded by piezoelectric transducers.

Since this type of detector records phase transitions, it performs as a threshold device and the energy threshold can be controlled by setting the temperature and/or the pressure. The dependence of the threshold energy on temperature was studied with mono-energetic neutron beams and the resulting calibration curve, which is shown in Fig. 1 (left) allows a precise description of recoil threshold energies ranging from 0.8 keV up to 800 keV [4].

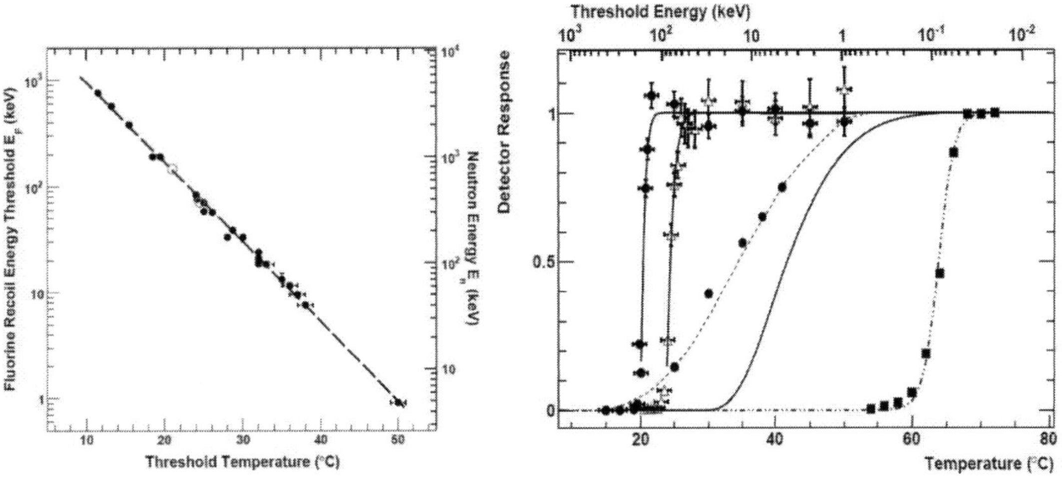

Figure 1. Left: Calibration curve for the energy threshold for ^{19}F recoils as a function of temperature obtained from measurements with mono-energetic neutrons; α- data are shown as open (red) circles. Right: Response to different kinds of particles in superheated C_4F_{10}. From left to right: ^{210}Pb recoil nuclei from ^{226}Ra spikes; α-particles at Bragg peak from ^{241}Am decays; poly-energetic neutrons from an AcBe source (dotted); ^{19}F recoils modeled assuming the scattering of a 50 GeV/c^2 WIMP; response to 1.75 MeV γ-rays and MIP's (dot-dashed).

3. Detector response to different particle species

Since each temperature at a given constant pressure corresponds to a defined recoil energy threshold, the spectrum of the particle induced energy depositions can be reconstructed by varying the threshold temperature. A summary is shown in Fig.1 (right). WIMP induced recoil energies of ^{19}F nuclei are expected to be smaller than 100 keV and therefore become detectable above 30° C. Particles which produce only low ionization densities, such as cosmic ray muons, γ- and β-rays, become detectable when they create sub-keV energy clusters within sub-nm sized regions and only become observable above 50^0 C. These particles are well separated from strongly ionizing neutron- or WIMP induced recoils, which allows an efficient suppression of such backgrounds at the level of 10^{-8} to 10^{-10}.

α-emitters produce a different response. In Fig. 1 (right) the α-curve with the lower threshold temperature (i.e. higher threshold energy) was obtained with ^{226}Ra spiked detectors. In this case the ^{226}Ra daughter, ^{222}Rn, diffuses into the droplets and the ^{210}Pb nucleus with the highest recoil energy in the decay chain (E_{rec} = 146 keV) defines the threshold (upper red circle in Fig.1 (left)). The α-threshold at higher temperature was recorded after spiking the inactive detector matrix with ^{241}Am such that only alpha particles entering the droplets from the bulk can induce nucleation. This threshold corresponds to a deposited energy of E_{dep}= 66 keV and only alpha particles with energy depositions at the Bragg peak are able to trigger nucleation (lower red circle in Fig. 1(left)).

With increasing temperature, in both cases the liquid becomes sensitive at threshold energies E_{th} < 66 keV to smaller dE/dx on the alpha track, but since the detector is already fully sensitive, the response levels off into a flat plateau. A more detailed discussion can be found in [4]. With the detector fully sensitive to alpha particles over the entire range of WIMP sensitivity, α- particles are the most important background for this kind of detector in dark matter searches. However the shapes of the WIMP response (essentially exponentially falling) and of the alpha response (constant) differ substantially, such that they can be separated by fitting the two contributions.

4. PICASSO at SNOLAB

The present PICASSO installation at SNOLAB accommodates 32 detector modules. A group of four detectors is installed in each of the 8 thermally and acoustically insulated boxes, serving as temperature control units, where the temperature can be controlled with a precision of ±0.1^{0}C in the range from 20^0 to 50^0 C. The entire installation is surrounded by a 50 cm thick water shield which serves as a neutron moderator and absorber; a dedicated ^{3}He- counter measurement established that the fast neutron flux from the surrounding rock is decreased by a factor 400.

The current detector generation consists of cylindrical modules of 14 cm diameter and 40 cm height as shown in Fig.2 (left). The containers are fabricated from acrylic and are closed on top by stainless steel lids sealed with polyurethane O-rings. Each detector is filled with 4.5 litres of polymerized emulsion loaded with 200 μm diameter droplets of C_4F_{10}. In the most recent detector generation the emulsion has glycerine and polyethylene glycol as its main ingredients. The best detectors have an α-background at the level of 20 cts/kg/d (kg of ^{19}F). The active mass of each detector is typically around 85 g of C_4F_{10} corresponding to 65g of ^{19}F. Each detector is read out by nine piezo-electric transducers. Details on detector operation and read-out are summarized in [5].

Figure 2. Left: A 4.5 L acrylic module with an active mass of 80g. Each module is equipped with 9 piezo-electric sensors for event localization. Right: Super structure and ultrapure water shielding of the 2.6 kg experiment, at its new SNOLAB location (2010). The set up accommodates 32 detectors installed in 8 temperature control units.

5. Data taking and analysis

Data have been taken with this set-up continuously from November 2008, with interspersed periods of maintenance, calibrations and a period of moving the experiment to a new location at SNOLAB in fall 2010. Roughly every three months neutron calibration data have been taken at several temperatures with a weak AmBe source (69 s^{-1}) to monitor the stability of the detectors and determine cut-efficiencies for discrimination variables. The count rates of all detectors show a flat plateau in the range from 1 to 60 keV 25^0 to 48 ^{0}C (or between 60 down to 1 keV), consistent with the presence of a pure alpha background (Fig.1(right)). An upper limit on the cross-section on ^{19}F is obtained by fitting the WIMP response curve and the flat alpha background. By fall 2011 the still ongoing analysis reached a sensitivity at the level of a few cts/kg/d. Given this sensitivity and the low detection threshold of ≈ 1 keV for the light ^{19}F recoil nuclei, PICASSO is particularly sensitive to low mass WIMPs below 15 GeV/c. This is interesting since it is in this mass region where the DAMA/LIBRA channeling results remained unchecked in the spin-dependent sector. Moreover for spin-independent interactions PICASSO is approaching the sensitivity to challenge or confirm the exciting recent claims of a positive effect by the CRESST, DAMA/LIBRA and CoGeNT collaborations [1,2,3].

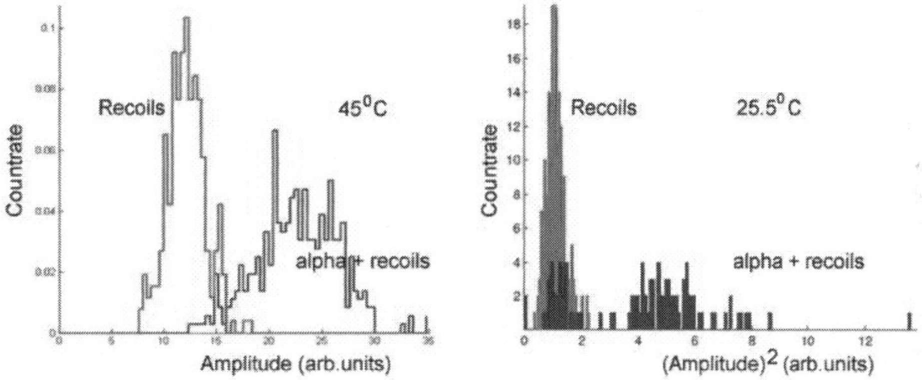

Figure 3. α- recoil discrimination. Left: α-background and exposure to n-source applying localisation and gain corrections. Right: same with improved electronics. At 25.5°C some α-decays trigger with nuclear recoils only and others with recoils and α-particles combined.

6. Next: event by event α-recoil discrimination

In previous studies the PICASSO collaboration showed that the acoustic signals contain information about the nature of the primary event: it was observed that the acoustic signals produced by α- emitters are a factor four more intense than signals of neutron or WIMP induced events [6]. This result has been confirmed by COUPP [7] and a theoretical model proposed by us is discussed in [4]. An important next step in PICASSO is the full implementation of this discrimination technique. Fig. 3 (left) shows results obtained with n-calibration and WIMP data and 400 kHz sampling. Events were reconstructed with a localization algorithm, which allowed for gain and solid angle corrections. The obtained amplitude resolution is 30% FWHM and for a nuclear recoil acceptance of 80% an alpha particle rejection of 99.34% could be established. Fig.3 (right) displays the same two categories of events but recorded with modified preamplifiers, adjusted filtering and 800 kHz sampling. Assuming the present exposure and an α-reduction by a factor 100 we expect an improvement of sensitivity to better than $\approx 3 \times 10^{-3}$ pb in the SD sector and $\approx 7 \times 10^{-6}$ pb in the SI sector.

References

[1] G. Angloher et al., arXiv:1109.0702
[2] R. Bernabei et al., Eur. Phys.J.C67 (2010) 39
[3] C.E. Aalseth et al., Phys. Lett.106 (2011) 131301
[4] S. Archambault et al., New J. Of Physics 13 (2011) 043006
[5] S. Archambault et al., Phys. Lett. B 682 (2009) 185
[6] F. Aubin et al., New Journal of Physics 10 (2008) 103017
[7] E. Behnke et al., arXiv:1008.3518v2; W.H. Lippincott, these proceedings

Journal of Physics: Conference Series **375** (2012) 012024 doi:10.1088/1742-6596/375/1/012024

EDELWEISS detectors: from R&D to dark matter search

Claudia Nones on behalf of the EDELWEISS collaboration
Service de Physique des Particules - CEA/DSM/IRFU/SPP, 91191 Gif-sur-Yvette, France
E-mail: claudia.nones@cea.fr

Abstract. Cryogenic Ge detectors used within the EDELWEISS experiment are subject to constant improvement with respect to the rejection capabilities against non WIMP interactions. These are driven by the performances obtained at a given step. A summary of the evolution of the "InterDigit" detectors with interleaved electrodes will be given with new possible solutions to improve the heat channel, with the ultimate goal of defining a building piece for the EURECA detector array.

1. Introduction

One of the key ingredients for the current cosmological concordance model is the existence of a dark, matter-like fluid ruling the dynamics of structures from the current galactic scales to the largest scales at early cosmic times. Weakly Interacting Massive Particles (WIMPs) are a generic class of dark matter particles with particularly appealing features. They appear in several extensions of the current Standard Model of particle physics, where thermal production mechanisms for such particles in the Big Bang naturally yield the order of magnitude for the observed cosmic abundance [1]. A vast effort is currently dedicated to the direct detection of WIMPs from the Milky Way halo through the coherent elastic scattering on nuclei constituting a terrestrial detector. A roughly exponential nuclear recoil spectrum is expected with typical energies of a few tens of keV. Theoretical models predict a wide range of WIMP-nucleon scattering cross sections. Current searches are approaching sensitivities for rates of a few 10^{-3} evts/kg/day, corresponding to cross sections for spin-independent interactions at the level of a few 10^{-8} pb. Since we are dealing with very rare event searches, the dedicated detectors must therefore achieve a low energy threshold combined with high background rejection capabilities.

2. The EDELWEISS experiment

EDELWEISS is a direct WIMP search experiment, where nuclear recoils induced by collisions with WIMPs from the galactic halo are detected using germanium detectors working at very low temperatures with the simultaneous measurement of ionization and phonon signals [2]. The comparison of the two signals makes possible to separate on an event-by-event basis the nuclear recoils from the electron recoils induced by β and γ radioactivity that constitute the major source of background in most present-day direct WIMP searches. The detectors are operated at the Underground Laboratory of Modane in the Frejus Tunnel under the French-Italian Alps. The 1780 rock overburden (4800 m water equivalent) results in a muon flux of about 4 $\mu/m^2/day$ in the experimental hall and the neutron flux above 1 MeV is $10^{-6} n/cm^2/s$. Each detector is a cylindrical absorber made of a HPGe single crystal equipped with sputtered Al electrodes. The thermal sensor is a Neutron Transmutation Doped Ge (NTD) thermistor glued on the absorber with Araldite epoxy. In the following sections we will review the different phases of this experiment from the detector point of view, underlining all the progresses done in terms of sensitivity, rejection capability and fiducial mass.

2.1. EDELWEISS-I: the starting point

The first phase of the experiment, EDELWEISS-I [3], has been characterized by three heat and ionization Ge bolometers (320 g each - 70 mm in diameter and 20 mm in height with edges beveled at an angle of 45°) operating simultaneously in a low background dilution cryostat running at a regulated temperature of 17 mK. Detectors were individually housed in separate 1 mm thick Cu casings, the distance between the Ge surfaces being 13 mm. For the heat measurement, a NTD-Ge thermometric sensor was glued on each detector. For the ionization measurement, detectors were equipped with two Al electrodes. One was segmented to define two regions, a central part and a guard ring, the applied collection voltage being of a few V/cm. The evaluated fiducial volume was $\sim$ 57%.

The final analysis of the EDELWEISS-I data covers a total fiducial exposure of 62 kg·d. The limits on the neutralino scattering cross-section were obtained from the observation of 40 nuclear recoil candidates with recoil energies between 15 and 200 keV, of which 3 are between 30 and 100 keV. The limits were obtained without the subtraction of any background, although the presence of a coincidence between two detectors and the study of charge collection distributions suggested that at least some of these events are due to a neutron background and surface interactions of electrons. In the preparation of the second phase of the experiment, it was necessary to perform a thorough investigation of all the events in EDELWEISS-I before the nuclear recoil selection, in order to better assess their origin and devise means to remove as many background components as possible for EDELWEISS-II [4]. It is clear that this first phase has been very useful to point out the main limiting factor for such a kind of experiment: the background coming from surface electron-recoil events. In fact for an interaction that takes place near the surface of the detector, the charge collection is incomplete and difficult to control. In particular, the local radioactivity from residual ^{210}Pb, a daughter of Radon which is present on all surfaces, generates such events that can create signals leaking down into the nuclear recoil band, where WIMP signals are expected.

2.2. EDELWEISS-II: ID detectors - a big improvement thanks to surface event rejection with interleaved electrodes

Given the experience acquired with EDELWEISS-I, a new design of a cryogenic germanium detector for dark matter search was required, taking advantage of the coplanar grid technique of event localization to improve the background discrimination and mainly to reject surface events [5]. The new-generation detectors called ID (InterDigit) method are essentially a variation of the coplanar grid technique, in which interleaved concentric strips are substituted for the classical disk-shaped collection electrodes. The depth of an event relative to the surfaces can be inferred from a comparison of the ionization signals on the different strips, making possible a rejection of energy deposits at the detector surfaces. This phase consisted of ten hyperpure germanium crystals of cylindrical shapes with a diameter of 70 mm and a height of 20 mm. Five of these detectors had their edges bevelled at an angle of 45° with a mass of 360 g. The mass of the other five detectors was 410 g. The detectors are in individual copper casings, stacked in towers of two to three ID detectors. During the entire data-taking periods, a dilution refrigerator maintains the detectors at a stabilized temperature of 18 mK. A total effective exposure of 384 kg·d has been achieved, mostly coming from fourteen months of continuous operation. Five nuclear recoil candidates are observed above 20 keV, while the estimated background is 3.0 events. The result is interpreted in terms of limits on the cross-section of spin-independent interactions of WIMPs and nucleons. A cross section of 4.4 $\times 10^{-8}$ pb is excluded at 90%CL for a WIMP mass of 85 GeV [6].

Here below the main detector properties are summarized:

- Detector mass: 400 g
- FWHM baseline: 1.2 keV for the heat signal
- FWHM baseline: 0.9 keV for the fiducial ionizing signal
- NTD sensitivity: 60 nV/keV
- Analysis threshold: 20 keV
- Fiducial mass: 160 g ($\sim$ 50% fiducial volume)

2.3. EDELWEISS-III: the FID800 detector - the new strategy
To go beyond the EDELWEISS-II sensitivity and to be competitive with other experiments, a third phase of this experiment is foreseen. The EDELWEISS-III project consists in an upgrade of both the current EDELWEISS setup and detectors to reach a sensitivity to WIMP-nucleon cross-section of the order of 5 $\times 10^{-9}$ pb in a short term requiring an exposure of 3000 kg·d. For the reduction of the β and γ backgrounds, the development of an improved detector design, named Full InterDigit (FID) has been pursued. 800 g crystals are equipped with two NTD heat sensors, and are covered by interleaved electrodes on all their surface. There is therefore no guard region anymore inside the crystal volume. Both the increase of crystal mass and the removal of guard regions increase strongly the fiducial mass for each individual detector. In addition, simulations have shown that Compton interactions spanning the fiducial and low-field guard regions may induce fake WIMP candidates. Large-statistics gamma-ray calibration of the first FID detectors have indeed shown improved rejection performances relative to ID detectors.The plan is to install 40 FID800 detectors in 2012. Infrastructure upgrades (cabling, cold electronics, cryogenics and acquisition) are necessary to host these new detectors, to reduce microphonic noise and hence the analysis threshold, and will also reduce the neutron budget within the cryostat.. In addition, an inner polyethylene shield will be installed to reduce the flux of fast neutrons coming from outside the cryostat. Fig.1 shows a picture of a FID detector: interleaved electrodes are visible on all its surface and also the NTD sensor is appreciable on the top surface. Fig.2 represents a sketch of the electric field lines inside the crystal resulting in a 75% of fiducial volume ($\sim$ 640g).

3. Conclusions and Perspectives

Important progresses have been done with cryogenic Ge detectors in the EDELWEISS experiment during last years. There is still a comfortable space for detector improvements including also larger masses. In addition, thanks to the big versatility of these detectors, they can be adapted to lower thresholds; at this purpose the collaboration is working on the heat channel with the development of new innovative heat sensors, called "NbSi superconductive resistive meanders", that have the potentiality to decrease the threshold improving also the energy resolution.

For the near future, new data in the coming year will contribute to clarify the present situation at low mass, a hot topic at nowadays, with the possibility to test the DAMA/CoGeNT/CRESST signals. In the next future, EURECA [7], a Ton scale Dark matter experiment, will search for WIMP interactions down to $\sigma \sim 10^{-10}$ pb corresponding to a rate of $\sim$1 event/ton/year. It is clear that to achieve this further step different present limitations should be overcome.

Figure 1. Picture of a FID detector of 800 g in its copper holder. Interleaved electrodes are visible on all its surface and also the NTD sensor is appreciable on the top surface.

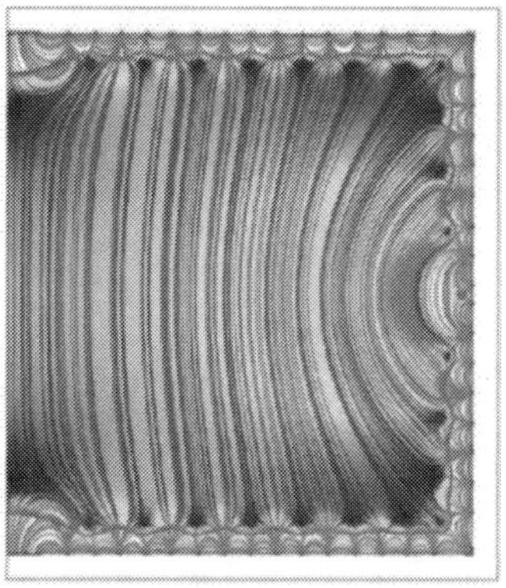

Figure 2. Sketch of the electric field lines inside the crystal.

Acknowledgments
This project is supported in part by the Agence Nationale pour la Recherche under contract ANR-06-BLAN-0376-01 and partly by the CEA - EUROTALENTS Programme (COW project) within the 7th European Community Framework Programme.

References

[1] Bertone G, Hooper D and Silk J 2005 *Phys. Rep.* **405** 279
[2] http://edelweiss.in2p3.fr/
[3] Sanglard V et al. 2005 Phys. Rev. D **71** 122002
[4] Fiorucci S et al. 2007 *Astropart. Phys.* **28** 143
[5] Broniatowski A et al. 2009 *Phys. Rev. Lett.* **681** 305
[6] Armengaud E et al. 2011 *Phys. Lett.* B **702** 329
[7] Kraus H et al. 2007 *Nucl. Phys.* B **173** 168

Scintillating bolometers for fast neutron spectroscopy in rare events searches

M. Martínez[1], N. Coron[1], C. Ginestra[2], J. Gironnet[1], V. Gressier[3], J. Leblanc[1], P. de Marcillac[1], T. Redon[1], P. Di Stefano[4], L. Torres[1], P. Veber[5], M. Velazquez[5] and O. Viraphong[5]

[1] Institut d'Astrophysique Spatiale (IAS), Bâtiment 121, Université Paris-Sud 11 and CNRS (UMR 8617), 91405 Orsay Cedex, France
[2] Grupo de Física Nuclear y Astropartículas. Universidad de Zaragoza, C/ Pedro Cerbuna 12, 50009 Zaragoza, Spain
[3] Institut de Radioprotection et de Sûreté Nucléaire, BP 3, 13115 Saint-Paul-lez-Durance, Cedex, France
[4] Department of Physics, Engineering Physics & Astronomy, Queen's University, Kingston, ON, Canada, K7L 3N6
[5] CNRS, Université de Bordeaux, ICMCB, 87 avenue du Dr. A. Schweitzer, 33608 PESSAC cedex, France

E-mail: mariam@unizar.es

Abstract. Neutrons are a relevant background in rare events physics. Detectors based on fast neutron-induced nuclear reactions are commonly used for fast neutron spectroscopy. In this subject, scintillating bolometers provide an excellent energy resolution and particle discrimination by the simultaneous measurement of the heat and emitted light. Our group has constructed several ^{6}Li and ^{10}B based massive scintillating bolometers (LiF, $Li_6Eu(BO_3)_3$, $Li_6Gd(BO_3)_3$), with energy resolutions ranging from 16 to 200 keV. First results of a 32 gr ^{6}LiF scintillating bolometer enriched at 95% in ^{6}Li operated at 20 mK are presented. The use of this material in a multi-target cryogenic dark matter experiment, like EURECA, would allow monitoring the incident neutron flux in the detector during the data-taking.

1. Introduction

Neutron induced nuclear recoils can mimic the WIMP signal in Dark Matter direct searches, so the determination and reduction of the neutron background is mandatory in any experiment of this kind. The expected fluxes in underground laboratories are very weak ($\sim 10^{-6}$ n/cm^2/s for neutrons coming form natural radioactivity in the laboratory rocks and $\sim 10^{-9}$ n/cm^2/s for μ-induced neutrons [1]) and can be strongly reduced with active and passive shieldings. Regarding the neutron background inside the detector, estimations based on Monte Carlo simulations are usually assumed, nevertheless there are always uncertainties (rock composition, veto efficiency, (α, n) reactions in the shielding or the detector itself...) that make it appealing a direct measurement. Between the most practical methods for very low neutron backgrounds spectroscopy are those based on neutron-induced nuclear reactions: the neutron energy is shared between the reaction products and the deposited energy gives directly the incident neutron energy, after subtracting the Q value of the reaction. ^{6}Li and ^{10}B are preferred isotopes due to their relatively high cross-section for neutrons in the range of several MeV (see table 1).

Table 1. Most used neutron-induced reactions for fast neutron spectroscopy.

Reaction (energy at thermal capture)	Q_{value} (MeV)	σ (E_n=1 MeV) (barns)
n+^{3}He→^{3}H (0.191 MeV) + p (0.574 MeV)	0.765	0.7
n+^{6}Li→α (2.050 MeV) + ^{3}H (2.730 MeV)	4.783	0.3
n+^{10}B→^{7}Li (1.015 MeV) + α (1.777 MeV) (6%)	2.792	0.04
n+^{10}B→^{7}Li* (0.840 MeV) + α (1.470 MeV) (94%) ↪^{7}Li + γ (0.48 MeV)		

The wide target choice and excellent energy resolution achievable with bolometric detectors make them very suitable for this issue. Moreover, in the case of scintillating materials, the particle discrimination power provided by the simultaneous measurement of light and heat produce an almost background-free spectrum, indispensable to measure very weak fluxes. In this work we present the performances of an enriched 32 gr ^{6}LiF for neutron detection and some preliminary results obtained with other interesting materials. Finally, prospects of using these materials in a multi-ton bolometric DM experiment, like the projected EURECA [2], are drawn.

2. LiF bolometers for neutron detector

LiF is an interesting material for bolometric neutron detection: its high Debye temperature makes it suitable as bolometer, and although its scintillation is negligible at room temperature the light emission increases as temperature goes down. The expected efficiency for neutron detection, calculated for a 800 gr cylindrical (h=ϕ) target, is ~20% at ^{6}Li resonance (240 keV), ~4% at 1-2 MeV and ~1% at 6 MeV [1] (see Figs 1 and 2).

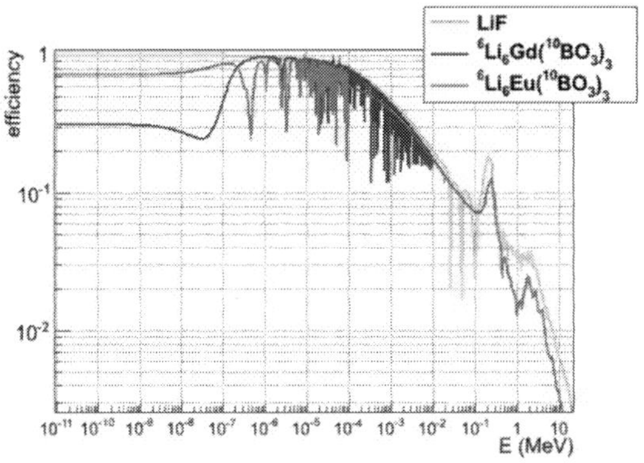

Figure 1. Calculated efficiency of ^{6}LiF, ^{6}Li$_6$Gd(^{10}BO$_3$)$_3$ and ^{6}Li$_6$Eu(^{10}BO$_3$)$_3$ (800 gr, cylindrical shape (h=ϕ), 95% enrichments).

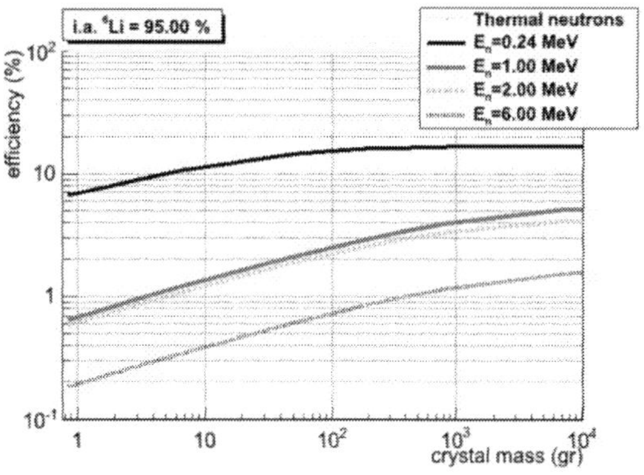

Figure 2. Expected efficiency of ^{6}LiF (95% enriched, cylindrical shape) as a function of the crystal mass.

Since 1993 our group has constructed several LiF bolometers for different purposes [3]. In particular, a 33 gr (not enriched) LiF was tested at the Canfranc Underground laboratory as neutron detector [4]. A portable neutron detector has been constructed, consisting of a small 534 mg ^{6}LiF (95%) (only heat channel) inside a ^{3}He cryostat (active charcoal pumping), with

[1] Multiple scattering is not considered in the calculation, so the given values correspond to lower limits.

an operating temperature of 300-400 mK and an autonomy of around 1 day. The measured energy resolution was 28.7 keV FWHM at the thermal neutrons peak. This detector allowed to demonstrate the energy recovery principle by a dedicated calibration in monoenergetic neutron fields at IRSN AMANDE facility, showing very good energy linearity between 50 keV and 17 MeV (deviation from linearity lower than 3%) [5].

A first prototype 32 gr ^{6}LiF (95%) cylindrical crystal (ϕ=25.1 mm, h=25.2 mm) was tested at sea level and preliminary results are shown in Figs 3 and 4. A lower than expected light yield (3 times lower than that of a 16 gr not enriched LiF [6]) was measured. The thermal/fast neutrons band is well separated from the $\mu/\gamma/\beta$ band down to 2 MeV but not from the α band (estimated with an ^{241}Am source facing the target), preventing a full discrimination of the α background. The energy resolution in the thermal neutrons peak is 50 keV FWHM. The fast neutron spectrum of the non-thermalized ^{252}Cf source is clearly visible in Fig 4 starting from Q_{val}=4.78 MeV.

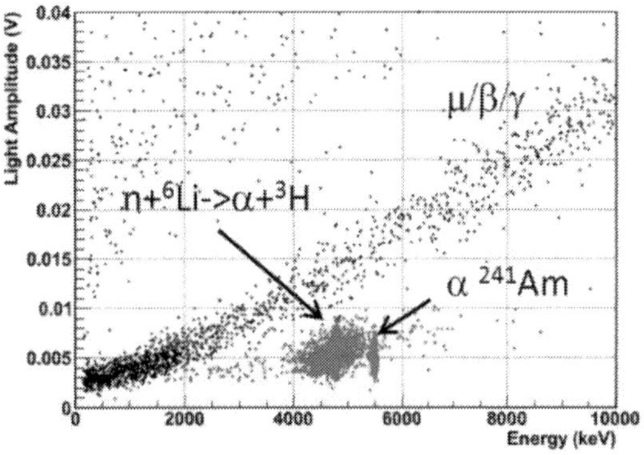

Figure 3. Scattering plot light vs heat of a 32 gr ^{6}LiF (95% enriched). The α/α+^{3}H band (in red) is well discriminated from the $\mu/\beta/\gamma$ band down to 2 MeV.

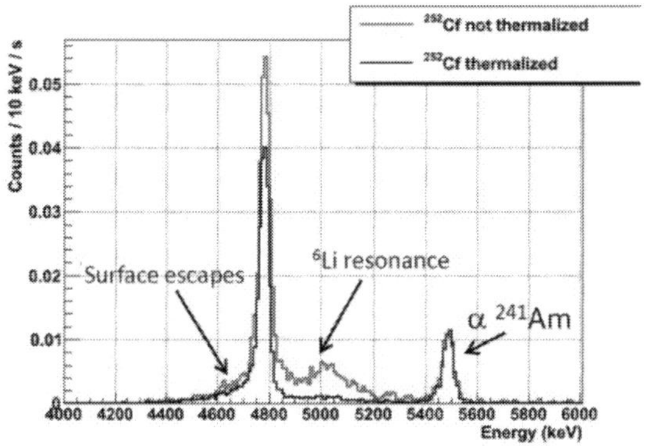

Figure 4. Energy spectra measured by a 32 gr ^{6}LiF (95% enriched) during a ^{252}Cf irradiation. neutron source, thermalized (black line) and non-thermalized (red line).

3. Other ^{6}Li and ^{10}B based materials

The rather poor light yield of LiF encouraged us to investigate other Li and B based scintillators, like the lithium borates LGBO ($Li_6Gd(BO_3)_3$) and $Li_6Eu(BO_3)_3$. Their calculated efficiency for neutron detection is shown in Fig 1 compared to that of LiF, being quite similar (but slightly inferior) in the fast neutron range (but in this case no ^{19}F elastic resonances degrade the efficiency between 10 and 200 keV). On the other hand, efficiency is lower than 1 for thermal neutrons, specially for LGBO (Gd having the higher cross section for thermal neutron capture (^{155}Gd, ^{157}Gd) and following a complicated gamma decay not easy to detect), a fact that could be interesting in certain applications (cleaner fast spectrum near the thermal peak). Only small (5x5x5 mm^3) lithium borate crystals have been tested so far and preliminary results are shown in Figs 5 and 6. Poor light yield and energy resolution were obtained in the case of LGBO (8% FWHM @ 5.49 MeV), but very promising results were found with $Li_6Eu(BO_3)_3$ (not enriched): a collected light yield 10 times better than that from our ^{6}LiF is reported, while 13 keV FWHM resolution is found on the ^{10}B neutron capture peak at 2.31 MeV.

4. Conclusions and prospects

Several ^{6}Li and ^{10}B based scintillating bolometers have been successfully mounted and tested as neutron detectors. Enriched ^{6}LiF has produced very good results, but the current light yield does

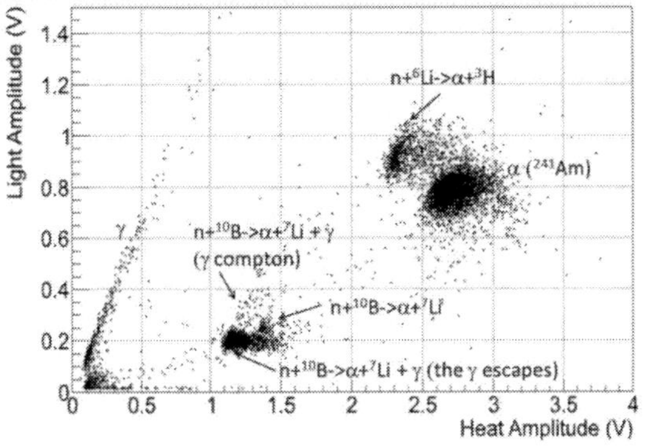

Figure 5. Scattering plot light vs heat of a 5x5x5 mm^3 $^6Li_6Gd(^{10}BO_3)_3$ (95% enriched) during a ^{252}Cf irradiation. Despite the poor resolution, the three reactions ($n+^{10}B \to \alpha+^7Li$, $n+^{10}B \to \alpha+^7Li+\gamma$ and $n+^6Li \to \alpha+^3H$) are distinguishable.

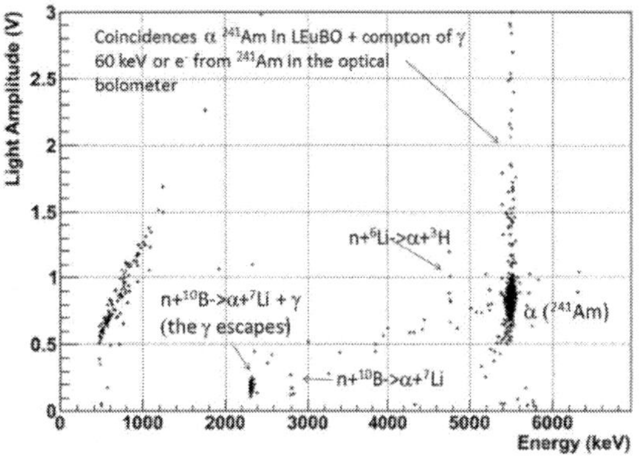

Figure 6. Same as Fig. 6 for a 5x5x5 mm^3 $Li_6Eu(BO_3)_3$ (not enriched). Also in this case the three reactions are well separated despite the poor statistics.

not allow for a full α background discrimination, so other alternatives are being investigated. The use of this kind of materials as targets in a multi-ton DM experiment could provide very interesting information, like the thermal neutron flux at the detectors location (a very valuable input/check for MC simulations), fast neutron detection during neutron calibrations, cross-check of the neutron shielding efficiency in background runs, and fast neutron detection during background runs. Due to the limited efficiency, a certain amount of material is needed to accomplish the last point. For example, to clarify the origin of the current backgrounds in bolometric Ge experiments[2][7] at least 10 kg of ^6LiF (95% enriched) would be needed.

Acknowledgments

This work has been supported by the French CNRS/INSU (MANOLIA and BOLERO projects) and by CNES (R&D MOB2). L. Torres is supported by a MICINN postdoctoral grant (Programa Nacional de Movilidad de Recursos Humanos 2008-2011), C. Ginestra by the MICINN Subprograma de Formación de Personal Investigador, and M. Martinez by a postdoctoral grant from the P2I program funded by Université de Paris Sud. We are grateful to Prof. F. Danevich from INR Kiev who kindly provided us with the LiEuBO sample. Partial funding has also been provided by Agence Nationale de la Recherche (ANR) grant SciCryo ANR-05-BLAN-0031.

References

[1] Kudryavtsev V A, Pandola L and Tomasello V 2008 *Eur. Phys. J. A* **36** 171
[2] Kraus H *et al* 2010 *PoS(IDM2010)* 109
[3] de Marcillac P *et al* 1993 *Nucl. Instr. Meth. A* **337** 95
[4] Coron N *et al* 2010 *J. Phys. Conf. Ser.* **203** 012139, Coron N *et al* (these proceedings)
[5] Gironnet J *et al*, *Proc.* of Fast Neutron Detectors and Applications, November 2011, Ein Gedi, Israel
[6] Coron N *et al* 2003 *Nucl. Instr. Meth. A* **520** 159
[7] CDMS and EDELWEISS collaborations 2011 *Phys. Rev. D* **84** 011102, arXiv:1105.3377v2

[2] Current limits: ~3x10^{-8} pb, corresponding to ~0.01 c/kg/d for energy threshold 10-20 keV (10-20% attributed to neutrons)[7]. Rescaling with Ge neutron elastic scattering cross section and the ^6LiF efficiency at ~1 MeV, around 0.001 c/kg/d are expected in ^6LiF. This value is a lower limit: elementary kinematics indicate that neutrons producing nuclear recoils under 10-20 keV energy threshold will be easily detected in the LiF detector.

12th International Conference on Topics in Astroparticle and Underground Physics (TAUP 2011) IOP Publishing
Journal of Physics: Conference Series **375** (2012) 012026 doi:10.1088/1742-6596/375/1/012026

Update on the ANAIS experiment. ANAIS-0 prototype results at the new Canfranc Underground Laboratory

J Amaré, S Cebrián, C Cuesta[1], D Fortuño, E García, C Ginestra, H Gómez, M Martínez, M A Oliván, Y Ortigoza, A Ortiz de Solórzano, C Pobes, J Puimedón, M L Sarsa and J A Villar

Grupo de Física Nuclear y Astropartículas, Universidad de Zaragoza

C/ Pedro Cerbuna 12, 50009, Zaragoza, Spain

ccuesta@unizar.es

Abstract. ANAIS experiment will look for dark matter annual modulation using NaI(Tl) scintillators at the Canfranc Underground Laboratory (LSC). Highly purified NaI(Tl) crystals are being developed to reach the required sensitivity. In a parallel way, the ANAIS-0 module (made with a low background St Gobain NaI(Tl) crystal) has been taking data at the LSC, testing different configurations: various photomultiplier tubes (PMTs) models with/without light guides. Low background PMTs with light guides and ultra low background PMTs without light guides have shown a similar contribution to the background. A complete simulation of the ANAIS-0 module with shielding in the different configurations tested has been carried out and compared with the experimental data, considering contributions to the background from NaI bulk contaminants, PMTs, light guides, quartz windows and shielding materials. A good understanding of the background above 500 keV can be reported.

1. The ANAIS experiment

ANAIS is a project aiming to set up, at the new facilities of the Canfranc Underground Laboratory, a large scale NaI(Tl) experiment in order to explore the DAMA/LIBRA [1] annual modulation positive result using the same target and technique. For this goal, some of the experimental features of ANAIS should be: energy threshold below 2 keVee in order to improve sensitivity for the annually modulated WIMP signal, background at low energy kept as low as possible and very stable operation conditions. The original proposal consisted of 100 kg NaI(Tl) crystals, but an enlargement up to 250 kg was recently proposed and funds allocated.

New NaI(Tl) crystals with very low content in potassium should be used in ANAIS to reach the required sensitivity for the dark matter signal. ANAIS collaboration signed an agreement to purify commercial NaI powder below 100 ppb and 1 kg samples are being characterized at our HP Ge test bench in the LSC. The terms of the crystal growing agreement are under discussion.

Different photomultipliers have been tested. Two options are still under consideration for ANAIS: Low background (LB) photomultipliers (Hamamatsu R6233-100 or Electron Tubes Limited 9302B) with 10-cm methacrylate light guides and ultralow background (ULB) photomultipliers (Hamamatsu R11065SEL) without light guides. Regarding the acquisition code and electronics, the new VME acquisition system is completed and first tests at the University of Zaragoza have been carried out. It

[1] Speaker

will be installed at LSC for further testing with ANAIS-0 in the next weeks. Stability conditions and environmental parameters are being monitored with the ANAIS-0 module at the new facilities of LSC.

2. The ANAIS-0 module

The ANAIS-0 module consists of an old NaI(Tl) ultrapure crystal (9.6 kg), made by Saint-Gobain and encapsulated by us in OFHC copper in the University of Zaragoza (figure 1). It was designed to characterize and fully understand ANAIS background at low energy, optimize NaI scintillation events selection, determine the calibration method and test the electronics. The detector encapsulation allowed for testing different configurations of light guides and photomultipliers.

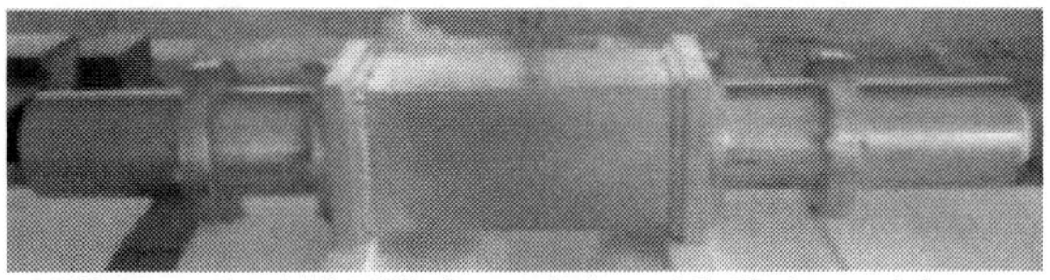

Figure 1. ANAIS-0 module with light guides.

Main results concerning background for ANAIS on site show that ultra low background photomultipliers (ULB PMTs) are radiopure enough to be a small contribution to the background of the ANAIS-0 module, and using light guides has not an important effect in the measured background (figure 2). ULB PMTs background without light guides is comparable to the background obtained with the low background photomultipliers (LB PMTs) with light guides. Work is in progress to establish the cuts to select of real bulk NaI scintillation events, mandatory before estimating the background at the very low energy region.

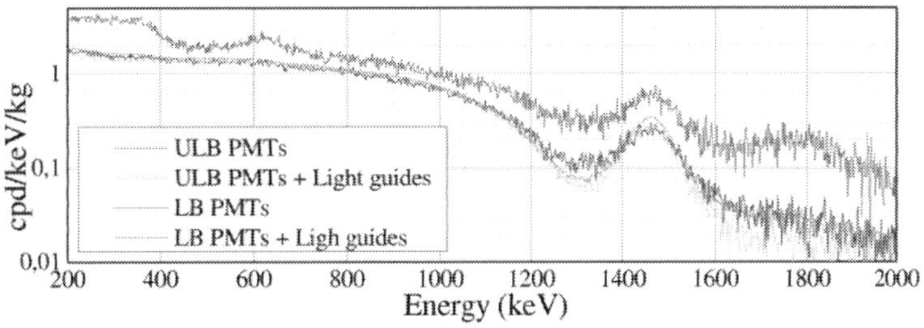

Figure 2. High energy raw background data for the different ANAIS-0 set-ups.

3. Background simulation

A complete simulation (using the GEANT4 package [2]) of the ANAIS-0 module with shielding in the different configurations tested has been carried out. Version Geant4.9.4.p01 has been used for all the simulations presented here. Physical processes and models commonly used in the context of underground experiments have been implemented for interactions of alpha, beta and gamma particles. The GEANT4 Radioactive Decay Model has been used for simulating radioactive contaminations, after checking carefully the energy conservation in the decay of all the isotopes taken into account; problems encountered in previous versions of GEANT4 code seem to be surpassed.

The simulated geometry (figure 3) includes the NaI crystal, Teflon and reflector lining, quartz windows, light guides (optional), photomultipliers, copper encapsulation as well as the shield made of both archaeological and standard low activity lead.

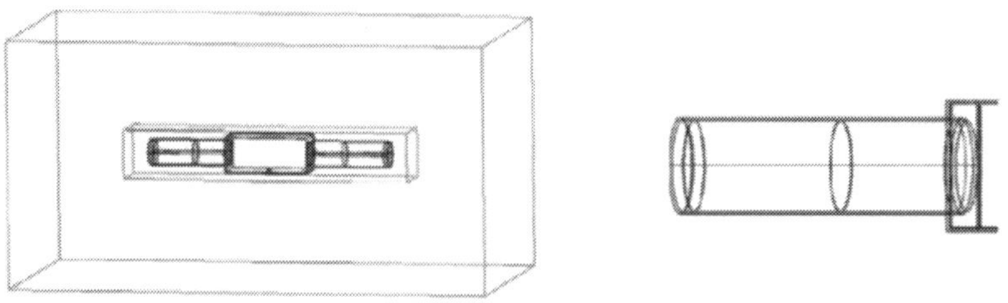

Figure 3. Geometry used in the ANAIS-0 module simulations (left). Detail of the PMT and light guide coupling (right).

Decays of radioactive impurities in the most relevant materials of the set-up, including mainly ^{238}U and ^{232}Th chains and ^{40}K, have been simulated assuming a uniformly distributed bulk contamination. Activities used to normalize simulation results come mostly from specific measurements made for ANAIS-0 building materials:

- For the ANAIS-0 module NaI crystal, the ^{40}K bulk content is determined through the coincidence method [2] to be 12.7 ± 0.6 mBq/kg. Activities of the different isotopes in the ^{238}U and ^{232}Th chains, assuming broken equilibrium, have been quantified after identifying their alpha emissions by means of Pulse Shape Analysis (PSA) and fitting the obtained alpha spectrum. The values obtained are: 0.075 ± 0.005 mBq/kg for ^{238}U/^{234}U, 0.023 ± 0.007 mBq/kg for ^{230}Th, 0.098 ± 0.004 mBq/kg for ^{226}Ra, 0.188 ± 0.005 mBq/kg for ^{210}Pb, 0.013 ± 0.005 mBq/kg for ^{232}Th and 0.035 ± 0.003 mBq/kg for ^{228}Th. Concerning the activity of ^{129}I, the value deduced for DAMA/LIBRA crystals [3], produced (as the ANAIS-0 module crystal) by Saint Gobain company, has been used (9.01 mBq/crystal).
- The activities for the several ETL and Hamamatsu photomultiplier models employed in the different ANAIS-0 set-ups were measured using our HP Ge detector operated at LSC [4].
- For quartz windows, light guides and copper encapsulation the upper limits on the activities derived in specific measurements using a HP Ge detector operated at LSC have been considered.
- For archaeological lead, an upper limit on ^{210}Pb activity quoted at [5] has been used.

The energy spectrum of alpha emissions identified in the data taken for a long background measurement of ANAIS-0 has been compared with the corresponding simulation (figure 4), assuming a volumetric distribution of ^{238}U and ^{232}Th, and daughters, isotopes in NaI crystal and the activity values deduced by PSA analysis. Effect of acquisition dead time on ^{214}Po emission and the summing of alpha and beta emissions for the ^{212}Bi-Po sequence have been properly taken into account.

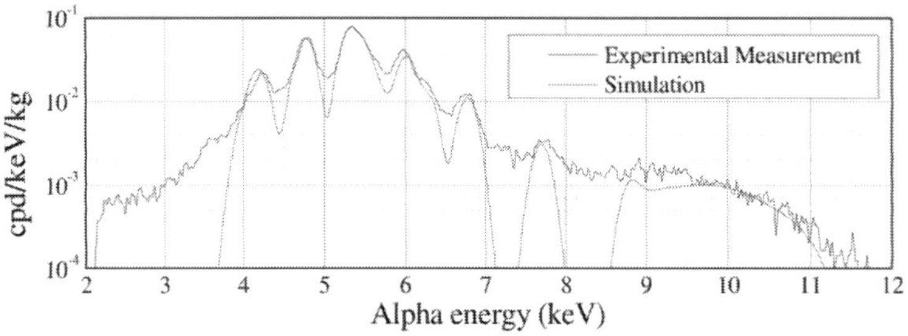

Figure 4. Simulation of the alpha spectrum compared with real data. Alpha+beta coincident events (Bi-Po sequence) are also shown.

The energy spectrum in ANAIS-0 due to beta-gamma events in a background measurement with the ULB PMTs without light guides has been compared with the simulated contribution of the different components (NaI crystal, copper encapsulation, photomultipliers, etc.) (figure 5). Looking at the high energy region, from 0.2 to 3 MeV, the main conclusion is that above 500 keV the measured background seems to be nicely reproduced by the simulated considered contributions; ^{40}K from the bulk NaI crystal is clearly the dominant background source from 500 keV to the 1460.8 keV peak.

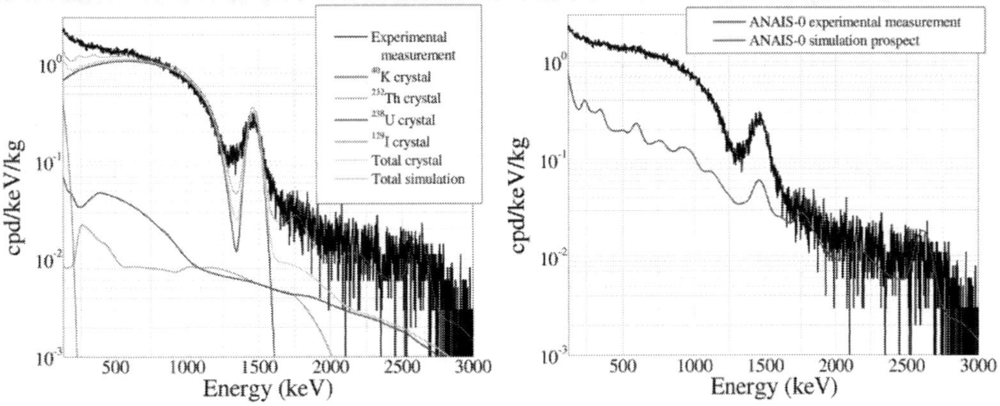

Figure 5. Simulation of the beta/gamma spectrum compared with real data. Contributions of bulk contaminations in NaI crystal are detailed in the plot and the total simulated spectrum is shown as red solid line (left). Prospect for the K contamination had as a goal for the new NaI crystals (20 ppb) keeping the same of all other contributions (right).

However, in the low energy region, below 0.2 MeV, agreement between the measured data and the total simulated contributions is worse, pointing to some non-explained component(s) in the background. Different possibilities are under study.

Drawn conclusions are equivalent for all the available configurations, either from the high or the low energy regions of the spectra.

Using the performed simulations, an estimate of the ANAIS new detectors achievable background levels has been attempted by reducing the K concentration in the NaI crystal down to 20 ppb, keeping activities for all the other emissions as presented before for the ANAIS-0 prototype (figure 5). It seems that the expected background will fulfill the ANAIS requirements provided if this K purity is effectively reached in the bulk crystal and the unexplained components (not included in the simulation for the moment) are kept under control.

4. Conclusions

New ultrapure NaI crystals are being developed, the NaI purification is on-going and crystal growing terms are under discussion. Besides, ANAIS-0 (old NaI crystal) is taking data at the LSC. Two configurations with ANAIS-0 have been tested: ultra low background photomultipliers and low background ones with light guides, both contribute equivalently to the background.

The ANAIS-0 background above 500 keV seems to be explained by the considered contributions. However, at low energy we find some non-explained components that are under study. Encapsulation materials and procedure allow us to reach the required background level if the new NaI crystals have a potassium content less than 20 ppb.

Acknowledgments

This work has been supported by the Spanish Ministerio de Ciencia e Innovación (MICINN) under project reference FPA2008–03228 and Consolider-Ingenio 2010 Programme under grant Multidark CSD2009–00064. C. Cuesta is supported by a Gobierno de Aragón predoctoral grant.

References

[1] Bernabei R et al 2008 *Eur. Phys. J.* C **56** 333-355 (arXiv:0804.2741)
[2] Agostinelli S et al 2003 *NIM A* **506** 250-303
[3] Amaré J et al 2010. *PoS* (IDM2010) 020
[4] Bernabei R et al 2008 *NIM A* **592** 297
[5] Cebrián S et al 1999 *Astropart. Phys.* **10** 361

DEAP-3600 Dark Matter Search at SNOLAB

M. G. Boulay for the DEAP Collaboration

Queen's University, Department of Physics, Engineering Physics and Astronomy, Kingston, Ontario, Canada, K7L 3N6

E-mail: mark.boulay@queensu.ca

Abstract. The DEAP-3600 detector, currently under construction at SNOLAB, has been designed to achieve extremely low background rates from all sources, including ^{39}Ar β-decays, neutron scatters (from internal and external sources), surface α-contamination and radon. An overview of the detector and its sensitivity are presented.

DEAP-3600 will perform a dark matter particle search on argon with sensitivity to the spin-independent WIMP-nucleon cross-section of 10^{-46} cm^2, a factor of approximately one hundred increase over current experiments. DEAP has been under development since 2005 and is a collaborative effort including over 60 researchers. The detector, shown in Fig. 1 is conceptually simple: a large spherical volume of liquid argon contained in a transparent acrylic vessel is viewed by 255 photomultiplier tubes that detect scintillation light pulses generated in the argon. DEAP-3600 will allow a three-year exposure of a one-thousand kilogram (fiducial) target of liquid argon in the SNOLAB Cube Hall, which itself has extremely suppressed backgrounds due its 6000 m.w.e. depth. Extreme care has gone into selection of detector materials, including their exposure to radon either during fabrication, or installation at SNOLAB, to ensure background target levels will be met. The acrylic vessel and lightguide acrylic have undergone extensive quality control to ensure radiopurity of the bulk acrylic; of particular concern was the diffusion of radon into the acrylic monomer leading to a buildup of long-lived ^{210}Pb, which would lead to a source of (α,n) neutron backgrounds in the inner detector. After the acrylic vessel has been fabricated, it will be sealed from the mine environment, and while purged with low-radon gas, a several-micron layer will be removed with a custom resurfacer to ensure the detector surface is as radiopure as the bulk acrylic. A custom large-area vacuum-deposition source developed by the DEAP collaboration will then be used to deposit tetraphenyl butadiene (TPB) wavelength shifter uniformly over the approximately 10 m^2 acrylic detector surface, before the vessel is cooled and filled with liquid argon. The detector design parameters and background targets are shown in Table 1.

Mitigation of the large ^{39}Ar background, approximately 1 Bq/kg of argon [2], will be accomplished using pulse-shape discrimination (PSD) of the scintillation signal, as described in [3, 4, 5]. All PMT signals will be digitized with CAEN v1720 digitizers. PSD in argon depends strongly on the number of detected photoelectrons (pe). Fig. 2 shows the expected PSD using the technique outlined in [5], for light yields of 4 or 8 pe/keV$_{ee}$. This strong dependence on light yield (a factor of approximately 5×10^4 difference in PSD between 4 and 8 pe/keV$_{ee}$) drove the design decision towards a single-phase detector, in order to pursue a very high photocathode coverage and thus maximize light collection and background discrimination. Given this PSD dependence on energy, a reduction of the β/γ background by a factor of ten allows a threshold

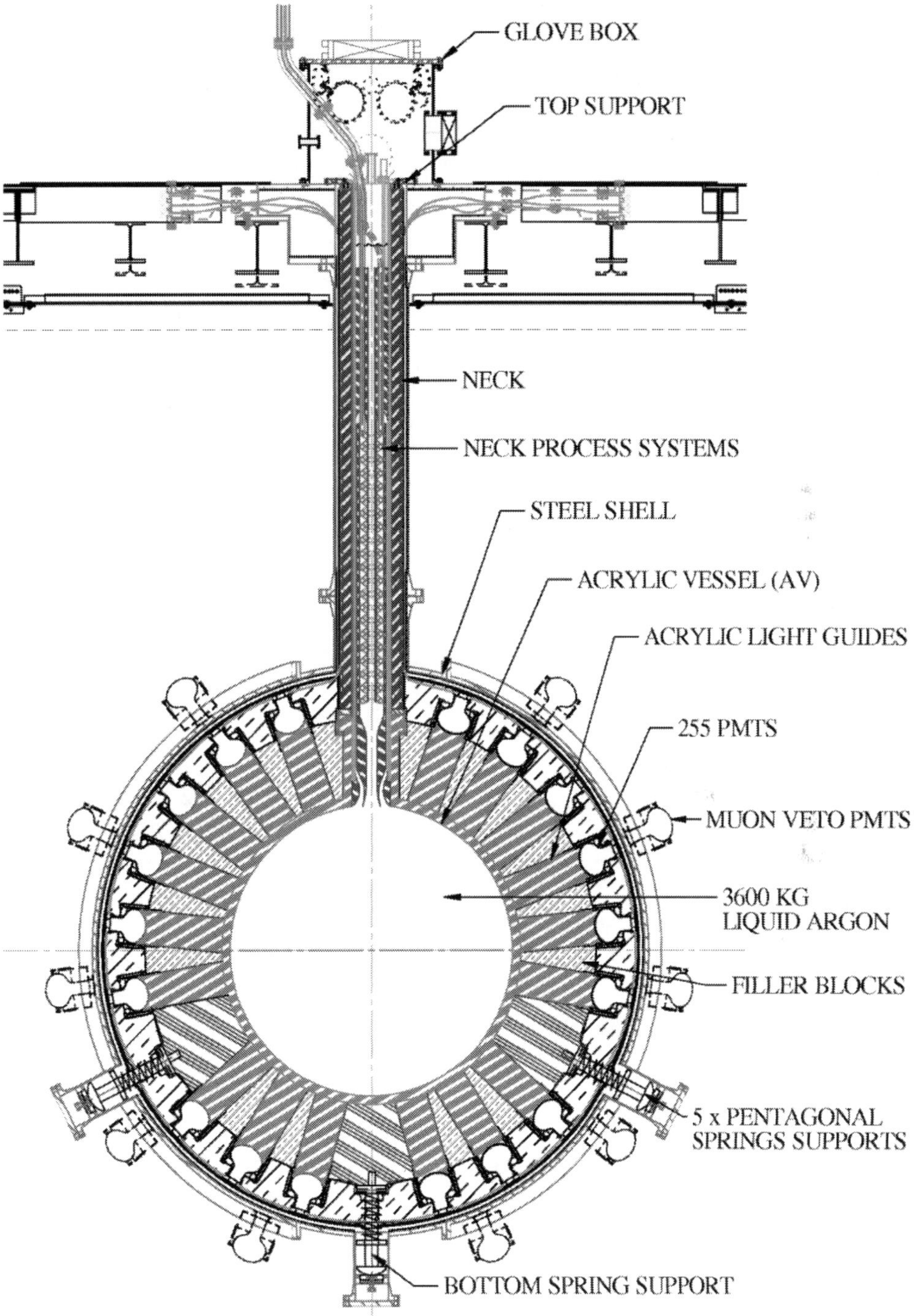

Figure 1. The DEAP-3600 detector. The acrylic vessel has in inner radius of 85 cm and holds 3600 kg of liquid argon, which is viewed by 255 8-inch high quantum efficiency photomultiplier tubes (PMTs) through 50-cm lightguides. The lightguides provide neutron shielding and thermal insulation between the cryogenic acrylic vessel and "warm" PMTs. The inner detector is housed in a large stainless steel spherical vessel, which itself is immersed in an 8 meter diameter water shielding tank.

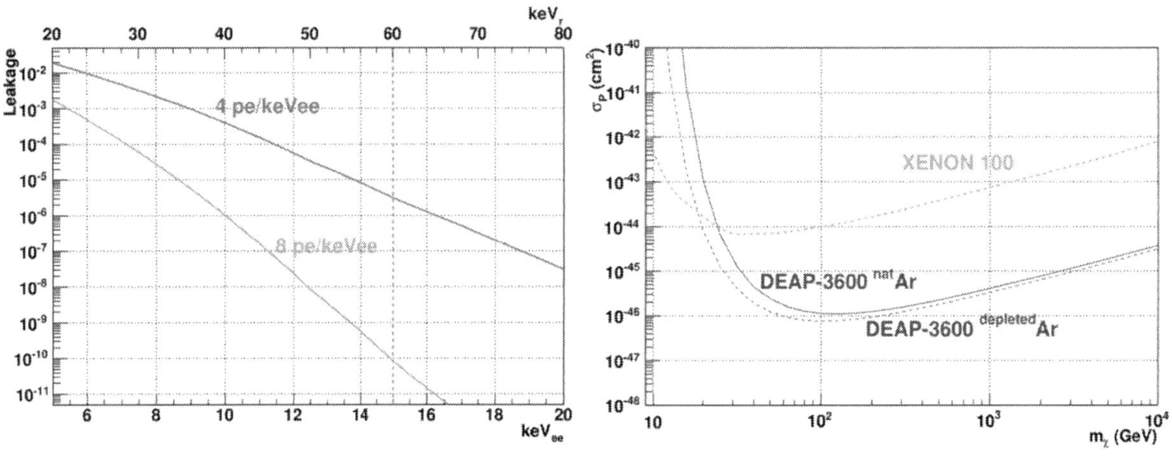

Figure 2. Pulse-shape discrimination and sensitivity with liquid argon. (Left) Discrimination of β events versus energy for light yields of 4 and 8 photoelectrons per keV$_{ee}$. The higher light yield improves discrimination by approximately 5×10^4 at the 15 keV$_{ee}$ threshold. (Right) 90% exclusion limit for DEAP-3600 with natural and depleted argon. Also shown for reference is the current XENON-100 exclusion limit [1].

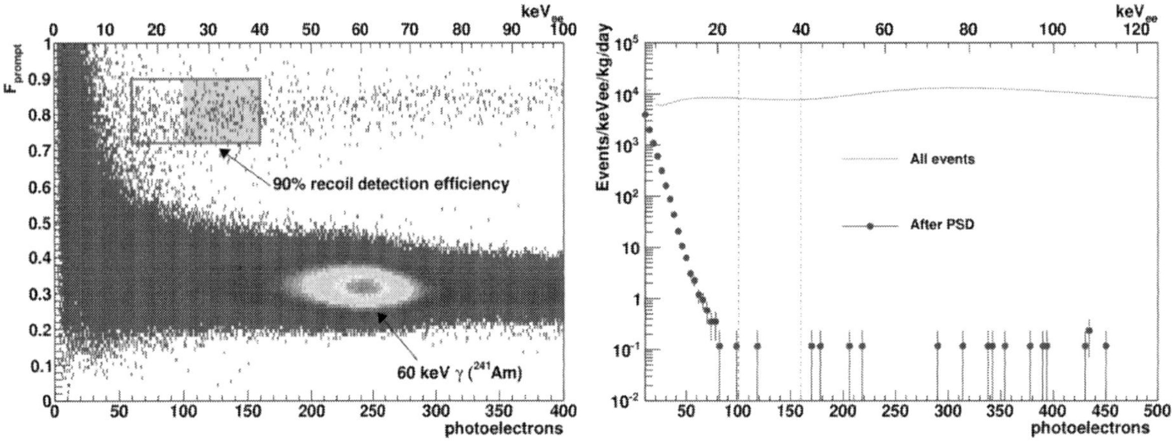

Figure 3. Results from the DEAP-1 7-kg liquid argon prototype detector at SNOLAB. (Left) Neutrons (high-F$_{prompt}$ band) and γ's (low-F$_{prompt}$ band) from an untagged Am-Be source. The boxes show F$_{prompt}$ regions with 90% detection efficiency, with DEAP-3600 and DEAP-1 thresholds. (Right) Backgrounds in the DEAP-1 prototype. The low-energy "wall" is from γ backgrounds that are not removed by PSD; high-energy events are from radon and surface contamination. In the energy region from 25 to 40 keV$_{ee}$, the backgrounds correspond to approximately 100 μBq/m^2.

reduction of 1.25 keV$_{ee}$. Acosta-Kane et al have demonstrated the possibility of argon with suppressed ^{39}Ar levels [6], and we are working in collaboration with the Princeton group to obtain depleted argon for DEAP-3600. 90% CL exclusion sensitivities are shown in Fig. 2, calculated using the standard assumptions outlined in [7] and with a galactic escape velocity of 544 km/s. The increased sensitivity obtained by reducing the threshold using depleted argon is also shown in Fig. 2; larger increases can be obtained for very low-threshold analyses.

Parameter	Value
Light Yield	8 photoelectrons per keV_{ee}
Nuclear Quenching Factor	0.25
Analysis Threshold	15 keV_{ee}, 60 keV_{r}
Total Argon Mass (Radius)	3600 kg (85 cm)
Fiducial Mass (Radius)	1000 kg (55 cm)
Position Resolution at threshold (conservative, design spec.)	10 cm
Position Resolution at threshold (ML fitter)	< 6.5 cm
Background Specifications	**Target**
Radon in argon	< 1.4 nBq/kg
Surface α's (tolerance using conservative pos. resolution)	< 0.2 $\mu Bq/m^2$
Surface α's (tolerance using ML pos. resolution)	< 100 $\mu Bq/m^2$
Neutrons (all sources, in fiducial volume)	< 2 pBq/kg
$\beta\gamma$ events, dominated by ^{39}Ar (after PSD)	< 2 pBq/kg
Total Backgrounds	< 0.6 events in 3 tonne-years

Table 1. Detector design parameters and background targets. The analysis threshold is set by the requirement for sufficient PSD of ^{39}Ar. The 10-cm position resolution is from a centroid charge resolution reconstruction algorithm. A more sophisticated maximum likelihood (ML) fitter improves the position resolution significantly, and allows an increased α surface background of approximately 100 $\mu Bq/m^2$ (a level already demonstrated with DEAP-1). The target background specifications correspond to approximately 1 event per 5 Gg-days for each source.

Fig. 3 shows data from the prototype DEAP-1 detector with 7 kg of liquid argon at SNOLAB. The inner detector has been prepared in a radon-reduced glovebox, with techniques similar to the surface preparation that will be used for DEAP-3600. Background levels of approximately 100 $\mu Bq/m^2$ have been demonstrated in DEAP-1; this is sufficiently low to achieve the target background from surface contamination in DEAP-3600 assuming the ML position reconstruction resolution shown in Table 1.

Most detector components have been fabricated. Acrylic sheets for the vessel production have been cast at Reynolds Polymer (Asia) and thermoformed into spherical sections (in Colorado). Light guides are being fabricated from acrylic (with measured attenuation lengths greater than 5 meters) cast at Spartech, Inc. Detector infrastructure and the shield tank have been installed at SNOLAB. Cooling for the experiment is provided by a distributed liquid nitrogen and cryocooler system. Assembly of cryogenic systems and inner detector components underground will be ongoing throughout 2012. Commissioning and first data are scheduled for late 2013.

References

[1] E. Aprile et al. Dark matter results from 100 live days of XENON100 data. *Phys. Rev. Lett.*, 107:131302, 2011.

[2] H. H. Loosli. A dating method with ^{39}Ar. *Earth and planetary science letters*, 63:51–62, 1983.

[3] M.G. Boulay and A. Hime. Technique for direct detection of weakly interacting massive particles using scintillation time discrimination in liquid argon. *Astroparticle Physics*, 25:179–182, 2006.

[4] W. H. Lippincott et al. Scintillation time dependence and pulse shape discrimination in liquid argon. *Phys. Rev.*, C78:035801, 2008.

[5] M.G. Boulay et al. Measurement of the scintillation time spectra and pulse-shape discrimination of low-energy beta and nuclear recoils in liquid argon with DEAP-1, 2009. arxiv.org:0904.2930.

[6] D. Acosta-Kane et al. Discovery of underground argon with low level of radioactive ^{39}Ar and possible applications to WIMP dark matter detectors. *Nucl. Instrum. Meth. A.*, 587:46–51, 2008.

[7] J. D. Lewin and P. F. Smith. Review of mathematics, numerical factors, and corrections for dark matter experiments based on elastic nuclear recoil. *Astropart. Phys.*, 6:87–112, 1996.

DARWIN
dark matter WIMP search with noble liquids

Laura Baudis, for the DARWIN consortium[1]
Physics Institute, University of Zürich, CH-8057 Zürich, Switzerland
E-mail: laura.baudis@physik.uzh.ch

Abstract. DARWIN (dark matter wimp search with noble liquids) is a design study for a next-generation, multi-ton dark matter detector in Europe. Liquid argon and/or liquid xenon are the target media for the direct detection of dark matter candidates in the form of weakly interacting massive particles (WIMPs). Light and charge signals created by particle interactions in the active detector volume are observed via the time projection chamber technique. DARWIN is to probe the spin-independent, WIMP-nucleon cross section down 10^{-48}cm^2 and to measure WIMP-induced nuclear recoil spectra with high-statistics, should they be discovered by an existing or near-future experiment. After a brief introduction, I will describe the project, selected R&D topics, expected backgrounds and the physics reach.

1. Introduction

Astrophysical observations show that 83% of gravitating matter in our universe is non-luminous and non-baryonic. The dark matter might be in the form of neutral, weakly interacting, stable or long-lived elementary particles, so-called WIMPs, which have eluded direct observation so far. Produced in the early universe, WIMPs would naturally lead to the observed dark matter abundance [1] and are predicted to exist in extensions to the Standard Model of particle physics [2]. Galactic WIMPs may be detected via scatters off atomic nuclei in deep underground experiments [3]. Since expected signal rates are below one interaction per kilogram of target material and year and momentum transfers are around 10 MeV - 100 MeV [4, 5], large detector masses, low energy thresholds and ultra-low backgrounds are essential experimental requirements to directly observe these hypothetical particles. Experiments using xenon [6, 7, 8, 9] and argon [10] as a homogeneous detection medium in a time projection chamber have reached sensitivities of $\sim 10^{-44}$cm^2 for the spin-independent scattering cross section on nucleons. While a factor of five improvement is predicted with data already in hand [8], ton-scale experiments under commissioning [11] or construction [12] will likely probe the cross section region down to $\sim 5\times 10^{-47}$cm^2. Notwithstading this remarkable leap in sensitivity, and assuming a local density and velocity distribution which are inferred from astronomical observations, significantly larger detectors are requisite to determine WIMP properties, such as its mass, scattering cross section

[1] The following institutions are members of the consortium (as of December 2011): ETH Zürich, University of Zürich (CH); Karlsruhe, Mainz, MPIK Heidelberg, Münster (DE); Nikhef (NL); Subatech (FR); Weizmann Institute of Science (IL); INFN (IT): Bologna, L'Aquila, LNGS, Milano, Milano Bicocca, Napoli, Padova, Pavia, Perugia, Torino; Associated groups: Columbia, Princeton, UCLA, Arizona State (USA).

and possibly spin [13]. To convincingly demonstrate the dark matter nature of a signal, a measurement of its interaction rate with multiple target materials is compulsory.

2. Technologies

The DARWIN study [14, 15, 16] is focused on a multi-ton liquid argon and/or xenon experiment rooted in the noble liquid time projection chamber (TPC) technique. The TPCs will record the prompt scintillation light[2] created when a particle interacts in the active detector volume along with the few liberated electrons after they are drifted in a strong electric field[3] and extracted into the vapor phase residing above the liquid. The prompt light signal will be observed by an array of photosensors immersed in the liquid, the electrons will be detected either directly, or indirectly via proportional scintillation in the gas phase with a second array of photosensors. The time difference between the prompt and delayed signals determines the z-position of an event, the spatial distribution of the delayed signal yields its $x - y$-position. The relative size of the charge and light signals, as well as their time structure will be used to distinguish nuclear recoils, as expected from WIMP scatters, from electronic recoils, which make the majority of the background. The spatial resolution allows to define an innermost, low-background volume and to reject fast neutrons, which – in contrast to WIMPs – tend to multiple scatter[4].

DARWIN will immensely benefit from the research and development, and from the construction and operation experience gained with XENON10 [17], XENON100 [18], XENON1T [12], WARP [10], ArDM [11], DarkSide [19], and much of the ongoing work is carried out within the framework of these projects. Here I mention a few studies only, some of these are specific to DARWIN. Other work deals with the cryogenic, gas purification, circulation, storage and recovery systems; with the external water Cerenkov shield and its potential extension with a liquid scintillator (depending on the depth of the underground laboratory – the Gran Sasso Laboratory and the Modane extension are under consideration); with material screening, selection and radon emanation measurements; with high-voltage systems, electrodes and field uniformity simulations; with low-noise, low-power electronics, cables and connectors, trigger schemes, data acquisition and treatment; with Monte Carlo simulations of the expected background noise, of the light collection efficiency and position reconstruction capability; with the design of the time projection chamber, of the cryostat and of the calibration system.

Light and charge response: the light and charge yields of noble liquids when exposed to low-energy nuclear recoils (from neutron or potential dark matter interactions) or electronic recoils (from γ- and β-interactions) are studied by several groups participating in DARWIN. A new measurement of the relative scintillation efficiency $\mathcal{L}_{\rm eff}$ in liquid xenon [20] shows an $\mathcal{L}_{\rm eff}$ behavior which is slowly decreasing with energy, with a non-zero value at 3 keV nuclear recoil energy, the lowest measured point. A similar measurement is ongoing for liquid argon [21]. A measurement of the liquid xenon scintillation efficiency for electronic recoils down to 2.3 keV is in progress [22]. A preliminary data analysis indicates that the scintillation yield falls with decreasing energy, as predicted by models of scintillation mechanisms in noble liquids [23]. Nonetheless, the scintillation response at 2.3 keV is observed to be non-zero, confirming that liquid xenon experiments will have a finite sensitivity at such low interaction energies. Measurements of the charge yields of LAr and LXe within the same energy regime are being planned.

Signal readout: the prompt scintillation light is to be observed either with conventional photomultiplier tubes (PMTs) which are low in radioactivity [24] and built to withstand low

[2] The peak photon emission from the transition of exited dimers into their dissociative ground state is centered around 128 nm and 178 nm in liquid argon and xenon, respectively [25].

[3] Typical drift fields are 0.5–1 kV/cm, with electron velocities around $\sim$2 mm/μs [25].

[4] The mean free path of $\sim$MeV neutrons is in the range of tens of cm.

temperatures and high pressures[5] or with a new, hybrid photodetector (QUPID [26]), which has an extremely low radioactivity content (<1 mBq/sensor for U/Th/K/Co) [24] and works both in liquid argon and xenon. The delayed signal can be observed directly, using detectors with single electron sensitivity and high spatial granularity (large electron multipliers [27]), or CMOS pixel detectors coupled to electron multipliers (GridPix [28]), or via proportional scintillation in the gas phase, using gaseous photomultipliers without dead zones (GPMs [29]), PMTs or QUPIDs.

3. Backgrounds, physics reach and timeline

DARWIN will be an "ultimate" argon and/or xenon dark matter experiment, before the solar and atmospheric neutrinos become the main, possibly irreducible background. It will directly probe WIMP-nucleon cross sections down to $\sim 10^{-48}\mathrm{cm}^2$. These cross sections are compatible with recent LHC results, should the dark matter particle turn out to be the neutralino [30, 31, 32]. The external background from gammas, muons and neutrons and the background from detector construction materials will be diminished to negligible levels by external shields, the self-shielding of the noble liquids[6], and the choice of fiducial volumes[7]. More difficult are intrinsic backgrounds from ^{85}Kr and ^{222}Rn decays in xenon and from ^{39}Ar decays in argon. In xenon, the natural krypton concentration is to be reduced by cryogenic distillation to <1 ppt and the radon level in the liquid is to be kept <1 μBq/kg. Argon gas that is extracted from deep underground wells is depleted in the radioactive ^{39}Ar [33]. Still, a background rejection by pulse-shape analysis of $>10^8$ is required in the case of a liquid argon detector [14, 15].

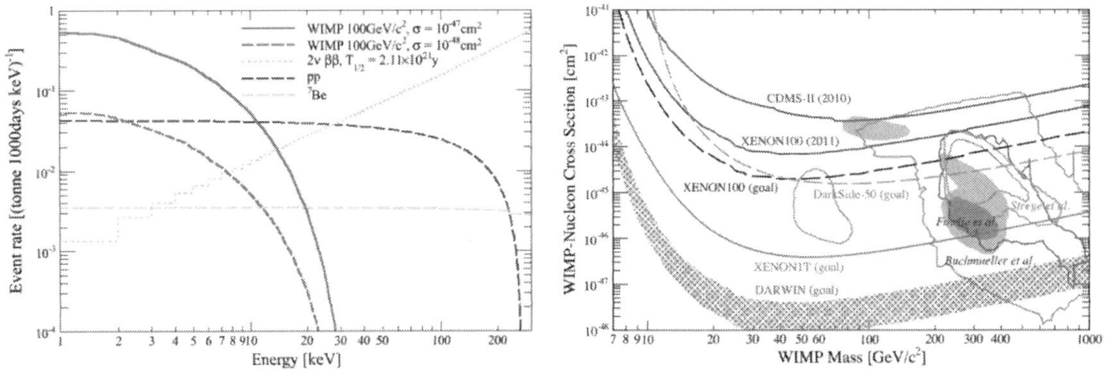

Figure 1. (Left): Expected nuclear recoil spectrum from WIMP scatters in LXe for a spin-independent WIMP-nucleon cross section of 10^{-47} cm^2 (red solid) and 10^{-48} cm^2 (red dashed) and a WIMP mass of 100 GeV/c^2 (using the standard halo model as in [13]), along with the differential energy spectrum for pp (blue) and ^{7}Be (cyan) neutrinos, and the electron recoil spectrum from the double beta decay of ^{136}Xe (green), assuming the natural abundance of 8.9% and the recently measured half life of 2.1×10^{21} yr [34]. Other assumptions are: 99.5% discrimination of electronic recoils, 50% acceptance of nuclear recoils, 80% flat analysis cuts acceptance. (Right): DARWIN's sensitivity goal for spin-independent WIMP nucleon cross sections, existing limits from XENON100 [8] and CDMS-II [35], future goals and updated theoretical predictions from supersymmetry (closed contours and shaded regions) [30, 31, 32].

The left side of figure 1 shows the expected nuclear recoil spectrum from WIMP scatters in xenon together with the background from neutrino-electron elastic scatters of solar neutrinos

[5] One example is the Hamamatsu R11410/R11065 3"-tube for LXe/LAr, currently tested for its performance, long-term stability and radioactivity levels at several DARWIN institutions.
[6] The mean free path of 3 MeV gammas is ~9 cm and ~20 cm in liquid xenon and argon, respectively.
[7] The final choice of the size and target materials are part of the outcome of the study, a baseline scenario is 20 t (10 t) total (fiducial) LAr/LXe mass.

and from the double beta decay of ^{136}Xe. The right side shows the aimed sensitivity of DARWIN, along with existing best upper limits on the WIMP-nucleon cross section, projections for the future and theoretically predicted regions from supersymmetric models.

DARWIN, which was endorsed in recent updates to the European and Swiss roadmaps for astroparticle and particle physics [36, 37], has officially started in 2010. A rough time schedule is the following: a technical design study is to be ready in spring 2013, leading to a letter of intent and engineering studies towards the proposal of a concrete facility in spring 2014, and a technical design report for detector construction by the end of 2014. The shield and detector construction phase is to start in 2015, commissioning in late 2016 with the start of the first physics run by mid 2017.

Acknowledgements: This work is supported through the first ASPERA common call, by SNF grant 20AS21-129329, by the DoE, and by the individual institutions participating in the study.

References

[1] B. W. Lee and S. W. Weinberg, *Phys. Rev. Lett.* **39**, 165 (1977).
[2] J. Feng, *Ann. Rev. Astron. Astrophys.* **48**, 495 (2010).
[3] M. W. Goodman and E. Witten, *Phys. Rev.* D **31**, 3059 (1985).
[4] G. Jungmann, M. Kamionkowski, K. Griest, *Phys. Rep.* **267**, 195 (1996).
[5] J. D. Lewin and P. F. Smith, *Astrop. Phys.* **6**, 87 (1996).
[6] J. Angle et al. (XENON10 Collaboration), *Phys. Rev. Lett.* **100**, 021303 (2008).
[7] E. Aprile et al. (XENON Collaboration), *Phys. Rev. Lett.* **105**, 131302 (2010).
[8] E. Aprile et al. (XENON Collaboration), *Phys. Rev. Lett.* **107**, 131302 (2011)
[9] V. N. Lebedenko et al. (ZEPLIN-III Collaboration), *Phys. Rev. D* **80**, 052010, (2009), V. N. Lebedenko et al., *Phys. Rev. Lett.* **103**, 151302, (2009), D.Yu Akimov et al., `arXiv:1110.4769` (2011).
[10] R. Brunetti et al. (WARP Collaboration), *Astropart. Phys.* **28**, 495 (2008).
[11] A. Rubbia et al. (ArDM Collaboration), *J. Phys. Conf. Ser.* **39**, 129 (2006).
[12] E. Aprile et al. (XENON Collaboration), XENON1T at LNGS, *Proposal*, April (2010) and *Technical Design Report*, October (2010).
[13] M. Pato, L. Baudis, G. Bertone, R. R. de Austri, L. E. Strigari and R. Trotta, *Phys. Rev. D* **83**, 083505 (2011).
[14] L. Baudis (DARWIN Consortium), PoS(IDM2010)122 (2010), `arXiv:1012.4764v1 [astro-ph.IM]`.
[15] M. Schumann (DARWIN Consortium), `arXiv:1111.6251v1 [astro-ph.IM]`.
[16] `http://darwin.physik.uzh.ch`
[17] E. Aprile et al. (XENON10 Collaboration), *Astroparticle Physics* **34**, 679-698 (2011).
[18] E. Aprile et al. (XENON Collaboration), `arXiv:1107.2155v1 [astro-ph.IM]`.
[19] A. Wright (DarkSide), `arXiv:1109.2979` (2011).
[20] G. Plante et al., *Phys. Rev. C* **84**, 045805 (2011).
[21] C. Regenfus, talk presented at TAUP2011 (2011).
[22] A. Manalaysay, talk presented at TAUP2011 (2011).
[23] M. Szydagis *et al.*, *JINST* **6** P10002 (2011).
[24] E. Aprile et al. (XENON Collaboration), *Astropart. Phys.* **35**, 43 (2011).
[25] E. Aprile, A.E. Bolotnikov, A.I. Bolozdynya, and T. Doke, *Noble Gas Detectors*, WILEY-VCH (2006).
[26] A. Teymourian, D. Aharoni, L. Baudis, P. Beltrame et al., *Nucl. Instr. Meth. A* **A 654**, 184 (2011).
[27] A. Badertscher et al, *Nucl. Instr. Meth. A* **641**, 48 (2011).
[28] V. Blanco Carballo et al., *JINST* **5**, P02002 (2010).
[29] S. Duval et al., `arXiv:1110.6053` (2011).
[30] A. Fowlie, A. Kalinowski, M. Kazana, L. Roszkowski, Y.-L. Sming Tsai, `arXiv:1111.6098v1 [hep-ph]`.
[31] O. Buchmueller, *et al.*, `arXiv:1110.3568v1 [hep-ph]`, `arXiv:1112.3564v1 [hep-ph]`.
[32] C. Strege, G. Bertone, D. G. Cerdeno, M. Fornasa, R. Ruiz de Austri, R. Trotta, `arXiv:1112.4192v1 [hep-ph]`.
[33] D. M. Mei, Z. B. Yin, J. Spaans, M. Koppang, et al., *Phys. Rev. C* **81**, 055802 (2010).
[34] N. Ackerman et al., *Phys Rev Lett.* **107**, 212501 (2011).
[35] Z. Ahmed et al., *Science*, 1186112 (2010).
[36] `http://www.aspera-eu.org/images/stories/roadmap/SAC-Roadmap-Nov-1-2011-final.pdf`
[37] `http://www.chipp.ch/documents/roadmap.pdf`

Status of indirect dark matter detection

Pasquale D. Serpico

LAPTh, Univ. de Savoie, CNRS, B.P.110, Annecy-le-Vieux F-74941, France

E-mail: serpico@lapp.in2p3.fr

Abstract. Indirect astrophysical channels (such as gamma rays, neutrinos, and cosmic ray antimatter) are a cornerstone in the dark matter particle identification program. A review of the present constraints is presented, together with some perspectives for the near future: we argue that the approach of "fitting any spectral feature to a dark matter model" is unlikely to lead to a convincing indirect detection of dark matter. Rather—if the WIMP paradigm for dark matter is correct—guidance from collider and direct detection programs is expected to allow soon for more fruitful *a priori* searches of correlated signatures in many indirect channels.

1. Introduction

The known matter (be it in visible form or not) cannot account but for a small fraction of the gravitational attraction binding together the largest structures of the universe, from galaxies to clusters and beyond. This is one of the most surprising results of 20th century astrophysics, that required evidence to be collected over many decades in order to be fully appreciated: from the first observations of the Coma cluster by Zwicky (summarized in [1]) to the puzzlingly flat rotation curves of spiral galaxies studied by V. Rubin and collaborators (see [2] for a review); from the indirect evidence from cosmology (like the baryometric measurement first obtained via primordial nucleosynthesis and the depth of the potential fluctuations inferred from CMB) to the spectacular recent inferences from colliding clusters thanks to lensing (a short account of most of these observations has been given in [3]). Yet, all these detections are gravitational: the universality of gravity prevents us to identify the particle physics nature of this (or these) "new species", dubbed dark matter (DM). A re-discovery of DM via other channels is clearly needed for that purpose. In this framework, indirect detection (ID) of DM means looking for consequences of DM interactions elsewhere than in the Lab: namely via putative decays, annihilations, and energy transfer to baryons. The motivations of this strategy are clear: i) It is an obvious thing to do, since DM has been seen only elsewhere than in the Lab, after all! ii) These features may imply a further impact of DM on cosmology or astrophysics, via non-gravitational effects on the baryonic fluids. iii) It provides an additional handle on properties one cannot probe otherwise in the Lab. That said, it is also worth recalling that

- The presence of indirect signatures is by no means guaranteed, rather is model-dependent.
- It need not be an electroweak (EW) scale signature, neither necessarily an annihilation one. Notable exceptions exist, two examples of which are keV mass scale sterile neutrinos possibly emitting a X-ray line via decay (see [4] for a review), or proposed annihilation signals of MeV DM impacting wildly different systems, from the Galactic Bulge soft gamma-ray emission [5] to indirect effects in primordial nucleosynthesis [6].

- There is no *astrophysical or cosmological evidence* for the electroweak scale being the right one for explaining the DM problem. Rather, it is a particle physics theory prejudice.

With these caveats in mind, if is fair to say that the most popular class of DM candidates are Weakly Interacting Massive Particles (WIMPs), based on the following three main arguments:

- The "WIMP miracle", the intriguing coincidence that a thermal relic with electroweak coupling and mass scale matches the cosmological measurement, $\Omega_{\rm DM} \approx 0.25$, within an order of magnitude or so.
- To ease agreement with EW observables (as well as to fulfill constraints such as the lower limit on the proton lifetime) in beyond the standard model (BSM) scenarios, model-builders often require some discrete symmetry differentiating SM particles from BSM ones, like R-parity in SUSY, K-parity in Extra-Dimensional modes or T-parity in Little Higgs. In turn, this implies the stability of the lightest particle of the BSM sector. More generally, this kind of arguments together with the first point above has brought some confidence that, whatever the BSM physics at the EW scale may be, it could be related to DM.
- Last but not least, EW-scale candidates have a rich phenomenology, allowing for more creativity from the model-building perspective, together with more detection strategies via collider, direct, and indirect techniques.

In the rest of this mini-review I will thus concentrate on WIMP-like candidates and signatures, but it is important to keep in mind other possibilities (among which some well motivated ones are axions, superheavy DM, and superWIMPs; see [7] for a review). Some of these have independent motivations and/or peculiar signatures, requiring ad hoc searches: at TAUP 2011 an example has been illustrated in the talk by M. Grefe on indirect searches for gravitino dark matter. When restricting oneself to WIMP candidates, the basic strategy can be summarized as follows:

- To demonstrate that astrophysical DM is made of particles: locally, via direct detection (DD) in underground labs; remotely, via ID.
- Possibly, create DM candidates in the controlled environments of accelerators, most notably the LHC.
- Find consistency between properties of the two classes of particles. Ideally, we would like to be able to calculate the relic abundance and DD and ID signatures on the basis of collider data, testing the link with cosmology and production mechanism.

For ID, one ca use different particles as probes, as reviewed in the following Secs. 2, 3, and 4. Section 5 is instead devoted to a discussion, and Sec. 6 to the conclusions. The emphasis is given to recent results and/or results mentioned at the TAUP 2011 conference.

2. Gamma rays

Gamma rays have several advantages as DM discovery tools: they retain directionality, so they map the DM distribution (squared, for annihilations) in the halo of the Milky Way as well as any other (sub)structures, either of our Galaxy or of the extragalactic sky. Another good feature is that gamma rays are generically found as byproducts of the DM, via the fragmentations/decays of the different SM particles created in the annihilation/decay process.

Theoretical predictions of spectra can be obtained via standard tools used in particle physics like PYTHIA, or via appropriate fittings/parameterizations to these results, see for example [8]. On the other hand, the photon line final states are suppressed by the very request of the DM being "dark": they must proceed trough loops and their intensity results both suppressed and model-dependent. This channel will be briefly commented upon in Sec. 5. As long as "vanilla WIMPs" are concerned, only a few qualitatively different spectra can be expected (e.g., τ leptons vs. softer quark-like ones) and quasi-universal spectra can be thus used as templates for searches.

Concerning the angular distribution of the signal, at least for the prompt photons (see Sec. 4.3 for a comment on non-prompt ones) it is completely specified by the DM distribution: for some expectations on amplitudes and anisotropies of the signal see e.g. [11]. Of course, the lack of detailed knowledge of the DM distribution translates into an uncertainty in the signal angular shape and normalization, which is particular important for the Galactic Center region and the extragalactic signal.

Perhaps most important, the signal prediction is often less troublesome than the presence of a significant background of astrophysical nature: the gamma-ray sky revealed by space telescopes like Fermi or ground-based Atmospheric Cherenkov Telescopes (ACTs) like HESS, MAGIC or VERITAS looks nothing like DM expectations, neither in spectra nor morphology. Not only backgrounds are important, but their nature and characteristics are revealed step by step, with the increased sensitivity of the instruments. The challenge might be compared with looking for BSM physics in LHC data while having only vague ideas of what the QCD and EW backgrounds look like. To different degrees of complexity, this is the basic problem of most ID searches. However, photons keep at least the advantage of directionality, so the search strategies can be "customized" for different targets, as illustrated in the following subsections.

2.1. Dwarf galaxies

These targets, which are known satellites of the Milky Way, are relatively rich in DM with respect to baryons, and are usually assumed to have no intrinsic background [1]. The dominating foreground uncertainty then comes from the uncertainty in the diffuse galactic background estimate and is of lesser concern. On the other hand the signal depends on the estimated distance and volume average of the squared DM density, and it is only as robust as the latter estimates are. Recent bounds from a *stacked* analysis by the Fermi collaboration—which rely on their own astrophysical data analysis for the dwarfs—nominally exclude "generic thermal" s-wave relic WIMPs annihilating into b-quarks with mass up to 27 GeV, and into τ's up to 37 GeV: see the Fermi-LAT collaboration analysis in [12] and the independent one in [13]. While ACTs have sensitivity to higher DM mass values and have considered dwarf galaxies as interesting targets, the bounds obtained are rather weak, orders of magnitude away from expectations. The recent MAGIC bounds from Segue 1 [14], of HESS from globular clusters M15 and NGC 6388 [15] or the 2010 ones by VERITAS [16] should give a feeling on how far the sensitivity is from theoretically interesting values of parameters.

2.2. Diffuse gamma-ray flux

The signal from the smooth Galactic DM halo constitutes a lower limit to the expected one, since it can be enhanced in case of clumpiness. Also, the signal is relatively large in terms of absolute numbers of photons. The major drawback is that the astrophysical background is also large and actually dominating. Basically, the strength of the constraints depend on how one deals with this background: one can take the hyper-conservative approach of using the observed number of photons and assume that the DM signal does not overshoot it, see e.g. [17]. Or one can vary the foregrounds within reasonable ranges of the parameters and use conservative assumptions to derive the bounds, as done in [18]. An independent analysis was performed in [19] based on the assumption of approximate cylindrical symmetry of signal (in galactic coordinates) and uncorrelated pixels, obtaining results essentially in agreement with [18]. It is also worth noting that the optimal strategy maximizing the signal/background is actually dependent not only on

[1] Needless to say, this assumption is going to break down at some level. For example, it is known that dwarfs can harbour globular clusters; based on Fermi data [9], globular clusters often host millisecond pulsars which are gamma-emitters. For the promising (nearby, DM-rich) Sagittarius dwarf, the presence of the globular cluster M54 alone makes one expect an astrophysical VHE signal two orders of magnitude larger than a typical thermal WIMPs one, and detectable with the future CTA [10].

the level of foregrounds, but also on the assumed profile of the inner halo (see the simplified example discussed in [20]).

At high Galactic latitutes, it is expected that both the Galactic Halo and the extragalactic flux contribute to the signal: actually, this is known to be the case at least for the background. In general, despite the relevant statistics, in this case it is even trickier to disentangle the signal from the numerous astrophysical foregrounds; a recent analysis has been presented by F. Donato at this conference (see [21]), while new strategies have been discussed e.g. in [22, 23]. Some effort is also going into refining and simplifying the theoretical predictions of the extragalactic signal, see for example [24].

2.3. Inner Galaxy

The Galactic Center region has been recognized long ago as a very promising target, given its potentially very high statistics and relatively narrow angular extent, which is suitable for detection also at ACTs [25]. Needless to say, it is subject to the usual problems in the foreground estimates, which involve in this case both discrete sources (point-like or extended) and the diffuse flux, see e.g. [26]. In general, in order to reduce the impact of the complex astrophysics at the Galactic Center, it is often more robust to look away from it to obtain reliable bounds. This strategy was followed for example to obtain the HESS bound in [27]: a *signal region* close to the Galactic Center but as much as possible free from backgrounds was selected; then a similar shape region where signal is expected to be smaller was selected for background estimate. The resulting analysis leads to the most stringent bound on DM from ACTs I am aware of, yet still one order of magnitude or larger away from expectations. Additionally, it is a halo-model dependent constraint, relaxing considerably for cored halo profiles.

Compared to early estimates based on pre-Fermi knowledge [28, 29], Fermi-LAT data have revealed a more complex structure of sources at the Galactic Center: intriguing "excesses" should thus be weighted against poorly understood backgrounds, see e.g. [30, 31, 32]. The Fermi-LAT team analysis is in fact still in progress: for preliminary results see [33].

3. Neutrinos

The same kind of scattering reactions probed in direct detection techniques may be responsible for the capture of halo DM particles by the Sun (or, to a minor extent, the Earth): after sinking to the core of the Sun, DM particles can annihilate at a significant rate, especially if equilibrium between capture and annihilation is reached. At steady state, the annihilation rate depends "almost exactly" (but see [34]) on the same product of local density and cross section entering DD experiments. Since the Sun is mostly made of protons, bounds on spin-dependent cross-sections are comparatively stronger than those on spin-independent ones, where terrestrial detectors can exploit the coherent scattering on massive target nuclei. Recently, Ref. [35] used Super-Kamiokande upper limits on additional flux on top of atmospheric neutrinos to probe light DM (order 10 GeV or less), of interest on the light of some direct detection results (DAMA/LIBRA, CoGeNT, CRESST). The analysis excludes for example spin-dependent interpretations of these data in terms of either s-wave or p-wave thermal relics, basically independent of the SM final state. Even for spin-independent interpretations significant tensions exist (especially for s-wave annihilation), unless the annihilation proceeds into light quark species [35, 36].

On the other hand, bigger (but higher-threshold) neutrino telescopes under water or ice have sensitivity to heavier DM candidates: recent bounds in interesting parts of parameter space have been derived by IceCube, see e.g. [37] or the contribution of M. Danninger. Encouraging progress has also been reported by ANTARES (see contribution to this conference by Vincent Bertin). Some less constraining IceCube bounds from the Inner Galaxy have been recently reported in [38].

4. Charged particles

Indirect DM searches via cosmic rays has been considered long ago, see e.g. [39]. Yet, this method is slightly more indirect since it relies not only on DM physics (like mass and cross sections) and astrophysics (halo distributions), but also on the plasma astrophysics processes which regulate the diffusive propagation in the Galaxy (see [40, 41, 42] for introductions to this topic). In general, antimatter channels are preferred with respect to matter CRs due to the lower astrophysical background expected (essentially secondaries from inelastic collisions in the interstellar medium). Here we briefly review antiproton and lepton constraints. We mention that $e^{\pm}$ can also produce synchroton signals in the Galactic magnetic field; this topic was covered in the talks by R. Lineros and M. Taoso, see also [43] and [44].

4.1. Antiproton flux

Antiproton constraints are particularly interesting for relatively light DM candidates. For example, recent analyses [45, 46] disfavour an s-wave annihilating thermal relic interpretation of some direct detection "signals". On the other hand, constraints on heavier DM candidates are less stringent. Yet, some of these are interesting in the context of models with high annihilation cross sections, such as some of those invoked a few years ago to explain the PAMELA positron fraction. In both cases, antiproton constraints tend to corroborate similar or stronger constraints coming from gamma ray results.

4.2. Electron and positron fluxes

Charged leptons are a relatively worse channel for indirect DM searches, at least as a discovery tool: Compared to antiprotons, for example, they are more sensitive to source and propagation inhomogeneities as well as to energy losses; also, the signal has to compete with many more potential astrophysical backgrounds: there are many more leptonic accelerator candidates than hadronic ones! For a recent review of astrophysical interpretations of positron data, see [47]. To use an analogy, it would be like starting to look for new physics at LHC in soft jets, where non-perturbative QCD plays an important role. Compared with the excitement of the past few years, at this conference there has been relatively little discussion of these topics. One exception has been the talk by E. Borriello, based on [48], on DM vs. astrophysical diagnostics via anisotropy constraints. Not surprisingly, a general trend in current research is focusing on possible ways to discriminate between DM signals and astrophysical foregrounds. One the other hand, recent results by Fermi [49] allowed for an indirect measurement of the separate electron and positron flux thanks to the geomagnetic "Earth shadow". They confirm the previous trend of rising positron fraction indicated by PAMELA [50] up to $E \simeq 200$ GeV, thus requiring higher minimum masses for DM to fit the rise. Since existing bounds are relatively stronger at higher DM masses, the parameter space for DM interpretations of positron data is narrowing down even further.

4.3. Other constraints and recent progress in phenomenology

Some activity has also been developed to investigate the impact of DM in a cosmological context. In particular, we mention here that energetic byproducts of DM annihilating during the *Dark Ages* (redshift $z \simeq 100$–1000) can ionize baryons. In turn, the CMB probes this phenomenon in particular via the optical depth τ. Detailed analyses such as [51] indicate that particles in the 5–10 GeV mass range annihilating into $\mu^{\pm}$ or $e^{\pm}$ can be constrained at the level of the thermal relic cross section already with present data. The forecast for Planck looks even more promising for light DM candidates.

Recent phenomenologically inspired model-building has also focused on heavy ($m \gtrsim$ TeV) "leptophilic" DM candidates. Indirect signatures of these candidates (for example gammas) are mostly observable at $E \ll m$. In this range, it is important to consider: i) *tertiary* signatures, like inverse Compton photons from the scattering on interstellar light of $e^{\pm}$ originating from DM; ii) multi-body final states, from sub-threshold virtual states [52, 53] or W,Z-strahlung [54, 55, 56, 57, 58, 59, 60, 61, 62]. A talk on electroweak effects has been given at this conference by A. De Simone.

5. A paradigm shift: towards indirect dark matter detection perspectives

When trying to move from the *constraint phase* to the *detection phase*, one immediately faces the trouble of identifying *reliably* genuine signatures of DM as opposed to astrophysical foregrounds. A first strategy is to look for (what are believed to be) *unique* signatures of DM, such as non-negligible fluxes of antideuterons in cosmic rays at relatively low energy, high energy neutrinos from the Sun or Earth Core, or gamma-ray lines. A discovery via these means, while possible, requires a great dose of luck for typical WIMP parameters: the signals are expected to be small and in some cases very special detector requirements are needed (as the mass/charge resolution in the case of antideuterons, or the size of neutrino detectors). Recently, Fermi-LAT data have allowed to derive new bounds on gamma-ray lines [63, 64], which are nonetheless orders of magnitude away from expectations. Despite some hope to investigate conceptually similar but more promising spectral features at future ACTs (see talk by C. Weniger and Ref. [65]), it is fair to say that not much could be learned from a negative result. These channels cannot be used as a crucial test for the WIMP paradigm.

What I believe to be a poor strategy is instead to ask oneself if a CR feature can be fitted with some DM model. While popular in the current literature, this approach can most likely lead to fool oneself. In fact, most DM models have enough free parameters to control: a) spectral shape (by adjusting the final state and/or propagation parameters) b) endpoint/energy scale (by adjusting the mass) c) normalization (by adjusting the mass, cross section or decay rate, as well as DM profile, the latter also controlling angular shape.) It is more likely than not that one will be able to find a reasonably good fit! The real issues for "claiming indirect detection" are however:

i) To find an explanation of *many phenomena* with particle physics *motivated models.*
ii) To predict indirect detection features *which cannot be understood by known astrophysics.*
iii) To have predictive links with direct search or collider signatures.

Needless to say, the number of key free parameters should be relatively small.

By looking at the near past, one immediately realizes that the DM community has probably been fooling itself in an ID several times: this is the case of the Galactic Center gamma-ray emission [66, 67], the 511 keV emission from positron annihilation in the Galaxy (for a review see [68]), or the e^+ fraction in cosmic rays [50], which have all been interpreted as possible if not likely DM signals. Later on, more mundane (but still astrophysically exciting!) origins have gained more and more plausibility. So, *what lesson should we draw?*

My opinion is that, barring a few hints still debated, most indirect signatures told us that *DM indirect signals are not dominant*, rather at the level or even well below astrophysical foregrounds, some of which have been discovered and investigated along the way. Experiments like AMS-02 [69] and ongoing as well as forthcoming gamma-ray detectors will provide further checks of the internal consistency of simple models of the cosmic rays. This field is being re-defined by high-quality data, extending over a larger dynamical range. A few years ago, the attitude was that the major uncertainties in ID was due to interstellar medium and propagation parameters, as well as nuclear physics input. A larger community now appreciates that a more challenging limitation to overcome is the lack of detailed knowledge of the sources. Fortunately for us, this is a wonderful time to do high-energy astrophysics, with plenty of data and excellent perspectives!

6. Conclusions

In summary, astrophysical and cosmological indirect probes have had a major impact on particle physics: they have told us that BSM physics does exist and contributes significantly to shape the Universe we live in!

On the other hand, they do not tell us *the scale* at which BSM physics should appear. More and more indirect detection channels are limited by systematic errors in the knowledge of foregrounds rather than by statistical errors. As a consequence, until a better understanding of high energy astrophysics will be achieved, blind searches for DM will become more and more challenging. To a large extent, it is likely that if the WIMP paradigm is correct, collider searches will collect the first unambiguous signals for BSM physics at the electroweak scale, perhaps accompanied by further evidence of DM at direct detectors underground. With the LHC finally up and running, it is advisable to go back to the standard, healthy practice in particle physics, when experimental results were paving the way to theoretical interpretations, which in turn would predict also independent and indirect signatures. We can just look forward to repeating this path to discovery, whose advantages for indirect detections cannot be underestimated: knowing what and where to look for DM would avoid the "look elsewhere effect" and increase the credibility of a signal hint. In fact, it should be appreciated that if a signal is found in other channels (collider/direct detection), we would still need ID:

- To confirm that whatever we find in the Lab is the same "dark stuff" responsible for astrophysical and cosmological observations.
- To access particle information not otherwise available in the Lab (annihilation cross section or decay time, for example).
- To infer cosmological properties of DM (e.g. power spectrum of DM at very small scales) not easily accessible otherwise.

In the meanwhile, ongoing or near future astrophysical observations will help gaining more sensitivity and precision as well as better understanding of astrophysical sources and propagation parameters. By the next edition of TAUP, we might be talking of the *recent detection of DM* or, perhaps, should start contemplating the need for a new paradigm in this fascinating quest.

References

[1] F. Zwicky, Astrophys. J. **86**, 217–246 (1937).
[2] Y. Sofue, V. Rubin, Ann. Rev. Astron. Astrophys. **39**, 137 (2001) [arXiv:astro-ph/0010594].
[3] M. Roos, arXiv:1001.0316 [astro-ph.CO].
[4] A. Boyarsky, O. Ruchayskiy, M. Shaposhnikov, Ann. Rev. Nucl. Part. Sci. **59**, 191–214 (2009) [arXiv:0901.0011 [hep-ph]].
[5] C. Boehm, D. Hooper, J. Silk, M. Casse, J. Paul, Phys. Rev. Lett. **92**, 101301 (2004) [astro-ph/0309686].
[6] P. D. Serpico, G. G. Raffelt, Phys. Rev. **D70**, 043526 (2004) [astro-ph/0403417].
[7] G. Bertone (ed.), "Particle dark matter: Observations, models and searches," Cambridge Univ. Pr. (2010).
[8] J. A. R. Cembranos, A. de la Cruz-Dombriz, A. Dobado, R. A. Lineros, A. L. Maroto, Phys. Rev. **D83**, 083507 (2011) [arXiv:1009.4936 [hep-ph]].
[9] A. A. Abdo *et al.* [Fermi-LAT Collaboration], Astron. Astrophys. 524, **A75** (2010).
[10] A. Viana *et al.*, arXiv:1103.2627 [astro-ph.HE].
[11] D. Hooper, P. D. Serpico, JCAP **0706**, 013 (2007) [astro-ph/0702328].
[12] M. Ackermann *et al.* [Fermi-LAT Collaboration], Phys. Rev. Lett. **107**, 241302 (2011) [arXiv:1108.3546 [astro-ph.HE]].
[13] A. Geringer-Sameth, S. M. Koushiappas, Phys. Rev. Lett. **107**, 241303 (2011) [arXiv:1108.2914 [astro-ph.CO]].
[14] J. Aleksic *et al.* [MAGIC Collaboration], JCAP **1106**, 035 (2011) [arXiv:1103.0477 [astro-ph]].
[15] A. Abramowski *et al.* [H.E.S.S. Collaboration], Astrophys. J. **735**, 12 (2011) [arXiv:1104.2548 [astro-ph.HE]].
[16] V. A. Acciari *et al.* [VERITAS Collaboration], Astrophys. J. **720**, 1174 (2010) [arXiv:1006.5955 [astro-ph.CO]].
[17] M. Cirelli, P. Panci, P. D. Serpico, Nucl. Phys. **B840**, 284–303 (2010) [arXiv:0912.0663 [astro-ph.CO]].

[18] G. Zaharijas, A. Cuoco, Z. Yang, J. Conrad [Fermi-LAT Collaboration], arXiv:1012.0588 [astro-ph.HE].
[19] E. J. Baxter, S. Dodelson, Phys. Rev. D **83**, 123516 (2011) [arXiv:1103.5779 [astro-ph.CO]].
[20] P. D. Serpico, G. Zaharijas, Astropart. Phys. **29**, 380-385 (2008) [arXiv:0802.3245 [astro-ph]].
[21] F. Calore, V. De Romeri, F. Donato, arXiv:1105.4230 [astro-ph.CO].
[22] S. Dodelson, A. V. Belikov, D. Hooper, P. Serpico, Phys. Rev. **D80**, 083504 (2009) [arXiv:0903.2829 [astro-ph.CO]].
[23] E. J. Baxter, S. Dodelson, S. M. Koushiappas, L. E. Strigari, Phys. Rev. **D82**, 123511 (2010) [arXiv:1006.2399 [astro-ph.GA]].
[24] P. D. Serpico, E. Sefusatti, M. Gustafsson, G. Zaharijas, Mon. Not. Roy. Astron. Soc. Letters, in press, [arXiv:1109.0095 [astro-ph.CO]].
[25] L. Bergström, P. Ullio, J. H. Buckley, Astropart. Phys. **9**, 137–162 (1998) [astro-ph/9712318].
[26] P. D. Serpico, D. Hooper, New J. Phys. **11**, 105010 (2009) [arXiv:0902.2539 [hep-ph]].
[27] A. Abramowski *et al.* [H.E.S.S. Collaboration], Phys. Rev. Lett. **106**, 161301 (2011) [arXiv:1103.3266 [astro-ph.HE]].
[28] S. Dodelson, D. Hooper, P. D. Serpico, Phys. Rev. **D77**, 063512 (2008) [arXiv:0711.4621 [astro-ph]].
[29] E. A. Baltz *et al.*, JCAP **0807**, 013 (2008) [arXiv:0806.2911 [astro-ph]].
[30] D. Hooper, T. Linden, arXiv:1110.0006 [astro-ph.HE].
[31] A. Boyarsky, D. Malyshev and O. Ruchayskiy, Phys. Lett. B **705**, 165 (2011) [arXiv:1012.5839 [hep-ph]].
[32] K. N. Abazajian, JCAP **1103**, 010 (2011) [arXiv:1011.4275 [astro-ph.HE]].
[33] A. Morselli, B. Canadas [Fermi-LAT Collaboration] arXiv:1012.2292 [astro-ph.HE].
[34] P. D. Serpico, G. Bertone, Phys. Rev. **D82**, 063505 (2010) [arXiv:1006.3268 [astro-ph.HE]].
[35] R. Kappl, M. W. Winkler, Nucl. Phys. **B850**, 505–521 (2011) [arXiv:1104.0679 [hep-ph]].
[36] S.-L. Chen, Y. Zhang, Phys. Rev. **D84**, 031301 (2011) [arXiv:1106.4044 [hep-ph]].
[37] K. Helbing [IceCube Collaboration], arXiv:1107.5227 [hep-ex].
[38] R. Abbasi *et al.* [IceCube Collaboration], Phys. Rev. **D84**, 022004 (2011) [arXiv:1101.3349 [astro-ph.HE]].
[39] J. Silk, M. Srednicki, Phys. Rev. Lett. **53**, 624 (1984).
[40] A. W. Strong, I. V. Moskalenko, V. S. Ptuskin, Ann. Rev. Nucl. Part. Sci. **57**, 285–327 (2007) [astro-ph/0701517].
[41] P. Mertsch, PhD Thesis Oxford 2010, arXiv:1012.4239 [astro-ph.HE].
[42] A. Castellina, F. Donato, arXiv:1110.2981 [astro-ph.GA].
[43] T. Bringmann, F. Donato, R. A. Lineros, arXiv:1106.4821 [astro-ph.GA].
[44] N. Fornengo, R. A. Lineros, M. Regis, M. Taoso, arXiv:1110.4337 [astro-ph.GA].
[45] J. Lavalle, Phys. Rev. **D82**, 081302 (2010) [arXiv:1007.5253 [astro-ph.HE]].
[46] C. Evoli, I. Cholis, D. Grasso, L. Maccione, P. Ullio, arXiv:1108.0664 [astro-ph.HE].
[47] P. D. Serpico, Astropart. Phys., in press [arXiv:1108.4827 [astro-ph.HE]].
[48] E. Borriello, L. Maccione, A. Cuoco, Astropart. Phys., in press [arXiv:1012.0041 [astro-ph.HE]].
[49] M. Ackermann *et al.* [Fermi LAT Collaboration], arXiv:1109.0521 [astro-ph.HE].
[50] O. Adriani *et al.* [PAMELA Collaboration], Nature **458**, 607–609 (2009) [arXiv:0810.4995 [astro-ph]].
[51] S. Galli, F. Iocco, G. Bertone, A. Melchiorri, Phys. Rev. D **84**, 027302 (2011) [arXiv:1106.1528 [astro-ph.CO]].
[52] C. E. Yaguna, Phys. Rev. **D81**, 075024 (2010) [arXiv:1003.2730].
[53] K.-Y. Choi, D. Restrepo, C. E. Yaguna, O. Zapata, JCAP **1010**, 033 (2010) [arXiv:1007.1728].
[54] M. Kachelrieß, P. D. Serpico, Phys. Rev. **D76**, 063516 (2007) [arXiv:0707.0209].
[55] N. F. Bell, J. B. Dent, T. D. Jacques, T. J. Weiler, Phys. Rev. **D78**, 083540 (2008) [arXiv:0805.3423].
[56] M. Kachelrieß, P. D. Serpico, M. A. Solberg, Phys. Rev. **D80**, 123533 (2009) [arXiv:0911.0001].
[57] P. Ciafaloni *et al.*, JCAP **1103**, 019 (2011) [arXiv:1009.0224].
[58] N. F. Bell, J. B. Dent, T. D. Jacques, T. J. Weiler, Phys. Rev. **D83**, 013001 (2011) [arXiv:1009.2584].
[59] N. F. Bell, J. B. Dent, T. D. Jacques, T. J. Weiler, Phys. Rev. D **84**, 103517 (2011) [arXiv:1101.3357 [hep-ph]].
[60] P. Ciafaloni *et al.*, JCAP **1106**, 018 (2011) [arXiv:1104.2996 [hep-ph]].
[61] N. F. Bell *et al.*, Phys. Lett. B **706**, 6 (2011) [arXiv:1104.3823 [hep-ph]].
[62] M. Garny, A. Ibarra, S. Vogl, JCAP **1107**, 028 (2011) [arXiv:1105.5367 [hep-ph]].
[63] A. A. Abdo *et al.*, Phys. Rev. Lett. **104**, 091302 (2010) [arXiv:1001.4836 [astro-ph.HE]].
[64] G. Vertongen, C. Weniger, JCAP **1105**, 027 (2011) [arXiv:1101.2610 [hep-ph]].
[65] T. Bringmann, F. Calore, G. Vertongen, C. Weniger, Phys. Rev. D **84**, 103525 (2011) [arXiv:1106.1874 [hep-ph]].
[66] H. Mayer-Hasselwander *et al.*, Astron. Astrophys. **335**, 161 (1998).
[67] F. Aharonian *et al.* [H.E.S.S. Collaboration], Astron. Astrophys. **425**, L13–L17 (2004) [astro-ph/0408145].
[68] N. Prantzos *et al.*, Rev. Mod. Phys. **83**, 1001–1056 (2011) [arXiv:1009.4620 [astro-ph.HE]].
[69] http://ams.cern.ch/

12th International Conference on Topics in Astroparticle and Underground Physics (TAUP 2011) IOP Publishing
Journal of Physics: Conference Series **375** (2012) 012030 doi:10.1088/1742-6596/375/1/012030

Downward-going tau neutrinos and Dark Matter

Viviana Niro

Dipartimento di Fisica Teorica, Università di Torino and INFN, Sezione di Torino
via P. Giuria 1, I-10125 Torino, Italy

E-mail: viviana.niro@gmail.com

Abstract. We discuss the possibility of detecting the downward–going tau neutrino flux from Dark Matter annihilation inside the Sun. In our analysis we focused on the hadronic showers produced by charged–current tau neutrino interactions followed by hadronic tau decay. We include the various sources of tau neutrino backgrounds as well as experimental sources of background due to misidentification of electron and muon events. We find that the downward–going tau neutrino signal has potentially very good prospects for Mton–scale Cherenkov detectors, if the level of misidentification of non–tau events is at the level of percent. In this case, within few years of exposure, a 5 σ significance discovery is potentially reachable for Dark Matter masses in the range from 20 to 300 GeV.

1. Introduction

Indirect Dark Matter (DM) searches look for the products of DM self–annihilation (or decay) occurring in the galactic halo, in the extragalactic environment or in those celestial bodies, like the Sun and the Earth, where they can be gravitationally trapped and accumulated. As a consequence of DM annihilations, a flux of high–energy neutrinos can emerge. In general, the flux is composed by all three neutrino flavours, but the typical signal which is looked at refers to the ν_μ component and consists of a flux of up–going muons in a neutrino detector. In this case, the major source of background is given by atmospheric muon neutrinos.

In this talk, which is based on Ref. [1], we propose, instead, a novel signature: the possibility of looking at the ν_τ component of the DM signal coming from the Sun. This is almost background–free in the downward–going direction, since the ν_τ amount in atmospheric neutrinos is negligible and atmospheric ν_μ do not sizably oscillate in the down–going baseline.

In Sect. 2 we discuss the neutrino fluxes coming from DM annihilation and the background fluxes: the atmospheric, solar corona and galactic neutrinos. The class of signals relevant for water Cherenkov detectors, *i.e.* the hadronic contained events, is then presented. We discuss both the signal events from tau decay as well as the background contributions due to misidentified electron and muon events. In Sect. 3 we show the level of sensitivity that can be achieved on the DM spin–independent cross section on protons σ_p. The capabilities of Mton–scale water Cherenkov detectors are also presented. Conclusions are given in Sect. 4.

2. Background and signal events

The neutrino flux at the detector, coming from DM annihilation inside the Sun, is calculated as described in Ref. [2], considering neutrino oscillations and neutrino neutral current ($\mathcal{NC}$) and charged current ($\mathcal{CC}$) interactions. We consider two DM annihilation channels: into neutrinos with flavor–blind branching ratios and into tau leptons.

The ν_τ coming from oscillation of atmospheric ν_μ [3] represents one source of background to the ν_τ DM signal. There are, however, three other forms of background: the intrinsic ν_τ contribution to the atmospheric flux [4], the neutrino flux produced in the solar corona [5] and the fluxes of tau neutrinos from the galactic plane [6].

For (GeV–TeV) energies, $\mathcal{CC}$ ν_τ interactions in water Cherenkov detectors will lead to multiple Cherenkov rings and the possibility of identifying these events is currently based only on statistical methods. Considering the hadronic decays of tau leptons, the misidentified events for the Super–Kamiokande detector is of the order of several percent [7, 8]. The experimental backgrounds that we consider are $\mathcal{NC}$ events from ν_e and ν_μ atmospheric neutrinos and $\mathcal{CC}$ events from ν_e.

The expression for the contained hadronic tau events is given by [9]

$$N_\tau^{\mathcal{CC}}\Big|_{\mathcal{S},\mathcal{B}} = M_{\rm det} N_y \times \int_{E_{\rm vis}^{\rm min}}^{E_{\rm vis}^{\rm max}} dE_{\rm vis} \int d\Omega\ \eta(\theta)\ \frac{d^2\mathcal{I}_\tau^{\mathcal{CC}}}{d\Omega\, dE_{\rm vis}}\Big|_{\mathcal{S},\mathcal{B}}, \qquad (1)$$

with $M_{\rm det}N_y$ being the detector exposure, $\eta(\theta)$ the on–source duty factor and

$$\frac{d^2\mathcal{I}_\tau^{\mathcal{CC}}}{d\Omega\, dE_{\rm vis}}\Big|_{\mathcal{S},\mathcal{B}} = \int dE_\nu \int dE_\tau\ \frac{d^2\phi_{\nu_\tau}}{d\Omega dE_\nu}\Big|_{\mathcal{S},\mathcal{B}} \Sigma_\tau^{\mathcal{CC}}(E_\tau, E_\nu)\ \frac{d\Gamma_h}{dE_{\rm vis}} + (\nu \to \bar{\nu}), \qquad (2)$$

where $\mathcal{S}$ and $\mathcal{B}$ denote signal and background and $d^2\phi_{\nu_\tau}/d\Omega dE_\nu$ the ν_τ flux. The function $\Sigma_\tau^{\mathcal{CC}}$ quantifies the number of interactions, while $d\Gamma_h/dE_{\rm vis}$ the decay rate of tau into hadrons [1].

In the left panel of Fig. 1, we show the number of downward–going signal events from hadronic tau decay as a function of the DM mass m_χ, for $M_{\rm det}N_y$ =1 Mton×year and for $\sigma_p = 10^{-41}$ cm^2. The number of signal events can reach the level of 50 or more, depending on m_χ. For the annihilation channels under study, this signal is most sensitive to DM masses in the range from 30 GeV up to 200–300 GeV. In the same figure, we also show the negligible number of $\mathcal{CC}$ events expected from the ν_τ background. The $\mathcal{NC}$ events from atmospheric ν_μ and ν_e as well as the $\mathcal{CC}$ events from ν_e are also shown. These classes of events pose a problem if they are not controlled at a level better than a few percent.

3. Detectability and statistical significance

To quantify the discovery reach of present and future water Cherenkov detectors, we use the statistical significance ς, defined as the signal–to–noise ratio:

$$\varsigma \equiv \frac{S}{\sqrt{S+B}}. \qquad (3)$$

In the ideal case, in which no misidentification is present and the detector efficiency for tau leptons ϵ_τ is 100%, the background contribution is given by $B_{\rm ideal} = N_\tau^{\mathcal{CC}}\big|_{\mathcal{B}}$. In a more realistic case, in which the misidentification and the detection efficiency for taus are considered, we have:

$$B_{\rm realistic} = \epsilon_\tau\ N_\tau^{\mathcal{CC}}\Big|_{\mathcal{B}} + \epsilon_\mu^{\rm mis}\ N_\mu^{\mathcal{NC}}\Big|_{\mathcal{B}} + \epsilon_e^{\rm mis}\left(N_e^{\mathcal{NC}}\Big|_{\mathcal{B}} + N_e^{\mathcal{CC}}\Big|_{\mathcal{B}}\right). \qquad (4)$$

We show in the right panel of Fig. 1 the contours for ς = 1.64 (which corresponds to a 90% C.L. upper bound) in the plane σ_p vs. m_χ, for an exposure $M_{\rm det}N_y$ =1 Mton×year. The dotted lines represent the limits without considering misidentification, while the solid lines correspond to ϵ_τ = 40% and $\epsilon_e^{\rm mis}$=$\epsilon_\mu^{\rm mis}$ = 4%. We show the allowed regions obtained from the DAMA [10],

Journal of Physics: Conference Series **375** (2012) 012030 doi:10.1088/1742-6596/375/1/012030

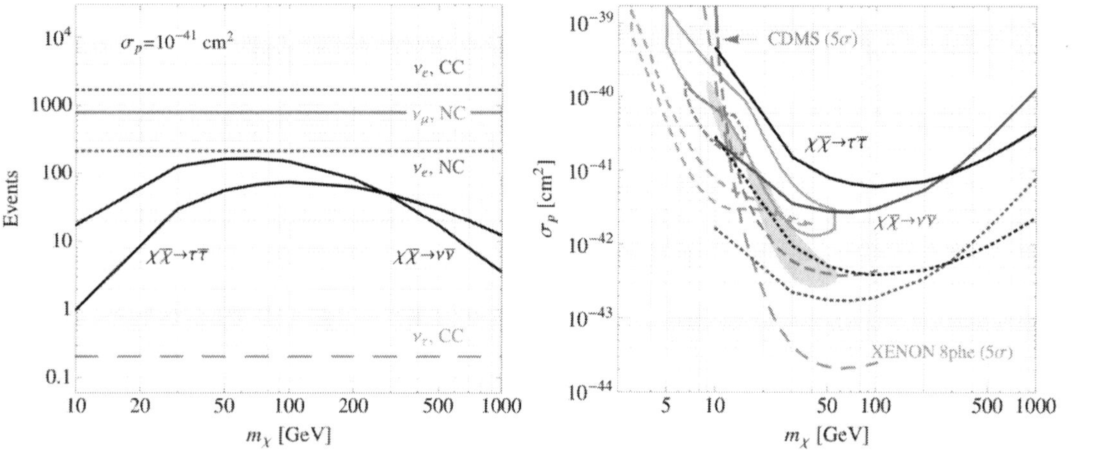

Figure 1. *Left*: Number of downward-going ν_τ hadronic events as a function of m_χ, for $\sigma_p = 10^{-41}$ cm^2 and $M_{\rm det}N_y$ =1 Mton×year. The horizontal lines represent the $\mathcal{NC}$ events expected from the atmospheric ν_e and ν_μ, and the $\mathcal{CC}$ events from atmospheric ν_e and background ν_τ. *Right*: Limits at 90% C.L. ($\varsigma = 1.64$) on σ_p as a function of m_χ, for $M_{\rm det}N_y$ =1 Mton×year. The dotted lines represent the limits without considering misidentification, while the solid lines correspond to $\epsilon_\tau = 40\%$ and $\epsilon_e^{\rm mis}$=$\epsilon_\mu^{\rm mis} = 4\%$. We show also the allowed regions from DAMA (orange solid line: without channeling, orange dashed line: with channeling), CoGeNT (dot-dashed red curve) and CRESST (cyan regions), and the limits from XENON 100 (green dashed line) and CDMS (gray dashed line) experiments.

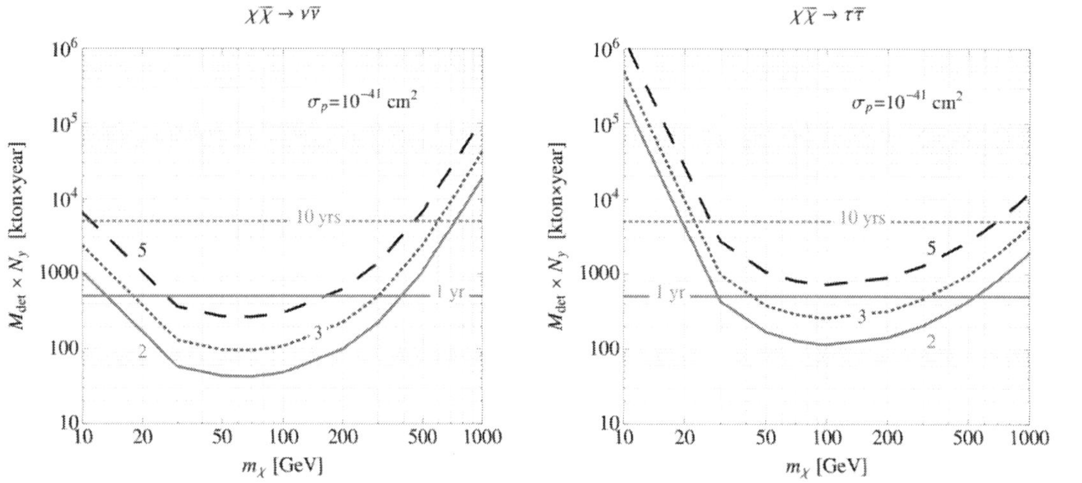

Figure 2. Iso-contours of statistical significance (ς = 2, 3 and 5 σ) for the detection of downward-going ν_τ hadronic events as a function of detector exposure (in kton × year) and DM mass m_χ, for $\sigma_p = 10^{-41}$ cm^2 (*left panel*: annihilation into neutrinos, *right panel*: annihilation into tau leptons). The two horizontal lines denote the exposures that can be reached by a 0.5 Mton detector, like HK, in 1 and 10 years. The detection efficiency is fixed to $\epsilon_\tau = 70\%$, while the misidentification to $\epsilon_e^{\rm mis} = \epsilon_\mu^{\rm mis} = 1\%$.

CoGeNT [11] and CRESST [12] positive results, see Ref. [13] for more details. The constraints from the XENON [14] and CDMS [15] experiments are also shown [16].

In Fig. 2 we present the iso–contours of statistical significance, ς =2, 3 and 5 σ, for the detection of downward–going ν_τ hadronic events as a function of m_χ and of the detector exposure. We consider $\epsilon_\tau = 70\%$, $\epsilon_e^{\rm mis} = \epsilon_\mu^{\rm mis} = 1\%$ and $\sigma_p = 10^{-41}$ cm^2. The horizontal lines denote the exposures that can be reached by a 0.5 Mton detector, like Hyper-Kamiokande (HK) [17], in 1 and 10 years. A few years of exposure would suffice to cover almost the whole DM mass range for our benchmark value of σ_p. The recent analyses of direct detection annual modulation effects observed by DAMA [10] and CoGeNT [11] (and the excess reported by CRESST [12]) point toward a DM candidate with $m_\chi \simeq$10 GeV and σ_p of the order of 10^{-42} cm^2 – 10^{-40} cm^2. For this type of particle, we would expect, for direct annihilation into neutrinos, between 9 and 900 hadronic events and a detection close to 5 σ with a 10 years exposure on HK (5 Mton×yr).

4. Conclusion

We propose a new channel for DM searches at neutrino telescopes: the downward–going hadronic tau events originated by the ν_τ signal produced in DM annihilations in the Sun. This specific signal potentially represents a very good opportunity for DM detection, since the background of atmospheric downward–going ν_τ is extremely reduced with respect to the upward–going ν_μ case commonly considered.

At water Cherenkov detectors, unfortunately, the hadronic tau events cannot be easily distinguished from $\mathcal{NC}$ events, mostly coming from atmospheric ν_e and ν_μ, and by $\mathcal{CC}$ e–like events. We found that the misidentification of non–tau events needs to be kept at the level of percent to have potentially good prospects for detecting DM through the downward–going tau neutrinos signal at future Mton–size Cherenkov detectors.

We showed that several tens of events per year (depending on the DM mass and annihilation channel) are potentially collectible in a Mton–scale detector. In the case of $\epsilon_\tau = 70\%$ and of $\epsilon_e^{\rm mis} = \epsilon_\mu^{\rm mis} = 1\%$, a 5 σ significance discovery is potentially reachable for DM masses in the range from 20 to 300 GeV with a few years of exposure, and for $\sigma_p = 10^{-41}$ cm^{-2}.

Acknowledgments

We acknowledge Research Grants funded jointly by MIUR, Università di Torino and INFN within the *Astroparticle Physics Project* (MIUR: PRIN 2008NR3EBK; INFN: FA51).

References

[1] Fornengo N and Niro V 2011 *JHEP* **11** 133 (*Preprint* 1108.2630)
[2] Cirelli M *et al.* 2005 *Nucl. Phys.* **B727** 99–138 (*Preprint* hep-ph/0506298)
[3] Honda M, Kajita T, Kasahara K and Midorikawa S 2011 (*Preprint* 1102.2688)
[4] Pasquali L and Reno M 1999 *Phys. Rev.* **D59** 093003 (*Preprint* hep-ph/9811268)
[5] Ingelman G and Thunman M 1996 *Phys. Rev.* **D54** 4385–4392 (*Preprint* hep-ph/9604288)
[6] Athar H, Lee F F and Lin G L 2005 *Phys. Rev.* **D71** 103008 (*Preprint* hep-ph/0407183)
[7] Abe K *et al.* 2006 *Phys. Rev. Lett.* **97** 171801 (*Preprint* hep-ex/0607059)
[8] Kato T 2007 Ph.D. thesis Stony Brook University
[9] Giordano G, Mena O and Mocioiu I 2010 *Phys. Rev.* **D81** 113008 (*Preprint* 1004.3519)
[10] Bernabei R *et al.* (DAMA Collaboration) 2010 *Eur. Phys. J.* **C67** 39–49 (*Preprint* 1002.1028)
[11] Aalseth C *et al.* (CoGeNT Collaboration) 2011 (*Preprint* 1106.0650)
[12] Angloher G *et al.* (CRESST-II Collaboration) 2011 (*Preprint* 1109.0702)
[13] Belli P *et al.* 2011 *Phys. Rev.* **D84** 055014 (*Preprint* 1106.4667)
[14] Aprile E *et al.* (XENON100 Collaboration) 2011 *Phys. Rev. Lett.* (*Preprint* 1104.2549)
[15] Ahmed Z *et al.* (CDMS-II Collaboration) 2010 *Science* **327** 1619–1621 (*Preprint* 0912.3592)
[16] Fornengo N, Panci P and Regis M 2011 *Phys. Rev.* **D84** 115002 (*Preprint* 1108.4661)
[17] Nakamura K 2003 *Int. J. Mod. Phys.* **A18** 4053–4063

Electron anisotropy: A tool to discriminate dark matter in cosmic rays

Enrico Borriello[1], **Luca Maccione**[2], **and Alessandro Cuoco**[3]

[1]Università di Napoli "Federico II" and INFN, Via Cintia, Napoli, Italy
[1]II. Institut für Theoretische Physik, Universität Hamburg, Germany
[2]DESY, Theory Group, Notkestraße 85, D-22607 Hamburg, Germany
[3]The Oskar Klein Centre for Cosmoparticle Physics, AlbaNova, SE-106 91 Stockholm, Sweden

E-mail: enrico.borriello@desy.de

Abstract. Indirect searches of particle Dark Matter (DM) with high energy Cosmic Rays (CR) are typically affected by large uncertainties. We show that, on the contrary, the DM intrinsic degree of anisotropy in the arrival directions of high energy CR electrons and positrons (CRE) is basically model independent and offers a straightforward criterion to discriminate among CRE from DM or from local discrete sources, like e.g. pulsars. In particular, in absence of the latter, DM sets the maximum degree of total anisotropy. As a consequence, if a larger anisotropy is detected, this would constitute an unambiguous evidence for the presence of astrophysical local discrete CRE sources. The Fermi-LAT will be able to probe such scenarios in the next years.

1. Introduction

High energy Cosmic Ray (CR) positrons are promising targets for indirect searches of Galactic particle Dark Matter (DM) [1]. The recent results reported by the PAMELA [2] and Fermi collaborations [3] on the positron fraction $e^+/(e^+ + e^-)$ and on the $(e^+ + e^-)$ (CRE) spectra in the energy range few GeV $\div$ 1 TeV show large discrepancies with standard astrophysical model predictions and have indeed received a large number of interpretations in terms of DM. In particular, it has been shown that a good fit of the PAMELA/Fermi data can be achieved with e^+e^- produced by a DM particle of ~TeV mass annihilating or decaying predominantly via leptonic channels, and several models realizing this scenario have been proposed [4]. However, interpretations based on astrophysical extra sources (like e.g. pulsars) [5, 6] have been shown to work equally well.

It is, however, very unlikely to distinguish the two scenarios using as observables only the fluxes, even with the larger statistics expected in the future [7]. It is thus mandatory to find other observables accessible to experiments, that are as much model independent as possible and can provide a clear discrimination between a DM dominated scenario and an "astrophysically" dominated one. As we showed in [8], the intrinsic degree of dipole anisotropy in the arrival directions of high energy CREs expected from a DM scenario, $\delta_{\rm DM}$, is indeed insensitive to many uncertainties, and constitutes a universal characteristics of galactic DM, offering a straightforward criterion to discriminate among CRE from DM or from local sources. Indeed, we showed that if future experiments will find an anisotropy larger than the maximum DM anisotropy we derive here, then a dominant DM contribution to the CRE anisotropy can be excluded in a basically model independent way, pointing instead to local discrete astrophysical CRE sources as the main source of anisotropy.

The reason why the dipole anisotropy has a very weak dependence on the various unknowns is, on the one hand, the very short electron path above ~ 100 GeV (~ 1 kpc) which makes this quantity very local in origin, and on the other hand, the fact that it is a flux ratio, so that most of the uncertainties cancel each other in the ratio. Furthermore, we find that the anisotropy signal from DM is intrinsically different from the one due to local discrete sources. The key point here is that the number of galactic DM substructures is $\mathcal{O}(10^{17})$ and a very nearby clump is always accompanied by the large, dominant and almost isotropic flux from the whole population of clumps, which washes out the single clump anisotropy. This has to be compared with the case e.g. of pulsars, which can produce a similar amount of CREs as DM, but concentrated in only 10^5 or less objects. In this scenario pulsars are rare and powerful enough that a few nearby objects can indeed dominate the flux and the anisotropy.

On the experimental side, the Fermi telescope recently placed the first upper limits on the integrated dipole anisotropy of the arrival directions of CRE with $E > 60$ GeV [9], and there are prospects for its actual observation after a few years of data taking [6].

2. Electron anisotropy

In the diffusive approach, the dipole anisotropy can be written as $\vec{\delta} = -(3D/\beta c)\vec{\nabla}\phi/\phi$ (see e.g. [10]), where D is the isotropic diffusion coefficient, βc and ϕ are the CRE velocity and flux respectively. We considered three different models of diffusion: with Kolmogorov-like turbulence, Kraichnan-like turbulence, and a last one with high turbulence equal to 0.7. The diffusion coefficient is always normalized in order to be in agreement with CR nuclei observations [6].

The total DM contribution to the e^+e^- fluxes is in general the sum of two components, $\phi_{\rm DM} = \phi_h + \phi_s$, where ϕ_h is the contribution from the smooth halo while ϕ_s is the contribution from the substructures. Current highest resolution N-body simulations roughly agree on the mass distribution of substructures, predicting a number density scaling like m^{-2} (Via Lactea II [11]) or $m^{-1.9}$ (Aquarius [12]). How substructures are distributed in the smooth halo is however more uncertain. We considered the two extreme cases of an unbiased distribution where substructures follow the main halo and an anti-biased case as suggested by the Via Lactea II simulation [13]. The internal concentration of substructures and the effects of tidal disruption were parameterized as in [14]. We chose a clump mass range $10^{-6} \div 10^{10}\ M_\odot$. For the spatial distribution of the smooth component and for the DM distribution inside the substructures we considered Navarro-Frenk-White (NFW) [15] and Burkert [16] profiles. Finally, we considered annihilation in μ, τ and quark pairs, for values of the DM mass: 100, 316, 1000, and 3162 GeV.

The sum over the substructure distribution in principle should run over a sizable fraction of the $\mathcal{O}(10^{17})$ substructures lying within the diffusive region. This is computationally prohibitive at present. We then computed analytically the average contribution from substructures with $10^{-6} < M/M_\odot < 10^2$, while we computed explicitly the contribution of each clump with $M > 10^2\ M_\odot$. We sampled the distribution of $M > 10^2\ M_\odot$ substructures via a MonteCarlo procedure as described in [17]. Further details can be found in [8].

3. Results and discussion

Both the analytic calculation and the MonteCarlo simulation give a mean anisotropy which is only slightly dependent by the mass decade. Fluctuations in the distribution of the high mass substructures however lead to large variability of $|\vec{\delta}|$, which is more relevant at the highest energies, because higher energy electrons probe smaller volumes than lower energy ones, thereby being more sensitive to fluctuations in the clump distribution.

The first two panels in fig. 1 show the results for the mean and maximum $\delta_{\rm DM} = |\vec{\delta}_{\rm DM}|$ for the considered annihilation channels and DM masses for our chosen DM density profiles and

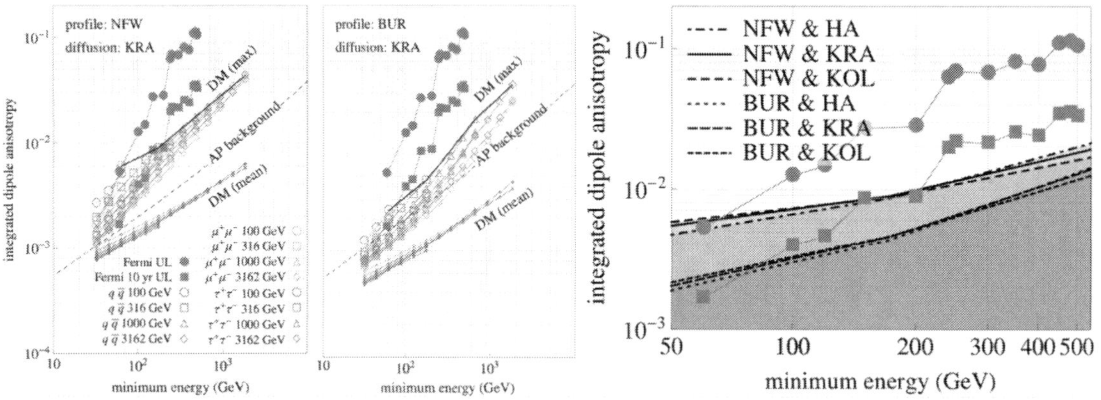

Figure 1. (First two panels) Intrinsic integrated dipole anisotropy of DM, compared to the standard astrophysical background, current Fermi upper limits at 95% CL and the sensitivity expected after 10 years of data taking. The points correspond to different annihilation channels and masses of the DM particle. Both average values and maxima are shown. The thick black line shows the upper envelope of the maxima. (Third panel) Upper limits on the intrinsic DM anisotropy for all the combinations of DM density profiles and propagation models under scrutiny in [8] and current and expected Fermi UL.

Kraichnan propagation setup in the unbiased case. Results for the other propagation setups are shown in [8]. The anti-biased case (not shown) gives results smaller by a factor of ~ 5 at 500 GeV. The maximum over our 100 realizations sample roughly corresponds to a 99% CL. The last panel in fig. 1 shows, instead, only the envelope of the maxima for the various propagation setups. Both the mean and maximum anisotropy increase with energy, as expected from the fact that at higher energies smaller and smaller propagation volumes are probed by the CREs and the role of fluctuations is more relevant.

The main property emerging from the above plots is that $\delta_{\rm DM}$ is almost independent of the detailed characteristics of the DM models and distributions in substructures and, in this sense, $\delta_{\rm DM}$ is a universal property of DM. Being a ratio, it is very little sensitive to integrated quantities, like the annihilation spectrum. Moreover, because CREs propagate only a few kpc distance in the Galaxy, $\delta_{\rm DM}$ is also little sensitive to the DM spatial profile, in particular on whether it is peaked or cored. The anisotropy is also not strongly sensitive to the internal concentration of the subhaloes, because diffusion over kpc scales smooths out the effect of a possible cusped over-density region. For these reasons, as we checked, the case of decaying DM gives similar results as the case of annihilating DM. Finally, remarkably, also the fluctuations in $\delta_{\rm DM}$ are basically model independent. The third panel in fig. 1, in particular, shows that the maximal anisotropies are almost independent of the diffusion setup, even in the extreme case of HA diffusion, and only slightly change for different DM profiles.

We now consider the role of the background. If the total flux is given by the contribution of an astrophysical (AP) flux and a DM originated one, $\phi = \phi_{\rm AP} + \phi_{\rm DM}$, the degree of anisotropy is given by a composition of the AP and the DM intrinsic anisotropies. It is easy to see that the bound $|\vec{\delta}| \leq \delta_{max}$ holds, with $\delta_{max} = x\,\delta_{\rm DM} + (1-x)\delta_{\rm AP}$, being $x = \phi_{\rm DM}/\phi$. In a specific scenario, δ_{max} is determined by x, the relative contributions of $\phi_{\rm DM}$ to the total flux ϕ, although, clearly, $\delta_{max} \leq \max(\delta_{\rm DM}, \delta_{\rm AP})$. If DM is distributed in substructures, and if local discrete CRE astrophysical sources are neglected, then we can have $\delta_{\rm DM} > \delta_{\rm AP}$ (fig. 1). In this case, $\delta_{\rm DM}$ sets the maximum anisotropy we can expect, with $\delta \simeq \delta_{\rm DM}$ when $\phi_{\rm DM} \gg \phi_{\rm AP}$.

Being the maximum of $\delta_{\rm DM}$ insensitive to the specific DM scenario, and to the details of the

CRE propagation, this upper limit is robust and universal. If a positive detection of anisotropy will occur in the future, and the anisotropy will be found larger than $\delta_{\rm DM}$, we can then exclude the presence of a substantial DM contribution, and therefore we have to demand $\delta_{\rm AP} > \delta_{\rm DM}$. This would point then unambiguously to a scenario dominated by local, discrete astrophysical sources, such as pulsars, as source of high energy CRE. However, this argument does not exclude that a subdominant contribution from DM annihilation in substructures can still be present [18].

Although our arguments are very natural, our findings result from a MonteCarlo computation of the local distribution of DM substructures. A possible bias of this approach is that we might have missed configurations whose probability is less than 1%, in which, e.g., a large mass clump emerges isolated and very close to the Earth. This could in principle produce an anisotropy larger than what we quote as a "maximum". We checked, however, that this configuration cannot produce a high degree of anisotropy. In fact, even in the unlikely case of a 10^8 $M_\odot$ clump at 100 pc from Earth (whose probability is much less then 1%) the anisotropy is strongly suppressed by the nearly isotropic flux of the much more abundant smaller substructures and it is thus always diluted below $\max(\delta_{\rm DM})$. This feature makes the DM signal intrinsically different from the one expected from pulsars. Indeed, while there might be a close-by, isolated pulsar, that can possibly lead to a large anisotropy [6], it is not possible to reproduce this configuration with DM.

In summary, our result is robust and can be used as a criterion to reject a DM dominated scenario in the framework of high energy CREs, in the case of detection of a large anisotropy.

Acknowledgments

EB acknowledges financial support by the Deutsche Forschungsgemeinschaft through SFB 676 "Particles, Strings and the Early Universe: The Structure of Matter and Space-Time", by the INFN I.S. FA51, and by the PRIN 2008"Fisica Astroparticellare: Neutrini ed Universo Primordiale" of the MIUR. LM acknowledge support from the State of Hamburg, through the Collaborative Research program "Connecting Particles with the Cosmos" within the framework of the LandesExzellenzInitiative (LEXI).

References

[1] G. Bertone, D. Hooper, J. Silk, Phys. Rept. **405** (2005) 279.

[2] O. Adriani *et al.* [PAMELA Collaboration], Nature **458** (2009) 607-609. [arXiv:0810.4995 [astro-ph]].

[3] A. A. Abdo *et al.* [Fermi LAT Collaboration], Phys. Rev. Lett. **102** (2009) 181101. M. Ackermann *et al.* [Fermi LAT Collaboration], arXiv:1008.3999 [astro-ph.HE].

[4] L. Bergström, T. Bringmann, J. Edsjo, Phys. Rev. D **78** (2008) 103520. I. Cholis *et al.* Phys. Rev. D **80**, 123511 (2009). M. Cirelli *et al.* Nucl. Phys. B **813**, 1 (2009). N. Arkani-Hamed, D. P. Finkbeiner, T. R. Slatyer, N. Weiner, Phys. Rev. D **79** (2009) 015014. L. Bergström, J. Edsjo, G. Zaharijas, Phys. Rev. Lett. **103** (2009) 031103.

[5] P. Blasi, Phys. Rev. Lett. **103** (2009) 051104. P. Blasi and E. Amato, arXiv:1007.4745 [astro-ph.HE]. P. D. Serpico, Phys. Rev. D **79**, 021302 (2009). D. Grasso *et al.*, Astropart. Phys. **32** (2009) 140. S. Profumo, Central Eur. J. Phys. **10** (2011) 1.

[6] G. Di Bernardo *et al.*, Astropart. Phys. **34** (2011) 528-538.

[7] M. Pato, M. Lattanzi, G. Bertone, JCAP **1012** (2010) 020.

[8] E. Borriello, L. Maccione, A. Cuoco, arXiv:1012.0041 [astro-ph.HE].

[9] M. Ackermann *et al.* [Fermi-LAT collaboration] arXiv:1008.5119 [astro-ph.HE].

[10] Berezinsky V S *et al.*, Astrophysics of Cosmic Rays, North-Holland, 1990.

[11] J. Diemand *et al.* Nature **454** (2008) 735.

[12] V. Springel *et al.*, Mon. Not. Roy. Astron. Soc. **391** (2008) 1685.

[13] M. Kuhlen, J. Diemand, P. Madau, [arXiv:0805.4416].

[14] L. Pieri, J. Lavalle, G. Bertone, E. Branchini, arXiv:0908.0195 [astro-ph.HE].

[15] J. F. Navarro, C. S. Frenk, S. D. M. White, Astrophys. J. **490** (1997) 493.

[16] A. Burkert, Astrophys. J. **447** (1995) L25.

[17] E. Borriello, A. Cuoco, G. Miele, Phys. Rev. **D79** (2009) 023518.

[18] I. Cernuda, Astropart. Phys. **34** (2010) 59.

Journal of Physics: Conference Series **375** (2012) 012032 doi:10.1088/1742-6596/375/1/012032

Cosmic-ray antiproton constraints on light dark matter candidates

Julien Lavalle[⋆,†,1]

⋆ (perm. addr.) Lab. Univers et Particules de Montpellier (LUPM), CNRS-IN2P3 & Univ. Montpellier II (UMR 5299), Place Eugène Bataillon, F-34095 Montpellier cedex 05 — France
† Instituto de Física Teórica (UAM/CSIC) & Dep[to] de Física Teórica (UAM), Univ. Autónoma de Madrid, Cantoblanco, E-28049 Madrid — Spain

E-mail: `lavalle@in2p3.fr`

Abstract. Some direct detection experiments have recently collected excess events that could be interpreted as a dark matter (DM) signal, pointing to particles in the ~10 GeV mass range. We show that scenarios in which DM can self-annihilate with significant couplings to quarks are likely excluded by the cosmic-ray (CR) antiproton data, provided the annihilation is S-wave dominated when DM decouples in the early universe. These limits applies to most of supersymmetric candidates, *e.g.* in the minimal supersymmetric standard model (MSSM) and in the next-to-MSSM (NMSSM), and more generally to any thermal DM particle with hadronizing annihilation final states.

1. Introduction

After the finding of an annual modulation consistent with a DM signal in the DAMA/Libra data [1, 2], some skepticism had arisen due to the failure of other direct detection experiments to confirm it in the preferred phase space corner. Interestingly enough, this initial caution has been recently tickled by the results of a few other experiments with different setups. CoGeNT [3] and CRESST [4] have indeed found excess events; CoGeNT has even released hints for an annual modulation [5]. Yet, XENON-10/100 [6, 7], CDMS [8] and SIMPLE [9] do not validate such excesses and therefore challenge the classical DM interpretation of the aforementioned positive results — *i.e.* the elastic collisions of weakly massive interacting particles (WIMPs) off target nuclei [10]. It is still important to recall that these excesses — and bounds — rely on data analyses led very close to the current experimental thresholds, where systematic errors can in principle get large and remain to be much better controlled. Nevertheless, these measurements have brought some excitement in the DM community, and many authors have tried to fit the available data with some phenomenological models (*e.g.* [11]). The preferred mass range is around 10 GeV, while difficulties are generally found when trying to get good fits to the signals while avoiding the constraints from other direct detection experiments [10]. Some ideas have been proposed to explain such discrepancies (*e.g.* [12, 13]), but what people usually do is to forget about either the bounds or the signals, depending on whether they want to promote their models or challenge them.

[1] former Multidark fellow

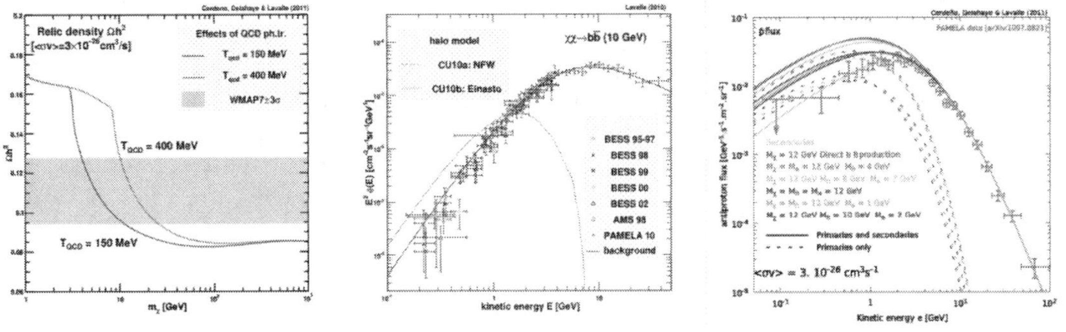

Figure 1. Left: Impact of considering 2 different temperatures for the quark-hadron transition on the relic density calculation as a function of the WIMP mass, for a given annihilation cross section [16]. Middle: antiproton flux prediction for a 10 GeV WIMP annihilating into $b\bar{b}$ [15]. Right: antiproton flux predictions for a 12 GeV WIMP annihilating into different mass combinations of an intermediate two-boson state which further decays into quarks [16].

Beside direct detection, indirect detection, that looks for DM annihilation or decay traces originating in high concentration astrophysical sites, provides an interesting way to check the relevance of any phenomenological proposal in this context. Among indirect signatures, CR antiprotons have long been known to potentially be an interesting target to look at [14]. In this proceedings' contribution, we show that they indeed provide quite strong limits on quarkophilic DM candidates, and are even likely excluding those models which have an S-wave dominated annihilation cross section. The following discussion is mostly based upon two articles, Refs. [15] and [16], to which we refer the reader for more details.

2. Ingredients and results

Before discussing the Galactic ingredients, it might be useful to recall a source of uncertainty in the annihilation cross section that comes from the calculation of the relic density, though commonly assumed to be under control. For typical thermal cold DM candidates, the chemical decoupling that sets the DM abundance occurs in the early universe at a temperature $T_f \approx m_\chi/20$. In the 10 GeV mass range, this translates into temperatures around 500 MeV, *i.e.* in the temperature range in which the quark-hadron transition is expected [17, 18]. A decoupling occurring before or after this transition induces a factor of 2 uncertainty in the annihilation cross section for a given cosmological abundance, since the relativistic degrees of freedom, which characterize the energy density in the radiation era (*i.e.* the squared expansion rate), vary by a factor of 4 meanwhile. We therefore note as a preliminary remark that the required annihilation cross section may be twice larger than the canonical value in this mass range, the transition mass being as uncertain as the quark-hadron transition temperature. This was discussed in detail in [16], and is illustrated in the left panel of Fig. 1. We now return to the prediction of the CR antiproton flux, for which the involved physics is a little more complex than for gamma-ray calculations. The additional steps come from that charged CRs scatter off magnetic turbulences in the Galaxy (which drives their motion to a diffusion pattern), are convected away from the disk by Galactic winds, and suffer nuclear reactions (spallation) with the interstellar medium (ISM) as well as energy redistribution due to scattering interactions and diffusive reacceleration. For a complete theoretical view on CR propagation, see Ref. [19]. Thus, a full prediction relies on solving a diffusion equation that reads (steady state and stable species):

$$-\vec{\nabla}\{K_x\vec{\nabla}\mathcal{N} - \vec{V}_c\mathcal{N}\} + \partial_E\left\{\widehat{\mathcal{K}}_E\mathcal{N}\right\} + \Gamma_s\mathcal{N} = \mathcal{Q}(\vec{x}, E)\ . \quad (1)$$

In this equation, $\mathcal{N} = dn/dE$ is the CR density per unit energy, K_x is the spatial diffusion coefficient, V_c is the convective wind, Γ_s is the rate of spallation with the ISM gas and $\widehat{\mathcal{K}}_E$ is an energy redistribution (differential) operator. The rhs term is the source term. For DM annihilation, it is proportional to the product of the annihilation cross section and the DM number density (the squared DM mass density to the squared WIMP mass), $\langle\sigma v\rangle\,(\rho/m_\chi)^2$. Semi-analytic solutions to the previous equation do exist in certain regimes, especially when one fixes the geometry of the diffusion zone to a slab inside which the diffusion coefficient is set homogeneous. Further reasonably fixing the spatial extent of this slab, *i.e.* its radius $R \approx 20$ kpc and half-width $L \approx 4$ kpc, closes the full definition of the propagation model. All these parameters are globally referred to as transport parameters. They are usually constrained with CR nuclei data — basically, one predicts the fluxes of those secondary species arising from the spallation of some primary CR nuclei, the source of which can be rather well modeled (the *a priori* unknown source parameters cancel out at zeroth order when using the secondary-to-primary ratios — *e.g.* [20]). Predictions of the CR antiproton flux due to DM annihilation have been derived for years (*e.g.* [21]), but they were often inconclusive because of uncertainties on the diffusion halo size, a diffusion zone down to $L \approx 1$ kpc inducing an unobservable flux. Nevertheless, the constraints on the diffusion volume have been strongly improved since then, and it is now expected to extend to $L \gtrsim 3$ kpc [22] — it is also interesting to note that taking $L \approx 1$ kpc inevitably leads to a strong excess of secondary astrophysical positrons below 5 GeV [23, 24]. The results to be presented below were essentially derived using $L = 4$ kpc, and a diffusion coefficient normalization of $K_0 = K_x(p/q = 1GV) \simeq 0.01\,\mathrm{kpc}^2/\mathrm{Myr}$.

Predictions of the DM-induced antiproton flux are poorly sensitive to the distribution of DM close to the Galactic center — as long as $\rho \propto r^{-\gamma}$ with $\gamma \lesssim 1.5$ — in contrast to gamma-ray predictions. This is due to diffusion itself, that tends to smear out far away inhomogeneities. Actually, the antiproton flux is much more controlled by the local DM density $\rho_\odot$. There has been some recent attempts to bracket it, and it is noteworthy that quite different methods have led to similar results, $\rho_\odot \simeq 0.4$, with a $\sim 20\%$ accuracy (*e.g.*[25, 26, 27]). The coming results can be derived by using the best-fit Einasto [28] or Navarro-Frenk-White [29] profiles provided in Refs. [25] and [27], indifferently.

As to DM models, we are going to consider two template cases that are likely representative of almost all quarkophilic configurations. In the first one [15], annihilation fully goes into b-quark pairs, which is conservatively representative of annihilation into any quark pair, up to an antiproton multiplicity factor typically $\lesssim 2$ in the interesting $\sim 0.1 - 5$ GeV kinetic energy range, unfavorable to heavy quarks (smaller dynamical range). In the second one [16], annihilation goes into two unstable scalar (or pseudo-scalar) particles, with any mass above 1 GeV allowed by kinematics, and which may further decay into quarks. These (pseudo-)scalar particles could be thought of as the light Higgs bosons arising in the NMSSM, or more generally in any singlet extension of the MSSM. Except for spin considerations, this annihilation channel is kinematically representative of any two-body intermediate state made of unstable bosons. The following analysis can therefore be extrapolated, for instance, to models that involve couplings to new light gauge fields [30].

The antiproton constraints merely come from the fact that the expected secondary astrophysical background already saturates the existing data, *e.g.* from BESS [31] or PAMELA [32]. Consequently, any other significant astrophysical yield quickly leads to an excess prediction. The important point is that the theoretical errors affecting the astrophysical background estimates are small, at the 15% level below a few GeV [33]. It turns out that the DM-induced antiproton flux generically exceeds the background expectation for WIMP masses between ~5-15 GeV and $\langle\sigma v\rangle \geq 10^{-26}\,\mathrm{cm}^3/\mathrm{s}$, for the two configurations discussed above. Therefore, if the annihilation cross setion is S-wave dominated at the decoupling time in the early universe, then requiring the correct DM relic abundance, which translates into having

$2.5 \times 10^{-26} \lesssim \langle \sigma v \rangle \lesssim 5 \times 10^{-26}\,\mathrm{cm}^3/\mathrm{s}$, generically leads to an antiproton excess in this mass range — a smaller cross section is difficult to achieve, since leading to an over-abundance of DM. Examples are given in the middle and right panels of Fig. 1. The only ways left to escape these antiproton bounds are (i) to consider the second configuration, *i.e.* the annihilation into a two-boson intermediate state where the decaying bosons would both have masses lower than the (anti)proton mass, or (ii) to be at the b-quark pair threshold production at which antiprotons can also barely be produced. The first case might be easy to achieve phenomenologically, but the second one is clearly very fine-tuned.

3. Conclusion

We have discussed how CR antiprotons can put limits on light DM candidates. They are generically strong and exclude quarkophilic DM particles in the 5-15 GeV mass range relevant to the direct detection signals, if they annihilate through S-waves with $\langle \sigma v \rangle \geq 10^{-26}\,\mathrm{cm}^3/\mathrm{s}$. The only constraint at the origin of this result is the upper bound on the cosmological DM abundance, which translates into a lower limit on the annihilation cross section at the decoupling time, $\langle \sigma v \rangle \gtrsim 2.5 \times 10^{-26}\,\mathrm{cm}^3/\mathrm{s}$. Light neutralinos arising in the MSSM with non-unified gaugino masses and in the NMSSM are therefore excluded as potential DM candidates if they fall in the quarkophilic categories discussed above, which is quite often the case. These antiproton limits are complementary to other studies using *e.g.* gamma-rays [34], CMB photons [35], and solar neutrinos [36], which concern more annihilation final states than only quarks.

[1] Bernabei R *et al* 2000 *Physics Letters B* **480** 23–31
[2] Bernabei R *et al* 2008 *European Physical Journal C* 167–+ (*Preprint* arXiv:0804.2741)
[3] Aalseth C E *et al* 2011 *Physical Review Letters* **106** 131301–+ (*Preprint* arXiv:1002.4703)
[4] Angloher G *et al* 2011 *ArXiv e-prints* (*Preprint* arXiv:1109.0702)
[5] Aalseth C E *et al* 2011 *Physical Review Letters* **107** 141301 (*Preprint* arXiv:1106.0650)
[6] Angle J *et al* 2011 *Physical Review Letters* **107** 051301 (*Preprint* arXiv:1104.3088)
[7] Aprile E *et al* 2011 *Physical Review Letters* **107** 131302 (*Preprint* arXiv:1104.2549)
[8] Ahmed Z *et al* 2011 *Physical Review Letters* **106** 131302–+ (*Preprint* 1011.2482)
[9] Felizardo M 2011 *ArXiv e-prints* (*Preprint* arXiv:1106.3014)
[10] Kopp J, Schwetz T and Zupan J 2011 *ArXiv e-prints* (*Preprint* arXiv:1110.2721)
[11] Bottino A, Donato F, Fornengo N and Scopel S 2010 *Phys. Rev. D* **81** 107302–+ (*Preprint* arXiv:0912.4025)
[12] Arina C, Hamann J and Wong Y Y Y 2011 *JCAP* **9** 22 (*Preprint* arXiv:1105.5121)
[13] Frandsen M T *et al* 2011 *Phys. Rev. D* **84** 041301 (*Preprint* arXiv:1105.3734)
[14] Silk J and Srednicki M 1984 *Physical Review Letters* **53** 624–627
[15] Lavalle J 2010 *Phys. Rev. D* **82** 081302–+ (*Preprint* arXiv:1007.5253)
[16] Cerdeño D G, Delahaye T and Lavalle J 2012 *Nuclear Physics B* **854** 738–779 (*Preprint* arXiv:1108.1128)
[17] Srednicki M, Watkins R and Olive K A 1988 *Nuclear Physics B* **310** 693–713
[18] Gondolo P and Gelmini G 1991 *Nuclear Physics B* **360** 145–179
[19] Berezinskii V S *et al* *Astrophysics of cosmic rays* (Amsterdam: North-Holland, edited by Ginzburg, V.L.)
[20] Maurin D *et al* 2001 *Astrophys. J.* **555** 585–596 (*Preprint* arXiv:astro-ph/0101231)
[21] Bottino A *et al* 2005 *Phys. Rev. D* **72** 083518–+ (*Preprint* arXiv:hep-ph/0507086)
[22] Putze A, Derome L and Maurin D 2010 *Astron. and Astrophys.* **516** A66+ (*Preprint* arXiv:1001.0551)
[23] Delahaye T *et al* 2009 *Astron. and Astrophys.* **501** 821–833 (*Preprint* arXiv:0809.5268)
[24] Lavalle J 2011 *MNRAS* **414** 985–991 (*Preprint* arXiv:1011.3063)
[25] Catena R and Ullio P 2010 *JCAP* **8** 4–+ (*Preprint* arXiv:0907.0018)
[26] Salucci P *et al* 2010 *Astron. and Astrophys.* **523** A83+ (*Preprint* arXiv:1003.3101)
[27] McMillan P J 2011 *MNRAS* 553–+ (*Preprint* arXiv:1102.4340)
[28] Einasto J 1969 *Astrofizika* **5** 137–+
[29] Navarro J F, Frenk C S and White S D M 1997 *Astrophys. J.* **490** 493–+ (*Preprint* arXiv:astro-ph/9611107)
[30] Cline J M and Frey A R 2011 *ArXiv e-prints* (*Preprint* arXiv:1108.1391)
[31] Abe K *et al* 2011 *ArXiv e-prints* (*Preprint* arXiv:1107.6000)
[32] Adriani O *et al* 2010 *Physical Review Letters* **105** 121101–+ (*Preprint* arXiv:1007.0821)
[33] Donato F, Maurin D, Salati P, Barrau A, Boudoul G and Taillet R 2001 *Astrophys. J.* **563** 172–184
[34] Geringer-Sameth A and Koushiappas S M 2011 *ArXiv e-prints* (*Preprint* arXiv:1108.2914)
[35] Galli S, Iocco F, Bertone G and Melchiorri A 2011 *Phys. Rev. D* **84** 027302 (*Preprint* arXiv:1106.1528)
[36] Kappl R and Winkler M W 2011 *Nuclear Physics B* **850** 505–521 (*Preprint* arXiv:1104.0679)

Indirect Search for Dark Matter with the ANTARES Neutrino Telescope

V. Bertin[1] on behalf of the ANTARES Collaboration

[1]CPPM, Aix-Marseille Université, CNRS/IN2P3,
163 avenue de Luminy – Case 902, 13288 Marseille cedex 9, France

E-mail : bertin@cppm.in2p3.fr

Abstract. Indirect search for Dark Matter trapped inside celestial bodies is one of the main physics goals of neutrino telescopes. The expected sensitivity of ANTARES with its data recorded in 2007 and 2008 to detect the flux of neutrinos originating from Dark Matter annihilations inside the Sun is presented. A comparison to current limits coming from direct detection experiments and other indirect detection experiments is shown with regards to the predictions of popular models such as the CMSSM and the Minimal Universal Dimension model.

1. Introduction

The most popular paradigm of modern cosmology considers the Dark Matter as a population of stable weakly interacting massive particles (WIMPs) relic from the Big Bang, although not yet discovered. Those particles would gravitationally accumulate in the core of massive celestial bodies such as stars or to a lesser extend planets as the Earth, where they could self-annihilate into ordinary matter and eventually produce significant high energy neutrino fluxes. Indirect search for Dark Matter looking at such neutrino fluxes coming from the core of the Sun, the Earth or the Galactic Centre is thus one of the main physics goals of the current and future neutrino telescopes.

2. The ANTARES neutrino telescope

The ANTARES detector [1] is the first undersea neutrino telescope and the largest one of the Northern hemisphere. It is composed of 12 mooring lines, each holding 75 photomultipliers distributed on 25 storeys (the titanium structure holding a triplet of photodetectors), installed at a depth of about 2500 metres off shore the Provençal coast of France, in order to form a 3D-matrix of ~900 photodetectors. The main goal of the experiment is to look for the Cherenkov light induced by high energy muons during their travel in the sea water throughout the detector. The trajectory of the muon track is reconstructed from the detection time of the Cherenkov photons as well as from the positions of the photodetectors. An indirect search for neutrinos can then be performed by selecting the upward-going muons produced by neutrinos which have passed through the entire planet and interacted in the vicinity of the detector. The direction of the incoming neutrino, being almost collinear with the secondary muon, can then be determined with an accuracy reaching 0.2° for high energy neutrinos above 10 TeV. Due to its size and the spacing of the photomultipliers, the ANTARES detector has a low energy threshold of ~20 GeV for reconstructed neutrinos and an effective area of ~10^{-3} m² for neutrinos with an energy of 500 GeV. The effective area increases strongly with the neutrino energy and reaches ~1 m² for PeV energy neutrinos. Its location in the Northern hemisphere makes it complementary in sky coverage with the South Pole neutrino telescope IceCube. In addition, a large fraction of the full sky can be observed with ANTARES thanks to the rotation of the Earth, including the central part of the Galaxy which is believed to be the host of many high energy phenomena.

The building of the ANTARES detector started in 2006 with the installation and the operation of the first line, and was completed in May 2008. The current analysis presented here is based on the data recorded in 2007 with a 5-line detector and in 2008 with a 9-to-12-line detector. The event reconstruction is performed by a χ^2 fit of the photodetector hit times as a function of their positions assuming that the light originates from the Cherenkov cone of a muon track passing through the detector [2]. Although the photomultipliers point at 45° downwards, the vast majority of reconstructed events are due to down-going atmospheric muons. The neutrino candidates are obtained by looking for upward-going tracks selected by a set of quality cuts on the reconstruction parameters in order to reject the background of badly reconstructed down-going atmospheric muons. After selection, the event sample contains about 1000 neutrino candidates recorded in about 295 effective days of data taking.

3. Indirect search for Dark Matter annihilations in the Sun with ANTARES

A search for neutrinos produced by Dark Matter annihilations into the Sun has been carried out in the data sample collected by ANTARES in 2007 and 2008. The analysis is based on a binned search strategy looking for an excess of neutrino events in a cone centred towards the direction of the Sun over the background of atmospheric neutrinos. Although a good agreement in the number and the distribution of events between data and Monte Carlo simulation is observed after selection, the background coming from atmospheric muon and atmospheric neutrino events has been estimated by scrambling the time of the data events in order to generate a fake Sun. This allows to supress the systematic errors coming from the uncertainties on the fluxes of atmospheric muon and neutrino events.

The estimation of the neutrino signal induced by Dark Matter annihilations in the core of the Sun has been estimated by using the WIMPSIM package [3] which generates the neutrino spectrum originating from the annihilations in a model independent way. For a given WIMP mass, this Monte Carlo simulation program calculates the capture rate and the annihilation rate in the Sun at equilibrium and generates the neutrino spectrum resulting from all possible self-annihilation channels. The propagation of the neutrinos within the Sun and in vacuum up to the Earth is simulated taking into account neutrinos interactions and regeneration of the tau leptons in the Sun medium, as well as neutrino oscillations in a full three-flavour framework.

The sensitivity to Dark Matter signal has been estimated firstly in a model independent way by considering an extreme "soft" neutrino spectrum corresponding to self-annihilations into b-quarks and a "hard" neutrino spectrum corresponding to self-annihilations into W/Z boson pairs or tau-leptons. These channels are well representative of a WIMP in the form of neutralinos in the framework of Minimal Supersymmetric extensions of the Standard Model. The specific case of the minimal Universal Extra Dimension model (mUED) in which the WIMP is the first excitation of the hypercharge gauge boson $B^{(1)}$ and thus the lightest Kaluza-Klein (LKP) particle [4] has also been studied explicitly. This model provides a highly predictive phenomenology with mainly two free parameters: the LKP mass $M_B{}^1$ and the relative mass splitting between the LKP and the first quark excitation $\Delta = (M_Q{}^1 - M_B{}^1)/M_B{}^1$. In particular, the branching ratios of LKP annihilations into the different final states are fixed in the model and dominated by annihilations into $\tau^+\tau^-$. In all cases, a full sensitivity study has been performed for twelve different values of the WIMP mass ranging from 10 GeV to 1 TeV.

For a WIMP with a given mass and a given neutrino annihilation spectrum, the number of signal events is obtained by convoluting the neutrino flux with the detector efficiency, the so-called effective area, determined for a given set of the selection parameters, mainly the values of the track fit quality parameter and of the half-opening angle of the cone around the Sun. The sensitivity to a given WIMP model is thus derived as the ratio between the average upper limit on the number of background events estimated from the scrambled data, considering a Poisson statistics in the Feldman-Cousins approach [5], and the number of signal events predicted for the corresponding WIMP mass and decay channel spectrum, and for the lifetime of the data taking. Following the Model Rejection Factor (MRF) technique [6], the values of the cuts on the track fit quality parameter and on the half-cone angle are optimized for each considered model in order to minimize the sensitivity. The optimization leads to a selection cone around the Sun of 3°-5° of half-opening angle for the various models corresponding to about 1-3 events of background inside the cone.

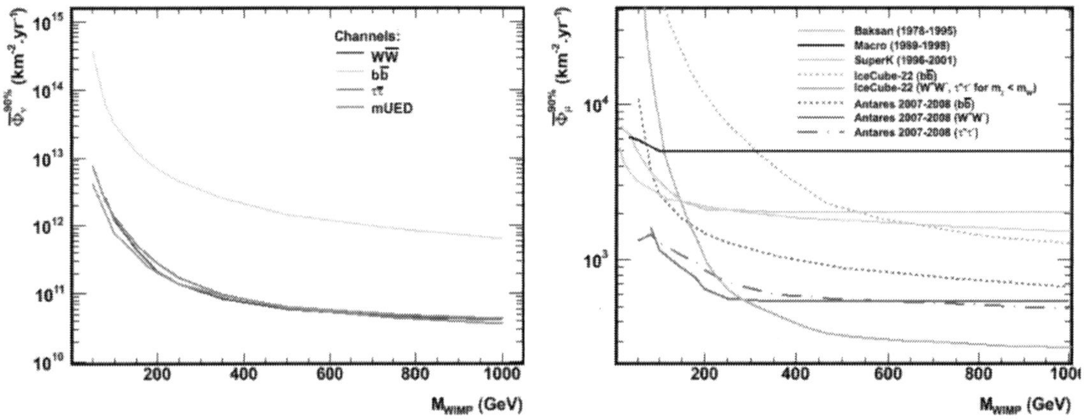

Figure 1. Sensitivity of ANTARES using the data recorded in 2007-2008 to the flux of neutrinos (left) and neutrino-induced muons (right) originating from Dark Matter annihilations inside the Sun as function of the WIMP mass for various decay channels and for the mUED model.

4. Results and perspectives

Figure 1 (left) shows the expected sensitivity of ANTARES with the data collected in 2007-2008 to the flux of neutrinos produced by annihilations of Dark Matter particles in the core of the Sun as function of the mass of the WIMP for various decay channels. In this calculation, a minimal threshold of 10 GeV has been considered for the energy of the neutrinos. As expected, the fluxes of neutrinos coming from Dark Matter annihilations into W^+W^- or $\tau^+\tau^-$ are much better constraint than annihilations into b-quarks which give a much softer neutrino spectrum. A high sensitivity is also observed for the specific case of the mUED model for which the theoretical branching ratios are taken into account for the calculation of the neutrino spectrum.

In order to compare our sensitivity to current limits set by other experiments, it is more convenient to present it in terms of neutrino-induced muon flux into the detector taking into account the interaction rate of neutrinos into the Earth, the muon mean range in matter and the nucleon density in the vicinity of the detector. Figure 1 (right) presents the expected sensitivity of ANTARES with the data collected in 2007-2008 to the flux of neutrino-induced muons produced by annihilations of Dark Matter inside the Sun integrated above a muon energy of 1 GeV. One can see that although the size of the ANTARES detector is limited compared to IceCube, its capability to detect neutrinos as low in energy as about 10 GeV leads to a very promising probability to detect a signal of Dark Matter annihilations inside the Sun in the WIMP mass range between 50 and 300 GeV.

The flux of neutrinos can also be related to the annihilation rate of Dark Matter into the Sun and then to the capture rate of the WIMP by elastic scattering, and thus to the spin-dependent cross section of the WIMP on protons, with the reasonable hypothesis of equilibrium between the annihilation rate and the capture rate. Figure 2 presents the sensitivity of ANTARES with the data collected in 2007-2008 to the spin-dependent cross section of the WIMP to protons as function to the WIMP mass compared to predictions of allowed regions of the parameter space of the CMSSM (left) and mUED (right)

models. In both cases, the scans of the model parameter spaces have been performed by using the SuperBayes program [7] taking into account the experimental constraints coming from the determination of the relic density or from collider observables. This demonstrates the complementary as well as the great sensitivity of neutrino telescopes with respect to direct detection experiments in the hunting quest of the Dark Matter of the Universe. In particular, ANTARES and IceCube are now starting to probe interesting regions of the parameter space in the case of SUSY models such as the CMSSM, while their sensitivity is still one to two orders of magnitudes above the predicted domain for the mUED model.
Thanks to the data taking performed in 2009 and 2010 with a complete detector, ANTARES has now accumulated more than 3000 neutrino candidates corresponding to about three times the statistic used in the analysis presented here. In addition, other potential Dark Matter sources such as the Earth, the Galactic Centre and Dwarf Spheroidal galaxies are being studied by dedicated analyses. The hunt for Dark Matter with ANTARES is therefore only starting and the near future will certainly be exciting.

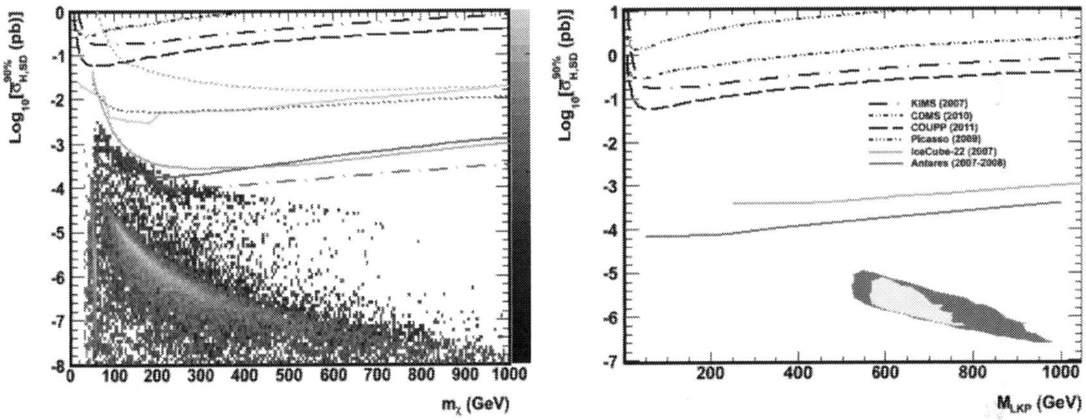

Figure 2. Sensitivity of ANTARES using the data recorded in 2007-2008 to the spin-dependent cross section of WIMP to proton as function of the WIMP mass compared to the predictions obtained for the CMSSM (left) and for the mUED model (right). In the left plot, the sensitivity of ANTARES in the "hard" channel (red line) and in the "soft" channel (red dashed dotted line) is compared to existing limits of direct detection experiments (black curves, see the legend details in the right plot) and of Super-Kamiokande (yellow) and IceCube-22 (green line for the "hard" channel and green dotted line for the "soft" channel). In the right plot, the light green region corresponds to models satisfying the observables input in the SuperBayes scan within one sigma while the purple zone highlights the two sigma region of the mUED parameter space.

References

[1] M. Ageron et al., ANTARES Collaboration, *Nucl. Inst. Meth. A* **656,** 11 (2011).
[2] J.A. Aguilar et al., ANTARES Collaboration, *Astropart. Phys.* **34**, 652 (2011).
[3] M. Blennow, J. Edsjo, T. Ohlsson, *JCAP* **0801**, 021 (2008).
[4] G. Servant and T.M.P. Tait, *Nucl. Phys. B* **650**, 391 (2003).
[5] G.L. Feldman and R.D. Cousins, *Phys. Rev. D* **57**, 3873 (1998).
[6] G.C. Hill and K. Rawlins, *Astropart. Phys.* **19**, 393 (2003).
[7] R. Trotta et al., *JHEP* 0812:024 (2008).

Spectral cutoffs in indirect dark matter searches

Christoph Weniger[1,*], Torsten Bringmann[2], Francesca Calore[2] and Gilles Vertongen[3]

[1] Max Planck Institute for Physics, Föhringer Ring 6, 80805 Munich, Germany
[2] II. Institute for Theoretical Physics, University of Hamburg, Luruper Chaussee 149, 22761 Hamburg, Germany
[3] Paris Institute of Astrophysics, 98 bis boulevard Arago, 75014 Paris, France
[*] presenting author

E-mail: weniger@mppmu.mpg.de

Abstract. Indirect searches for dark matter annihilation or decay products in the cosmic-ray spectrum are plagued by the question of how to disentangle a dark matter signal from the omnipresent astrophysical background. One of the practically background-free smoking-gun signatures for dark matter would be the observation of a sharp cutoff or a pronounced bump in the gamma-ray energy spectrum. Such features are generically produced in many dark matter models by internal Bremsstrahlung, and they can be treated in a similar manner as the traditionally looked-for gamma-ray lines. Here, we discuss prospects for seeing such features with present and future Atmospheric Cherenkov Telescopes.

1. Introduction

Indirect dark matter (DM) searches aim at seeing an excess in cosmic rays from the annihilation or decay of DM in the Galactic halo [1]. Very often, indirect searches focus on *secondary photons* from the fragmentation of hadronic annihilation products. The corresponding spectra are rather broad and peak at energies much lower than the DM mass m_χ, which generically makes a convincing claim of a DM detection above the astrophysical backgrounds difficult. In many models, however, *pronounced spectral features* are expected at the kinematic endpoint $E_\gamma = m_\chi$; they include monochromatic gamma-ray lines [2], sharp steps or cutoffs [3, 4] as well as pronounced bumps [5]. The type and strength of these features are intricately linked to the particle nature of DM; a detection would thus not only allow for a convincing discrimination from astrophysical backgrounds but also to determine important DM model parameters like the value of m_χ. So far, only line-signals have explicitly been searched for [6]—despite the fact that they are loop-suppressed and thus generically subdominant compared to other spectral signatures [5]. Here, we discuss a general method to search for sharp spectral features in gamma-ray observations. Concentrating on DM models with a large internal Bremsstrahlung (IB) component, and on observations with Imaging Atmospheric Cherenkov Telescopes (IACTs), we derive projected limits and prospects to see such signatures with current and future instruments.

2. Methods and targets

The defining aspect of the above-mentioned spectral features is an abrupt change of the gamma-ray flux as function of energy. It is therefore possible to concentrate the search for spectral

	DM particle	m_χ^{th} [TeV]	$\langle\sigma v\rangle^{\text{th}}$ [cm^{3}s^{-1}]	relevant channel	spectral feature
$\gamma\gamma$	any WIMP	$\mathcal{O}(0.1\text{–}10)$	$\mathcal{O}(10^{-30})$	$\gamma\gamma$	line
KK	$B^{(1)}$	1.3	$1\cdot10^{-26}$	$\ell^+\ell^-\gamma$	FSR step
BM3	neutralino	0.23	$9\cdot10^{-29}$	$\ell^+\ell^-\gamma$	IB bump
BM4	neutralino	1.9	$3\cdot10^{-27}$	$W^+W^-\gamma$	IB bump

Table 1. DM benchmark models used in our analysis as examples for the typical spectral endpoint features to be expected in WIMP annihilations. For these particular models, we also state the annihilation channel that is most important in this context, as well as mass and total annihilation rate for thermally produced DM. See text and Ref. [10] for further details about the DM models and Fig. 1 for the corresponding photon spectra.

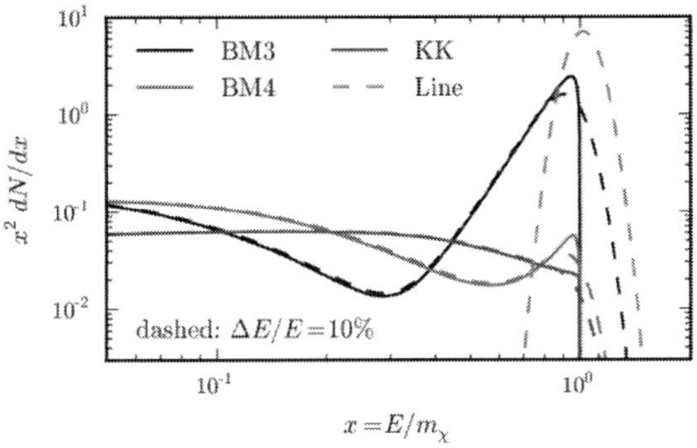

Figure 1. Photon spectra for the DM benchmark models of Tab. 1. Dashed lines show the same spectra smeared with a Gaussian of width $\Delta x/x = 0.1$. From Ref. [10].

features on a small sliding energy window $[E_0, E_1]$, with $E_0 < m_\chi < E_1$, and window sizes of the order of a few times the energy resolution of the instrument. An important advantage of considering only small window sizes is that astrophysical gamma-ray fluxes can be typically very well described by a simple power-law. Limits on and the significance for dark matter induced fluxes can then be derived by standard statistical methods like the profile likelihood method.

We are interested in deriving projected limits on the dark matter annihilation cross-section that follow from the spectral end-point features of different dark matter scenarios *alone*, *i.e.* without taking into account secondary photons, using the above method. We will focus on observations of the Galactic center region with IACTs, considering benchmark scenarios that roughly correspond to the telescope characteristics of the currently operating H.E.S.S. (IACT1) [7], the future CTA (IACT2) [8] and—as the most optimistic choice for indirect DM searches—the proposed Dark Matter Array (DMA, IACT3) [9]. For the background, we take into account cosmic-ray fluxes of electrons and protons, the diffuse gamma-ray flux and the source HESS J1745-290 at (or very close to) the Galactic center. We adopt an Einasto dark matter profile, and a relatively small target region $\Delta\Omega = 2° \times 2°$ around the Galactic center. The energy window size is chosen such that the impact of a nonzero background curvature on our final results is less than 50% (see Ref. [10] for details).

We will discuss three types of typical endpoint features that arise from radiative corrections to the tree-level annihilation process: (1) The most striking spectral signature, in terms of a possible discrimination from a power-law background, is a *gamma-ray line*, which would result from the direct annihilation of DM into $\gamma\gamma$, $Z\gamma$ or $H\gamma$. Generically, the annihilation rate is loop-suppressed and expected to be of the order of $\langle\sigma v\rangle_{\text{line}} \sim \alpha_{\text{em}}^2 \times \langle\sigma v\rangle_{\text{tree}} \sim 10^{-30}\text{cm}^3\text{s}^{-1}$

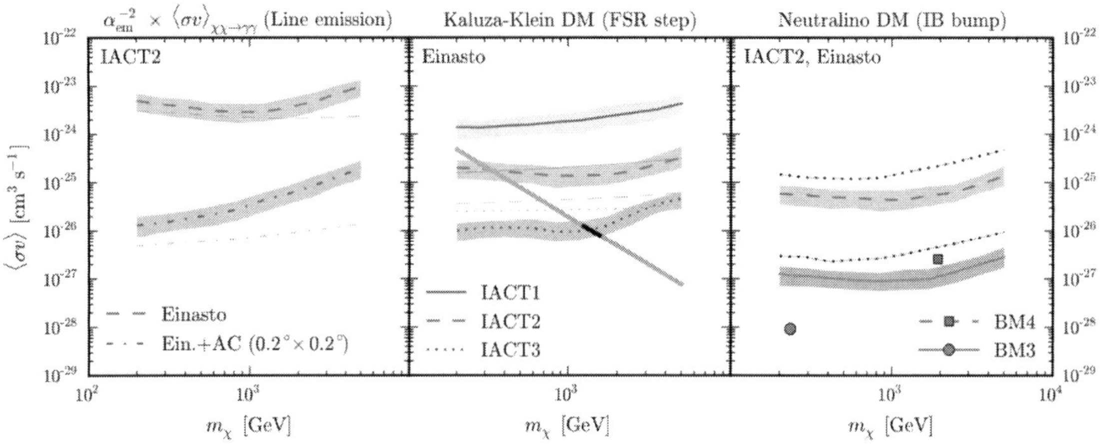

Figure 2. *Thick lines:* Expected 2σ upper limits on $\langle\sigma v\rangle$ for selected DM models, DM profiles and observational scenarios; bands indicate the variance of these limits. *Thin lines:* Spectral feature of DM signal has $\mathcal{S}/\mathcal{B} \approx 1\%$ (after convolution with energy dispersion). The *left panel* shows limits on gamma-ray lines, rescaled by a loop-factor of α_{em}^{-2} for better comparison. In the *central panel*, the gray band indicates the expected $\langle\sigma v\rangle$ for KK DM, the black part being compatible with the observed relic density. In the *right panel*, we indicate the adopted neutralino benchmark points, and the *dotted lines* show the projected 5σ sensitivity. The benchmark experiments IACT1, IACT2 and IACT3 correspond roughly to H.E.S.S., CTA and DMA, respectively, see text for details. The figure is taken from Ref. [10].

(although in some cases much stronger line signals are possible). (2) As an example for a step-like feature we use the gamma-ray spectrum expected from annihilating *Kaluza-Klein (KK) DM* in models of universal extra dimensions [11]. Its total gamma-ray annihilation spectrum at high energies is dominated by final state radiation (FSR) off lepton final states, and the shape of the spectrum dN/dx, with $x = E_\gamma/m_\chi$, turns out to be essentially independent of m_χ and other model parameters [4]. (3) Pronounced bump-like features at $E \simeq m_\chi$ may arise from IB in the annihilation of *neutralino DM* [5]. Here, BM3 is a typical example for a neutralino in the stau co-annihilation region, where photon emission from virtual sleptons greatly enhances the photon spectrum at high energies; BM4 refers to a situation in which IB from $W^\pm$ final states dominates.

In Tab. 1, we shortly summarize the properties of the DM benchmark models described above, including for completeness the actual DM mass and total annihilation rate needed to obtain the observed relic density for thermally produced DM. Note, however, that we essentially treat these values as free parameters in our analysis and that we are rather interested in the spectral shape of the annihilation signal, represented by dN/dx; in Fig. 1 we show these spectra for a direct comparison.

3. Results and discussion

In Fig. 2 we show our results for the expected 2σ upper limits (thick lines) on the above DM models as well as the variance of these limits among the mock data sets. We find that in particular IB features in the spectrum (right panel) have the potential to constrain the annihilation rate at least down to values typically expected for thermal production, $\langle\sigma v\rangle \sim 3\cdot 10^{-26}\text{cm}^3\text{s}^{-1}$, already for modest assumptions about the DM distribution. For models with very large IB contributions like BM3, we find that our method would provide even stronger limits on $\langle\sigma v\rangle$ than what was obtained by the H.E.S.S. analysis of the Galactic center region assuming annihilation into $\bar{b}b$ [12].

For the case of not too strongly pronounced endpoint features (like line signals in most models or the step for Kaluza-Klein DM), secondary photons will usually be more powerful in constraining the total annihilation rate $\langle\sigma v\rangle$. However, in case of an adiabatically compressed profile our limits could improve by maybe two orders of magnitude, as demonstrated for gamma-ray lines in the left panel. As shown in the central panel of Fig. 2, the future CTA should be able to place limits about one order of magnitude stronger than currently possible, and the proposed DMA could further improve these by another factor of ten.

Finally, we would like to stress that our limits in general provide rather *complementary information* on the DM nature and cannot easily be compared with limits on secondary photons. In any case, from the point of view of indirect DM searches, the *detection* of the kinematic cutoff will be much more interesting than the detection of secondary photons, because they would provide rather unambiguous evidence for the DM nature of the signal as well as allow one to determine important parameters like the DM mass. An obvious extension of the approach presented here is to apply it to the *discrimination* of models [13].

Acknowledgments

T.B. and C.W. thank the organizers of the TAUP 2011 for an interesting and stimulating conference. T.B. and F.C. acknowledge support from the German Research Foundation (DFG) through Emmy Noether grant BR 3954/1-1.

References

[1] G. Bertone, D. Hooper, J. Silk, Phys. Rept. **405** (2005) 279.
[2] L. Bergström, P. Ullio and J. H. Buckley, Astropart. Phys. **9** (1998) 137 [arXiv:astro-ph/9712318].
[3] A. Birkedal, K. T. Matchev, M. Perelstein and A. Spray, arXiv:hep-ph/0507194.
[4] L. Bergström, T. Bringmann, M. Eriksson and M. Gustafsson, Phys. Rev. Lett. **94** (2005) 131301 [arXiv:astro-ph/0410359].
[5] T. Bringmann, L. Bergström and J. Edsjö, JHEP **0801** (2008) 049 [arXiv:0710.3169 [hep-ph]].
[6] A.A. Pullen, R.R. Chary, M. Kamionkowski, Phys. Rev. D **76** (2007) 063006 [Erratum-ibid. **83** (2011) 029904] [arXiv:astro-ph/0610295]; F. A. Aharonian *et al.* [HEGRA Collaboration], Astron. Astrophys. **400**, 153 (2003) [astro-ph/0302347]; A. A. Abdo *et al.*, Phys. Rev. Lett. **104** (2010) 091302 [arXiv:1001.4836 [astro-ph.HE]]; G. Vertongen and C. Weniger, JCAP **1105** (2011) 027 [arXiv:1101.2610 [hep-ph]].
[7] F. Aharonian *et al.* [H.E.S.S. Collaboration], Astron. Astrophys. **457** (2006) 899 [arXiv:astro-ph/0607333].
[8] The CTA Consortium, arXiv:1008.3703 [astro-ph.IM].
[9] L. Bergström, T. Bringmann and J. Edsjö, Phys. Rev. D **83** (2011) 045024 [arXiv:1011.4514 [hep-ph]].
[10] T. Bringmann, F. Calore, G. Vertongen and C. Weniger, Phys. Rev. D **84** (2011) 103525 [arXiv:1106.1874 [hep-ph]].
[11] G. Servant and T. M. P. Tait, Nucl. Phys. B **650** (2003) 391 [arXiv:hep-ph/0206071].
[12] A. Abramowski *et al.*, Phys. Rev. Lett. **106**, 161301 (2011) [arXiv:1103.3266 [astro-ph.HE]].
[13] T. Bringmann, F. Calore, G. Vertongen and C. Weniger, work in progress; see also, *e.g.*, M. Perelstein and B. Shakya, JCAP **1010** (2010) 016 [arXiv:1007.0018 [astro-ph.HE]].

12th International Conference on Topics in Astroparticle and Underground Physics (TAUP 2011) IOP Publishing
Journal of Physics: Conference Series **375** (2012) 012035 doi:10.1088/1742-6596/375/1/012035

Indirect searches for gravitino dark matter

Michael Grefe

Deutsches Elektronen-Synchrotron DESY, Notkestraße 85, D-22607 Hamburg, Germany

Departamento de Física Teórica and Instituto de Física Teórica UAM/CSIC,
Universidad Autónoma de Madrid, Cantoblanco, E-28049 Madrid, Spain

E-mail: `michael.grefe@uam.es`

Abstract. The gravitino in models with a small violation of R-parity is a well-motivated decaying dark matter candidate that leads to a cosmological scenario that is consistent with big bang nucleosynthesis and thermal leptogenesis. The gravitino lifetime is cosmologically long-lived since its decays are suppressed by the Planck-scale as well as the small R-parity violating parameter. We discuss the signals in different cosmic-ray species coming from the decay of gravitino dark matter, namely gamma rays, positrons, antiprotons, antideuterons and neutrinos. Comparison to cosmic-ray data can be used to constrain the parameters of the model.

1. Introduction

The existence of the gravitino in the particle spectrum of locally supersymmetric theories leads to problems in the standard cosmological scenario. One manifestation of this so-called cosmological gravitino problem appears in scenarios where the baryon asymmetry is generated via thermal leptogenesis. The late decays of the abundantly produced gravitino are in conflict with the successful predictions of big bang nucleosynthesis. A similar problem even persists in scenarios where the gravitino is the lightest superpartner and thus a candidate for the dark matter. In this case late decays of the next-to-lightest superpartner can spoil the predictions of big bang nucleosynthesis. Among several other possibilities a very attractive solution to this problem are theories with a slight violation of R-parity [1].

2. Gravitino dark matter with bilinear R-parity violation

Theories with bilinear R-parity violation only violate lepton number and are thus not affected by proton stability constraints. The size of the R-parity violating couplings are, however, constrained by cosmological considerations: They must be large enough to allow for the decay of the next-to-lightest superpartner before the time of big bang nucleosynthesis and they must be small enough that the baryon asymmetry of the universe is not washed out by these additional interactions. Interestingly, the allowed window for the couplings leads to a gravitino lifetime exceeding the age of the universe by several orders of magnitude, thus making the unstable gravitino a viable dark matter candidate [2].

An intriguing feature of this scenario is that it exhibits a rich phenomenology although the gravitino usually is thought to be one of the most elusive particles. For instance, the next-to-lightest superpartner decays mainly via the small R-parity violating couplings and is thus long-lived on collider scales [3]. In addition, the decay of gravitino dark matter in the galactic halo can lead to observable signals in the spectra of cosmic rays as we will discuss in the following.

The gravitino can decay into several two-body final states: $\psi_{3/2} \to \gamma\nu, Z\nu, h\nu$ and $W\ell$. We calculate analytically the branching ratios for these decay channels, while the spectra of stable final state particles – *i.e.* photons, electrons, protons, deuterons and neutrinos – are obtained by a PYTHIA simulation of the gravitino decay [4]. The purpose of this conference contribution is to briefly report on the phenomenological results obtained in [4].

3. Indirect detection of gravitino dark matter

In recent years several anomalies in the spectra of cosmic rays have been reported that cannot be explained by astrophysical secondary production and could be a hint for a signal from dark matter annihilation or decay. Indeed, the decay of gravitino dark matter leads to a hard spectrum of positrons (and electrons) that – depending on the gravitino mass and lifetime – could explain the rise in the positron fraction observed by the PAMELA experiment [5]. This contribution could also be observed as a subdominant component in the absolute electron flux (see Figure 1). However, gravitino decays unavoidably produce associated fluxes of antiprotons and gamma rays that are only compatible with observations if the gravitino lifetime is larger than needed for an explanation of the PAMELA data (see Figure 2). Therefore, a gravitino dark matter explanation of the PAMELA data is ruled out and additional astrophysical sources like pulsars are required to explain the observed positron fraction.

Also antideuterons can be formed in the fragmentation of gravitino decay products. The expected signal may exceed the astrophysical background especially for lower gravitino masses and can be tested with forthcoming experiments like AMS-02 and the GAPS balloon, thereby making antideuterons a valuable tool to probe light gravitinos (see Figure 3).

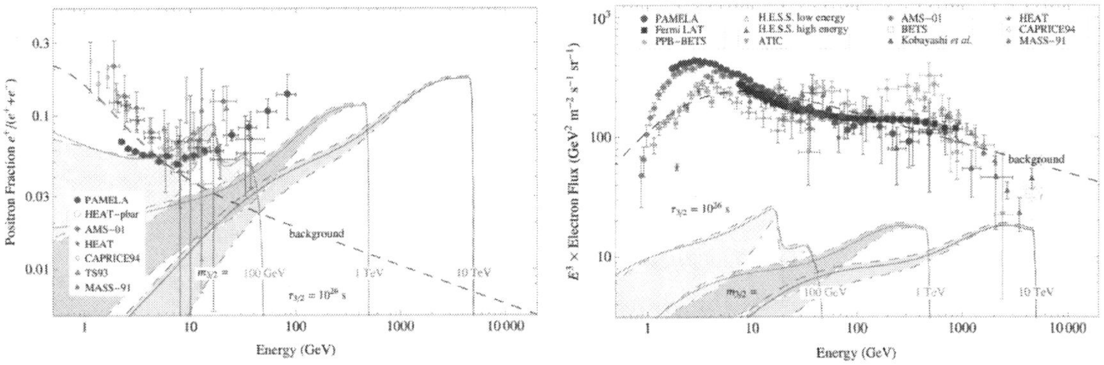

Figure 1. Positron fraction (*left*) and electron spectrum (*right*) expected from gravitino dark matter decays compared to data and the expected astrophysical background. Shaded bands show the uncertainty due to charged cosmic-ray propagation through the galaxy.

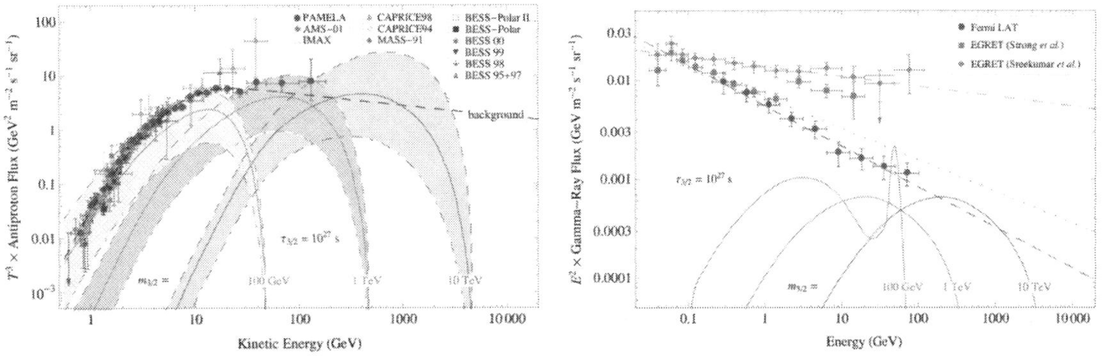

Figure 2. Antiproton spectrum (*left*) and gamma-ray spectrum (*right*) expected from gravitino dark matter decays compared to data and the expected astrophysical background. From [4].

Another valuable channel are neutrinos that – like gamma rays – provide angular information that can be used to discriminate gravitino decay from dark matter annihilation [6]. An intriguing feature of the expected signal is the line at the end of the spectrum that persists – in contrast to gamma rays – also for large gravitino masses (see Figure 4). In neutrino telescopes like IceCube neutrinos are mainly detected via the Cherenkov light of neutrino-induced upward through-going muons. Although the rather poor energy resolution of this channel obscures the expected spectral features, a statistical analysis might reveal a departure from the featureless spectrum of atmospheric neutrinos (see Figure 5). We find that a new detection channel, namely neutrino-induced showers, has two advantages compared to the muon signal: The dominant background of atmospheric muon neutrinos contributes less to this channel and its better energy resolution allows for a better discrimination of spectral features (see Figure 6). Although there are currently still problems to discriminate the shower signal from short muon tracks we conclude that this channel is the most promising for a detection of neutrinos from gravitino decays.

The non-observation of gravitino decay signals in the spectra of various cosmic-ray species allows to estimate lower limits on the gravitino lifetime (see Figure 7). For light gravitinos the strongest limits come from gamma-ray line searches, while at larger masses the antiproton and diffuse gamma-ray limits are the most important. For very heavy gravitinos also neutrino bounds become significant. We expect that forthcoming antideuteron observations and searches based on neutrino-induced showers allow to improve these bounds in the future.

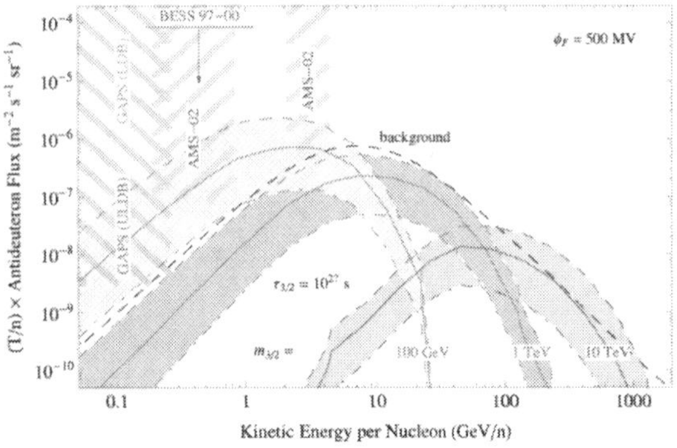

Figure 3. Antideuteron spectrum expected from gravitino dark matter decays compared to sensitivities of forthcoming experiments and the expected astrophysical background.

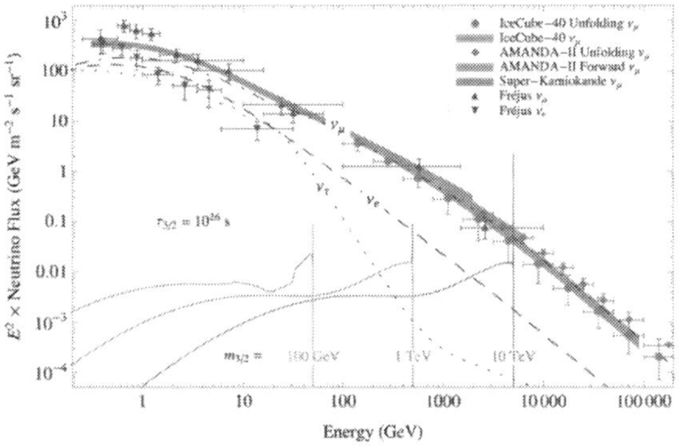

Figure 4. Neutrino spectrum expected from gravitino dark matter decays compared to data and the expected background of atmospheric neutrinos. Figure taken from [4].

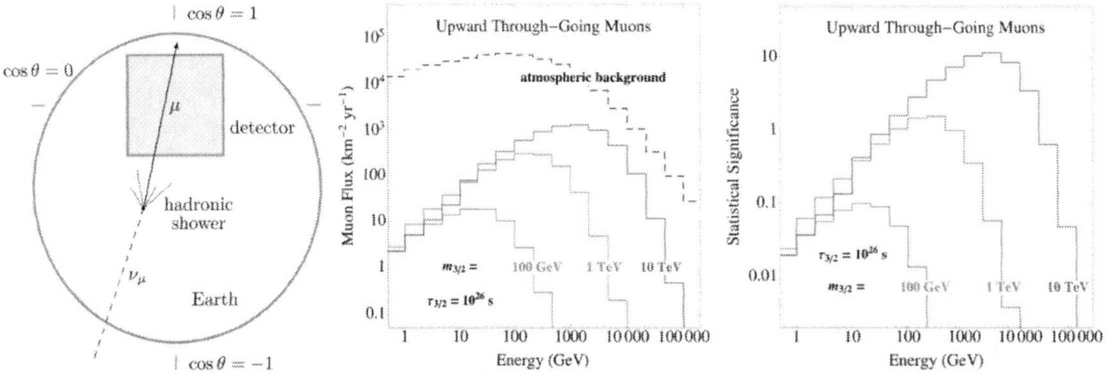

Figure 5. Event topology of neutrino-induced upward through-going muons (*left*), muon spectrum with finite energy resolution (*middle*) and statistical significance of the muon signal (*right*). Figures taken from [4].

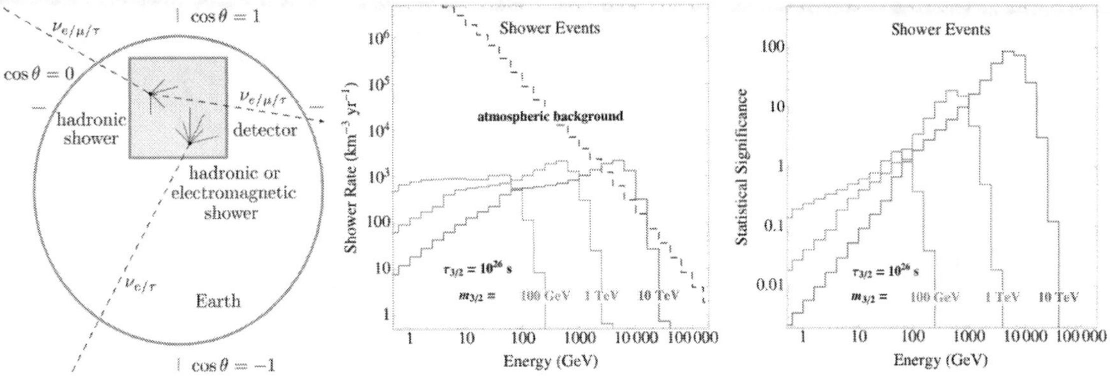

Figure 6. Event topology of neutrino-induced showers (*left*), spectrum of showers with finite energy resolution (*middle*) and statistical significance of the shower signal (*right*). From [4].

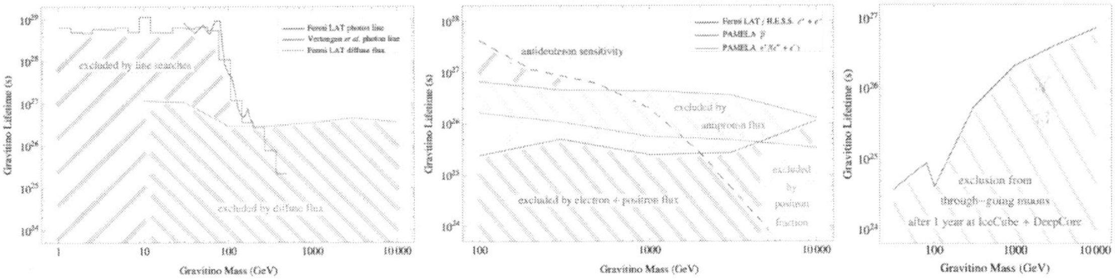

Figure 7. Estimated limits on the gravitino lifetime from gamma-ray observations (*left*), observations of charged cosmic rays (*middle*) and forthcoming observations of upward through-going muons at IceCube (*right*). Figures taken from [4].

4. Conclusions

Unstable gravitino dark matter with a lifetime exceeding the age of the universe by many orders of magnitude is well motivated from cosmology. Although gravitino decays produce cosmic-ray spectra that could explain the PAMELA excess, this possibility is ruled out due to strong constraints on the associated production of antiprotons and gamma rays. Forthcoming antideuteron searches will be a valuable channel to search for signals of light gravitinos while neutrino telescopes like IceCube have the capability to probe heavy gravitinos – in particular if new detection channels like shower events are employed.

In conclusion, we find that the gravitino lifetime can be effectively constrained over a large range of gravitino masses using a multi-messenger approach. For an exhaustive discussion of the presented results see [4].

Acknowledgments

I am grateful to Laura Covi and Gilles Vertongen for useful discussions and collaboration on part of the presented work. I acknowledge the support of the DFG within the Collaborative Research Center 676, the Marie Curie ITN "UNILHC" under grant number PITN-GA-2009-237920 and the grant HEPHACOS S2009/ESP-1473 of the Comunidad de Madrid.

References

[1] Buchmüller W *et al.* 2007 *J. High Energy Phys.* JHEP03(2007)037 (*Preprint* hep-ph/0702184)
[2] Takayama F and Yamaguchi M 2000 *Phys. Lett.* B **485** 388–92 (*Preprint* hep-ph/0005214)
[3] Bobrovskyi S *et al.* 2010 *J. High Energy Phys.* JHEP10(2010)061 (*Preprint* arXiv:1007.5007 [hep-ph])
[4] Grefe M 2011 *Unstable Gravitino Dark Matter – Prospects for Indirect and Direct Detection* DESY-THESIS-2011-039 (*Preprint* arXiv:1111.6779 [hep-ph])
[5] Adriani O *et al.* [PAMELA Collaboration] 2009 *Nature* **458** 607–9 (*Preprint* arXiv:0810.4995 [astro-ph])
[6] Covi L *et al.* 2010 *J. Cosmol. Astropart. Phys.* JCAP04(2010)017 (*Preprint* arXiv:0912.3521 [hep-ph])

Dark matter synchrotron emission and radio observations

Marco Taoso

IFIC, CSIC–Universidad de Valencia, Ed. Institutos, Apdo. Correos 22085, E–46071 Valencia, Spain, and MultiDark fellow

E-mail: taoso@ific.uv.es

Abstract. We compute the synchrotron emission induced by electrons produced by DM annihilations inside our galaxy. The signal is compared with observations in a large range of frequencies, from 22 MHz up to 1420 MHz. We set constraints on the DM mass and annihilation cross-section and highlight the impact of astrophysical uncertainties.

1. Introduction

The detection of a non-gravitational signal of Dark Matter (DM) would be one of the greatest pillars of modern physics, simultaneously confirming our views of cosmology, astrophysics and particle physics. This possibility might be not far ahead if DM is in the form of Weakly Interacting Massive Particles (WIMPs), which currently are the most investigated class of DM candidates. Signatures of this scenario include a multiwavelength spectrum associated to radiative emissions involving electrons and positrons generated in WIMP annihilations or decays. In this work, we focus our attention on the galactic synchrotron emission from DM annihilations and we concentrate on observations in the low-frequency range from 22 to 1420 MHz (see Ref. [1]). Low radio frequencies have not been addressed in the past in connection with DM studies, and we will show that they may be quite adequate in the search for DM, especially for low-mass WIMPs. Indeed, for a magnetic field of μG strength, as typical in the Galaxy, synchrotron emission below GHz-frequency is generated by electrons and positrons with energies well below 10 GeV. In particular, we will show that WIMP candidates annihilating into light leptons with a thermal annihilation cross-sections can be strongly constrained for masses $M_{DM} \leq 10$ GeV.

Radio emission arising from DM annihilation have instead been addressed in Refs. [2, 3] by using higher frequency surveys and by studying heavier dark matter candidates, with a particular focus on the galactic Center by Refs. [4, 5, 6, 7, 8] , or in order to reproduce the so called WMAP Haze in Refs. [9, 10, 11]. DM searches with radio observations can also be pursued with the cosmological diffuse radio background, using both intensity [12] and anisotropies informations [13].

In Sec. 2 we describe the formalism used to compute the radio emission induced by DM annihilations. In Sec. 3 we compare the predictions for DM radio fluxes with observations and we derive bounds on the DM annihilation cross sections. We conclude in Sec. 4.

Journal of Physics: Conference Series **375** (2012) 012036 doi:10.1088/1742-6596/375/1/012036

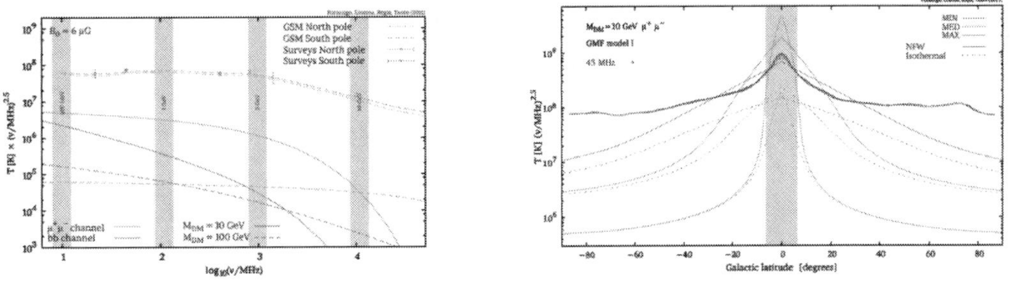

Figure 1. Left: Temperature versus frequency calculated for the galactic poles ($b = \pm 90°$) for different DM masses and annihilation channels. Here we adopt a NFW profile and the MED propagation model. The data points are the temperature at north and south galactic poles averaged in a 10° circle. Right: Temperature versus galactic latitude at 45 MHz. The data corresponds to a thin strip $|l| < 3°$, with l the galactic longitude. Lines are predictions for DM models for $l = 0°$. Blue band corresponds to $|b| < 10°$ and it denotes the directions towards the galactic center region.

2. Synchrotron emission from galactic DM annihilations

The rate and spectrum of electron and positrons injected in the galaxy by DM annihilations depends on the DM density distribution and on particle physics inputs: the DM mass M_{DM}, the annihilation cross-section (σv) and the energy spectrum of electrons/positrons per single annihilation, which in turn depends on the particular DM annihilation channel. In our analysis, we consider two DM density profiles, the popular NFW profile, which nicely fits the results of N-body simulations, and an isothermal profile which is instead used as a conservative case of a cored DM distribution. The propagation of cosmic rays electrons is then modeled following a semi-analytical approach [14]. Cosmic rays are assumed to be confined by the galactic magnetic field inside a propagation zone which is described by a cylinder centered at the galactic center with a radius $R_g = 20$ kpc. The number density of electrons per unit of energy $n_e(E, \mathbf{x})$ follows a transport equation which takes into account the spatial diffusion of the electrons and the energy losses due to the interactions with the galactic magnetic field and the interstellar radiation field. The parameters of the propagation model are constrained by cosmic-rays data, notably the Boron over Carbon ratio (B/C) and radioactive isotopes. In order to bracket the uncertainties on the electrons propagation we consider three propagation benchmarks, the so-called MIN, MED, and MAX models. These models are fully compatible with B/C and antiproton/proton ratio observations, describing the extreme behaviors (MIN and MAX) and best fit (MED) of these observables [15]. Finally, the synchrotron flux observed at the Earth results from the integration along the line of sight of the synchrotron emissivity j_ν:

$$j_\nu(\mathbf{x}) = \int dE\, n_e(E, \mathbf{x}) \frac{dw}{d\nu}(\mathbf{x}) \tag{1}$$

where $dw/d\nu$ denotes the emission power, which depends on the galactic magnetic field (GMF) strength $B(\mathbf{x})$ (see Appendix of Ref. [1] for details). The structure of the GMF is still not well understood. Mainly, it is composed by a regular and a turbulent part, the last one responsible for the diffusive behaviour of the cosmic rays. Here, the total magnetic field is assumed to have a cylindrical symmetry with an exponential dependence on the distance on the galactic center and on the vertical height (more details in Ref. [1]). The normalization at the Sun position is set to 6μG, which is well consistent with the measurements.

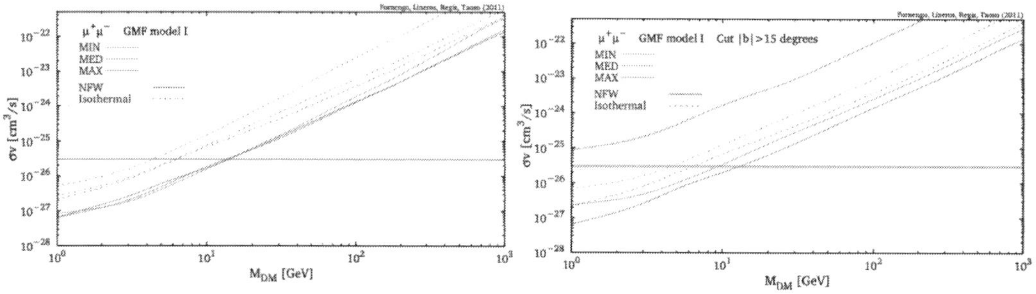

Figure 2. Upper bounds on the annihilation cross section (σv) as a function of the DM mass for the muonic channel. Left panel shows the constraints inferred using all the regions of sky available in each of the surveys considered. In the right panel the bounds are obtained excluding the strip along the galactic plane $|b| < 15°$.

3. Comparison with observations and constraints

Using the formalism explained in the previous section we have generated skymaps of the synchrotron radiation produced by DM annihilations at different frequencies and compared with different radio surveys available. We found that low ($\lesssim$GHz) radio frequencies are particularly suitable to search for light/intermediate DM masses $\mathcal{O}(100)$ GeV. Indeed, both WIMP models with dominant hadronic/bosonic annihilation final states and leptophilic light WIMP models induce a synchrotron spectrum which is softer than the observed galactic one, as shown in Fig. 1. In the right panel of Fig. 1 we compare observations and DM emissions at 45 MHz for a 10 GeV DM particle annihilating into muon pairs and with a thermal annihilation cross section, $(\sigma v) = 3 \times 10^{-26}$ cm^3 s^{-1}. The observational data correspond to a thin strip crossing the galactic center and perpendicular to the galactic plane ($|l| < 3°$). We notice that the emission is more extended for propagation models with larger diffuse height, like the MED and MAX models. This means that in these cases it is possible to search for the DM signal outside the inner galactic center region ($|b| < 10°$). While the observed radio emission is dominated by the astrophysical background, Fig. 1 suggests that DM could substantially contribute to the radio flux, especially close to the galactic center region. Certainly a possible DM detection is challenged by the large uncertainties which affect the determination of the background. In the following, in order to be conservative we just use the present observational data to constrain DM models, without attempting any background subtraction. In order to that we consider five surveys from 22 MHz to 1420 MHz (see Ref. [1] for more details about the method adopted to set the bounds.) In Fig. 2 we show the results for the muonic channel. Models with DM masses $M_{DM} \leq 10$ GeV and thermal value of the annihilation cross section are strongly constrained in the case of the NFW profile. Despite the morphology of the emission is quite different for the three propagation models, the derived bounds are instead similar. The constraints significantly weaken for the isothermal profile, which presents a much lower DM density in a large region around the galactic center. Then, we compute the bounds by cutting a large region around the galactic plane by imposing $|b| > 15°$ (right panel in Fig. 2). In general, with this conservative choice, we reduce the uncertainties on the propagation related to possible unaccounted astrophysical effects occurring on the galactic disk. Moreover, the fact that the radio astrophysical background and its uncertainties are maximal in the galactic plane, complicates the searches of DM signals in this region. Therefore, by avoiding the region $|b| < 15°$, we study the reach of current radio surveys in the DM parameter space focusing on cleaner targets of observations. Interestingly, we find that the bounds are almost unchanged for the MAX model and only slightly affected for the MED one. For the MIN model, they are more altered since the emission is more concentrated in the galactic center region.

4. Conclusions

We have argued that radio observations in the low-frequency regime are particularly suitable to search for DM annihilations in the galactic halo. To show that, we have compared available radio surveys with the synchrotron emission produced by DM annihilations for different DM setups and astrophysical assumptions. Assuming a NFW density profile and for realistic galactic magnetic fields, current radio surveys constraint thermal values of DM the annihilation cross section for DM masses $M_{DM} \leq 10$ GeV. We notice that this is the mass range suggested by the signals recently found in some DM direct detection experiments: DAMA [16], CoGeNT [17] and CRESST-II [18].

While we have conservatively focused on the bounds which can be inferred from observations, it could be possible that the present data already contain a significant DM contribution. In Ref. [1] we have tried to single out this DM component using the 408 MHz map as a spatial template to model the galactic CR synchrotron emission in the low-frequency maps. With this method, a DM-induced component softer than the astrophysical galactic synchrotron emission would appear as an excess in the central galactic center region with an approximately spherical shape. We have concluded that present data does not support any evidence for the presence of this additional synchrotron component. However, we expect that future radio surveys, in particular with the LOFAR telescope [19], will improve coverage, angular resolution, and sensitivity in low-frequency radio maps, which may allow to disentangle a faint DM contribution.

Acknowledgments
The work was supported by the EC contract UNILHC PITN-GA-2009-237920, by the Spanish grants FPA2008-00319, FPA2011-22975, MultiDark CSD2009-00064 (MICINN) and PROMETEO/2009/091 (Generalitat Valenciana).

References

[1] N. Fornengo, R. A. Lineros, M. Regis and M. Taoso, JCAP **1201** (2012) 005 [arXiv:1110.4337 [astro-ph.GA]].
[2] E. Borriello, A. Cuoco, G. Miele, Phys. Rev. **D79** (2009) 023518. [arXiv:0809.2990 [astro-ph]].
[3] T. Delahaye, C. Boehm, J. Silk, [arXiv:1105.4689 [astro-ph.GA]].
[4] G. Bertone, M. Cirelli, A. Strumia, M. Taoso, JCAP **0903** (2009) 009. [arXiv:0811.3744 [astro-ph]].
[5] C. Boehm, J. Silk, T. Ensslin, [arXiv:1008.5175 [astro-ph.GA]].
[6] R. M. Crocker, N. F. Bell, C. Balazs, D. I. Jones, Phys. Rev. **D81** (2010) 063516. [arXiv:1002.0229 [hep-ph]].
[7] M. Regis, P. Ullio, Phys. Rev. **D78** (2008) 043505. [arXiv:0802.0234 [hep-ph]].
[8] L. Bergström, G. Bertone, T. Bringmann, J. Edsjö, M. Taoso, Phys. Rev. **D79** (2009) 081303. [arXiv:0812.3895 [astro-ph]].
[9] G. Dobler and D. P. Finkbeiner, Astrophys. J. **680** (2008) 1222 [arXiv:0712.1038 [astro-ph]].
[10] D. T. Cumberbatch, J. Zuntz, H. K. K. Eriksen, J. Silk, [arXiv:0902.0039 [astro-ph.GA]
[11] T. Linden, S. Profumo, B. Anderson, Phys. Rev. **D82** (2010) 063529. [arXiv:1004.3998 [astro-ph.GA]].
[12] N. Fornengo, R. Lineros, M. Regis and M. Taoso, Phys. Rev. Lett. **107** (2011) 271302 [arXiv:1108.0569 [hep-ph]].
[13] L. Zhang, G. Sigl, JCAP **0809** (2008) 027. [arXiv:0807.3429 [astro-ph]].
[14] D. Maurin, F. Donato, R. Taillet, P. Salati, Astrophys. J. **555** (2001) 585. [astro-ph/0101231].
[15] F. Donato, D. Maurin, P. Salati, A. Barrau, G. Boudoul, R. Taillet, Astrophys. J. **563** (2001) 172. [astro-ph/0103150].
[16] R. Bernabei *et al.*, DAMA/LIBRA collaboration, Eur. Phys. J. **C67** (2010) 39. [arXiv:1002.1028 [astro-ph.GA]].
[17] C. E. Aalseth *et al.*, CoGeNT collaboration, Phys. Rev. Lett. **107** (2011) 141301. [arXiv:1106.0650 [astro-ph.CO]].
[18] G. Angloher *et al.*, CRESST-II collaboration, [arXiv:1109.0702 [astro-ph.CO]].
[19] http://www.lofar.org/

12th International Conference on Topics in Astroparticle and Underground Physics (TAUP 2011) IOP Publishing
Journal of Physics: Conference Series **375** (2012) 012037 doi:10.1088/1742-6596/375/1/012037

Galactic synchrotron emission from astrophysical electrons

Roberto A. Lineros[1,0], **Torsten Bringmann**[2], **Fiorenza Donato**[3]

[1]IFIC, CSIC-Universidad de Valencia, Ed. Institutos,Apdo. Correos 22085, E–46071 Valencia, Spain & MultiDark fellow

[2] II. Institute for Theoretical Physics, University of Hamburg, Luruper Chausse 149, D–22761 Hamburg, Germany

[3] Dipartimento di Fisica Teorica, Università di Torino, Istituto Nazionale di Fisica Nucleare, via P. Giuria 1, I–10125 Torino, Italy

E-mail: `rlineros@ific.uv.es`

Abstract. The interaction between the galactic magnetic field and the non-thermal population of electrons is responsible for a large part of the radio sky from 10 MHz up to several GHz. This population is mostly composed of electrons with primary and secondary origin. Cosmic ray propagation models describe their evolution in space and energy, and allow to study the impact on the radio sky in intensity and morphology at different frequencies. We consider different propagation models and test their compatibility with available radio maps. We find models highly consistent both with B/C data, the local electron flux and synchrotron emission observations. The resulting constraints on propagation models could significantly improve prospects for indirect dark matter searches in these channels and, even more so, in antiprotons.

Contribution for the proceedings of the 12th International Conference on Topics in Astroparticle and Underground Physics (TAUP).
5th – 9th September 2011, Munich, Germany.

1. Introduction

The emission of synchrotron radiation from a non-thermal population of electrons and positrons (from here simply refer to both as *electrons*) has given one of the first insights that Galactic cosmic rays (CR) propagate diffusively inside a magnetic halo which extends far beyond the distribution of gas and stars.

This evidence came from observations of extended radio haloes around edge-on galaxies. Current CR observations are nicely explained by semi–analytical propagation models which incorporate physical processes that typically happen in the Galactic environment. These models have been successfully tested and reproduce the observations with reasonible precision. Despite this precision, however, there is still a big degeneracy in the propagation parameters. These type of uncertainties highly affect possible dark matter searches based on CR observations. In this work, we study the possible effects of different propagation models on the radio sky. Further details can be found in Ref. [1].

[0] speaker

Model	L_z [kpc]	K_0 [kpc^2/Myr]	δ
min	1	0.0016	0.85
med	4	0.0112	0.70
max	15	0.0765	0.46

Table 1. Benchmark propagation models compatible with nuclear CR observations [6].

2. Electron propagation and synchrotron radiation

Electrons produced by Galactic sources travel across the Galaxy, interacting on their way with the radiation fields, the interstellar gas, and the Galactic Magnetic Field (GMF). The propagation of electrons is modeled consistently with the propagation of nuclear CR by using a successfully tested two–zone propagation model [?] in which electrons are confined by the GMF inhomogeneities to a cylinder, centered on the Galactic Center and oriented with the Galactic plane, with radius $R_g = 20$ kpc and vertical length L values between 1 and 15 kpc. Similar to nuclear CR, the propagation is described by a transport equation which for the case of electrons reads

$$-K_0(E/E_0)^\delta \nabla^2 \psi - \frac{\partial}{\partial E}(b(E)\psi) = q(\mathbf{x}, E). \tag{1}$$

Here, ψ is the number density of electrons per unit of energy, $b(E)$ is the energy loss term (for a further discussion of this term, see Sec. 3), $q(\mathbf{x}, E)$ is the *electron* source term, and E_0 is an energy scale with value of 1 GeV. We take into account that electrons have both a primary and a secondary origin [2] while positrons are only produced as secondaries [3]. The parameters K_0 and δ parameterize the diffusion constant; together with L they provide the most important parameters of the model. Different configurations produce different propagation models, each one with proper features in the predicted CR fluxes. The parameter space can be constrained by nuclear CR observations like boron over carbon ratio (B/C), radioactive isotopes (i.e. ^{10}Be/^{9}Be) and the antiproton to proton ratio ($\bar{p}/p$) [4, 5, 6]. More details may be found in Ref. [1] and references therein.

In our study, we consider three benchmark propagation models [6] *min*, *med*, and *max* (Tab. 1). The *min* and *max* models produce extreme behaviors (with respect to the *antiproton* signal expected from dark matter annihilation) which are still compatible with CR observations. The *med* model is very close to the best-fit point of the secondary/primary analysis. Let us comment that CR fluxes are measured with relatively small errors. The constrained space of parameters is still quite degenerate and allows, e.g., for rather big variations in the size L of the diffusion zone.

Together with the propagation, the interaction of CR *electrons* with the GMF will inevitably produce synchrotron radiation [7]. For an average magnetic field intensity of $\mathcal{O}(\mu\mathrm{G})$ [8] and *electron* energies in the MeV–GeV range, the emission will be in the MHz–GHz range, i.e. at radio frequencies. This provides a link on how radio data and CR electrons are related. In order to compare the observations with predictions from propagation models, we calculate the synchrotron emissivity and integrate it along the line of sight (details in Ref. [1]). We consider 11 sky-maps at frequencies 0.010, 0.022, 0.045, 0.408, 1.42, 2.326, 23, 33, 41, 61 and 94 GHz. These were generated using a public software [9], which includes a compilation of several radio surveys in the frequency range between 10 MHz and 94 GHz.

3. Discussion and summary

We compare propagation model predictions with radio data using different approaches: frequency spectrum, *electron* CR flux at Earth, and morphology. Each of those reveals different

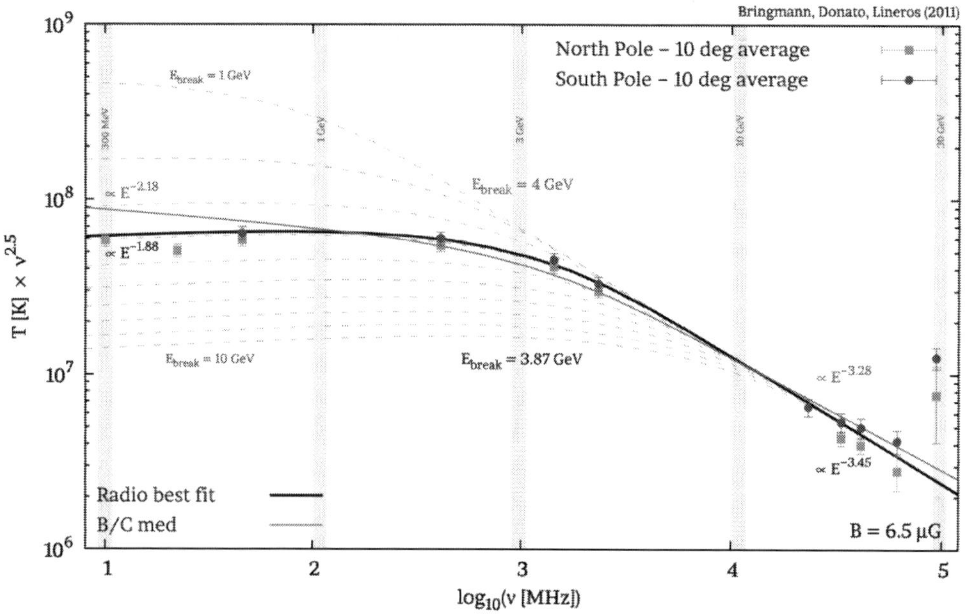

Figure 1. Temperature versus frequency (taken from Ref. [1]). The black line corresponds to the emission from an effective (propagated) electron distribution that follows a broken power-law with indices -1.88 and -3.45. The red line shows the emission from propagated electrons in the *med* model, with an energy loss term described by a broken power-law with indices 0.9 and 2 for low and high energies, respectively. The data points correspond to 10° angular averages in the direction of North and South poles and were calculated with the software described in Ref. [9].

features that help to understand the connection with the propagation model. In figure 1, we present the comparison in terms of the frequency spectrum. The observed frequency spectrum presents a shape resembling a broken power–law with indices ~ 2.5 below, and ~ 3.2 above, a break around 800 MHz. This shape in the frequency spectrum should be related to a similar behaviour in the (propagated) *electron* distribution. In fact, the frequency spectrum can be reproduced by a distribution of propagated *electrons* modeled by a broken power–law described by indices $(1.88, 3.45)$, with an energy break at 3.86 GeV (black line). Also, we calculate the prediction of the propagation model (red line). We find that a possible explanation of this frequency break could be that energy losses from bremsstrahlung and gas ionization start to dominate over inverse Compton scattering; these terms will induce a break in the energy loss term in eq. 1 that provides an alternative to the conjectured break in the injection spectrum of primary electrons [10]. A more thorough discussion is presented in [1] where the comparison with the *electrons* flux data is also addressed.

Another interesting prediction coming from different propagation models is the radio morphology. In figure 2, we calculate the emission at different latitudes (b), where the galactic anti-center is at $b = 0°$. We find that observations at 408 MHz and 1.42 GHz are in better agreement with a typical diffusive halo size of $L \sim 4$ kpc than with halo sizes as large as 15 kpc or as small as 1 kpc. This provides first indications that a complementary and/or independent analysis of the propagation model parameters can be done using radio data.

The connection between radio data and CR physics relates two rather different research fields. Making use of this connection would increase the possibilities to obtain a better global picture of Galactic diffusion processes. The reduction of the uncertainties in CR physics will affect

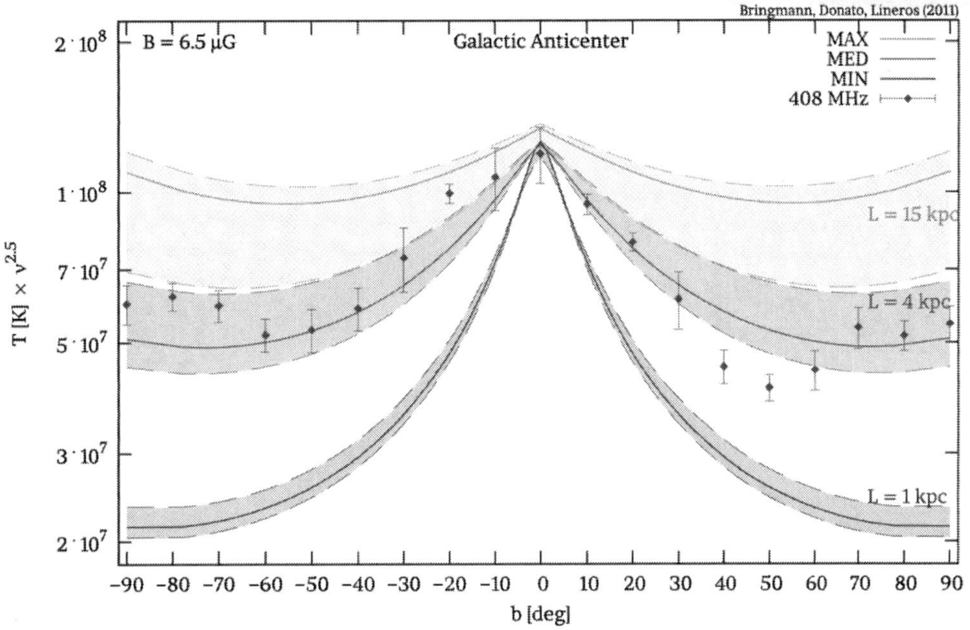

Figure 2. Temperature at 408 MHz versus galactic latitude (taken from Ref. [1]). The emission produced by the three propagation models of Tab. 1 (solid lines) and uncertainty bands for fixed L are shown (resulting from the variation of all other propagation parameters within the bounds allowed by B/C). Models with $L \lesssim 1$ kpc and $L \gtrsim 15$ kpc are clearly disfavoured by observations. Data points correspond to 10° averages.

tremendously Dark Matter searches due to a better control of the background and, for the case of antiprotons even more important, greatly reduced uncertainties in the predicted flux from dark matter annihilation.

Acknowledgments

We would like to thank the organizers of TAUP for a very successful conference. RL is supported by the EC contract UNILHC PITN–GA–2009–237920, by the Spanish grants FPA2008–00319, FPA2011–22975, MultiDark CSD2009–00064 (MICINN) and PROMETEO/2009/091 (Generalitat Valenciana). TB is supported by the German Research Foundation (DFG) through Emmy Noether grant BR 3954/1-1.

[1] Bringmann T, Donato F and Lineros R A 2012 *JCAP* **2012** 049 (*Preprint* `1106.4821`)
[2] Delahaye T, Lavalle J, Lineros R, Donato F and Fornengo N 2010 *A & A* **524** A51 (*Preprint* `1002.1910`)
[3] Delahaye T, Lineros R, Donato F, Fornengo N, Lavalle J, Salati P and Taillet R 2009 *A & A* **501** 821–833 (*Preprint* `0809.5268`)
[4] Maurin D, Donato F, Taillet R and Salati P 2001 *APJ* **555** 585–596 (*Preprint* `astro-ph/0101231`)
[5] Maurin D, Taillet R and Donato F 2002 *A & A* **394** 1039–1056 (*Preprint* `astro-ph/0206286`)
[6] Donato F, Fornengo N, Maurin D, Salati P and Taillet R 2004 *PRD* **69** 063501 (*Preprint* `astro-ph/0306207`)
[7] Berezinskii V, Bulanov S, Dogiel V, Ginzburg V and Ptuskin V 1990 *Astrophysics of Cosmic Rays* (North Holland)
[8] Han J L 2009 *IAU Symp.* **259** 455 (*Preprint* `0901.1165`)
[9] de Oliveira-Costa A, Tegmark M, Gaensler B M, Jonas J, Landecker T L and Reich P 2008 *MNRAS* **388** 247–260 ISSN 1365-2966
[10] Strong A W, Orlando E and Jaffe T R 2011 *A & A* **534** A54+ ISSN 1432-0746 (*Preprint* `1108.4822`)

Journal of Physics: Conference Series **375** (2012) 012038 doi:10.1088/1742-6596/375/1/012038

Searches for Dark Matter with the IceCube Detector

M Danninger[1] for the IceCube collaboration

[1] Department of Physics, Stockholm University, AlbaNova, S-10691 Stockholm, Sweden

E-mail: `danning@fysik.su.se`

Abstract. Construction of the IceCube neutrino observatory was recently completed, including the full DeepCore sub-array, a low-energy extension of the IceCube detector. We present recent results from the searches of dark matter candidates with IceCube, performed with the 22-and 40-string configurations using the Sun and the Galactic Center and Halo as possible dark matter sources. We also report on the search for dark matter annihilations with the IceCube neutrino detector in the 79-string configuration. Furthermore, a formalism for quickly and directly comparing event-level IceCube data with arbitrary annihilation spectra in detailed model scans is presented. We show an application of this formalism to both model exclusion and parameter estimation in models of supersymmetry.

1. Introduction

While the presence of dark matter (DM) in the universe has been inferred through its gravitational interactions, it has yet to be directly or indirectly observed. One of the most promising and experimentally accessible candidates for DM are so-called Weakly Interacting Massive Particles (WIMPs). In current models, WIMPs are predicted to have a mass in the range of a few tens of GeV to a few TeV. Neutrino telescopes, such as IceCube, can be used in the search for WIMPs and their underlying fundamental properties. IceCube searches for decaying or self-annihilating DM in the halo of galaxies and thereby can probe its livetime, or self-annihilation cross section [1]. Additionally, IceCube can directly test the WIMP-nucleon scattering cross section by looking for neutrino signals from self-annihilating DM captured in large celestial bodies [2], like the Sun or the Earth.

Here recent results are presented from the searches for DM candidates with AMANDA-II and IceCube in its 22- and 40-string configurations. First, we focus on the latest limits on $\langle\sigma_A v\rangle$ from the Galactic Center and Halo. Then, we discuss a multi-year search for DM annihilations in the Sun with the AMANDA-II and IceCube detectors. We conclude with a brief discussion on a new formalism for quickly and directly comparing event-level IceCube data with arbitrary annihilation spectra in detailed model scans and comments on prospects for the full IceCube 86-string detector.

For IceCube search strategies for self-annihilating DM in dwarf galaxies surrounding the Milky Way, the reader is referred to reference [3].

The IceCube detector

The IceCube Neutrino Telescope [4] records Cherenkov light in the ice from relativistic charged particles created in neutrino interactions in the vicinity of the detector. By recording the arrival times and intensities of these photons with optical sensors, the direction and energy of the muon and parent neutrino may be reconstructed. IceCube instruments 1 km^3 of glacial ice at the South Pole with 5160 Digital Optical Modules (DOMs) on 86 strings deployed between depths of 1450 m and 2450 m. Eight more densely instrumented strings optimized for low energies plus the 12 adjacent standard strings at the center of the detector geometry make up the DeepCore subarray, which increases the sensitivity at low energies and substantially lowers the energy threshold. In addition, by using the surrounding IceCube strings as a veto, DeepCore will enable searches at low energies in the southern hemisphere, transforming IceCube into a full sky observatory. IceCube's predecessor, AMANDA-II, was operational until May 2009. It consisted of an array of 677 optical modules deployed on 19 strings at depths between 1500 m and 2000 m.

2. Galactic Dark Matter searches with IceCube

The expected DM density distribution in the Milky Way can be approximated by a spherically symmetric function. Such a distribution is strongly motivated by observational evidence e.g., galaxy rotation curves, but also by large scale N-body simulations. While halos are relatively well understood in the outer regions, the inner structure currently cannot be directly measured or resolved in simulations. Therefore, halo models generally show good agreement in the outer halo, and differ significantly for the Galactic center (GC). The analyses discussed here uses the Navarro-Frenk-White, NFW, profile [5] as a benchmark model, while other profiles are used to estimate the impact on the results from the choice of halo model. The expected neutrino flux from DM self-annihilations is proportional to ρ^2_{DM} integrated along the line of sight, $J(\Psi)$, for a given angular distance to the GC, where ρ_{DM} is the local DM density, which is normalized to 0.3 GeV/cm^3 at a distance of 8.5 kpc from the GC.

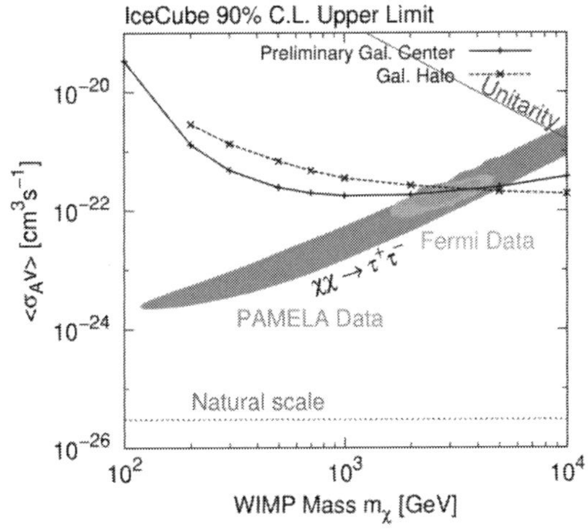

Figure 1. 90% C.L. upper limits for τ channel, compared to preferred phenomenological annihilation channels if PAMELA excess and Fermi e$^-$ are interpreted as DM. IceCube 40-string limits are preliminary, since they do not include signal acceptance systematic uncertainties.

The differential neutrino multiplicity per annihilation is highly model dependent. Therefore, various possible annihilation channels are investigated, assuming a branching ratio of 100% for each of them.

The highest neutrino flux from WIMP annihilation is expected to come from a relatively wide region, centered in the direction of the GC that is at the location of IceCube about 30° above the horizon. Data from this direction are dominated by atmospheric muons. Therefore, this analysis is based on the identification of events with an interaction vertex inside the detector, to distinguish signal from atmospheric muons that produce incoming tracks. The optimum on-source region both in right ascension and declination were found to be ±8°. This analysis is performed on IceCube 40-string data taken in 2008 and 2009.

In the Galactic halo analysis we search for a neutrino anisotropy by comparing fluxes from an on-source region centered around the same right ascension as the GC, to that from an equally sized off-source region shifted by 180° at the same declination. This analysis is performed on IceCube 22-string data taken during 2007 and 2008. Here, the final event selection is dominated by upwards going neutrino-induced muon events.

Observations in both analyses were consistent with background-only expectations. Upper limits on $\langle\sigma_A v\rangle$ are computed at the 90% confidence level as a function of WIMP mass and annihilation channel. Figure 1 shows the obtained limits for the τ channel, compared to preferred phenomenological annihilation channels if the PAMELA excess and Fermi electrons are interpreted as DM [6]. Details on both analyses can be found in Ref. [7] and [8].

3. Solar Dark Matter searches with IceCube

In theories of the Minimal Supersymmetric Standard Model [9], the WIMP can take the form of the lightest neutralino, χ, while in the framework of universal extra dimensions [10] it would be the lightest Kaluza-Klein particle (LKP). Whatever their underlying physics, WIMPs may be swept up by the Sun on its transit through the Galactic halo and become gravitationally bound by scattering weakly on solar nucleons. Over time, this leads to an accumulation of DM at the center of the Sun, exceeding the mean galactic density. Self-annihilation to standard model particles may result in a flux of high energy neutrinos that will be spectrally dependent on the annihilation channel and WIMP mass, and can be searched for as a point-like source in AMANDA-II and IceCube.

In this analysis, the two most extreme annihilation channels are studied; the softest ($\chi\chi \rightarrow b\bar{b}$) and hardest ($\chi\chi \rightarrow W^+W^-$) possible neutrino spectrum. The same event selection is also applied to search for the LKP. The dominant background for this search comes from downwards going muons created by cosmic ray interactions in the atmosphere. By restricting the search to the Austral winter period, when the Sun is below the horizon, the muon background can be significantly reduced. Background is further eliminated by a series of quality cuts, resulting in a final event selection that is dominated by neutrino-induced muon events. The analysis is performed in a blind manner such that the azimuth of the Sun is not revealed until the selection cuts are finalized. This analysis combines data from previous work on AMANDA-II and IceCube 22-string with the latest IceCube 40-string analysis (AMANDA-II is embedded within IceCube 40-string array and was used as a low energy extension to IceCube). As all three data samples are independent, we can combine them in a statistically suitable way in one likelihood analysis with a total livetime of 1065 days. Here, we use the shape of the space angle distribution of the final samples with respect to the Sun's position, to build a confidence interval for the number of signal events compatible with the observed data at the 90% C.L.. The full study is detailed in Ref. [11].

From the upper limit on the number of signal events, a limit on the annihilation rate in the Sun can be derived, which in turn can be used to set a limit on the WIMP capture rate in the Sun. Assuming equilibrium between WIMP capture and annihilation in the Sun, the capture rate is then directly proportional to the WIMP-proton scattering cross sections σ_{SI} and σ_{SD}. By assuming that the capture is dominated by spin-dependent scattering, it is possible to derive the limit on σ_{SD} at the 90% C.L. for this multi-year analysis (Fig. 2). Also shown in Fig. 2 is a projected sensitivity for the full IceCube and DeepCore detector and current experimental limits from direct searches. Corresponding limits on the LKP are discussed in Ref. [11] and [12].

4. Global SUSY analysis with IceCube

We will perform two specialized analyses of the impacts of IceCube observations of the Sun upon (SUSY) DM models. The primary difference between these analyses and the existing IceCube upper limits is that the new analyses would be based on explicit exploration of theoretical SUSY parameter spaces, including a model-by-model comparison of fluxes with the IceCube observations. This approach would also allow models to be tested against a range of complementary data, including accelerator constraints, the relic density of DM, limits on rare processes from B-physics and the anomalous magnetic moment of the muon.

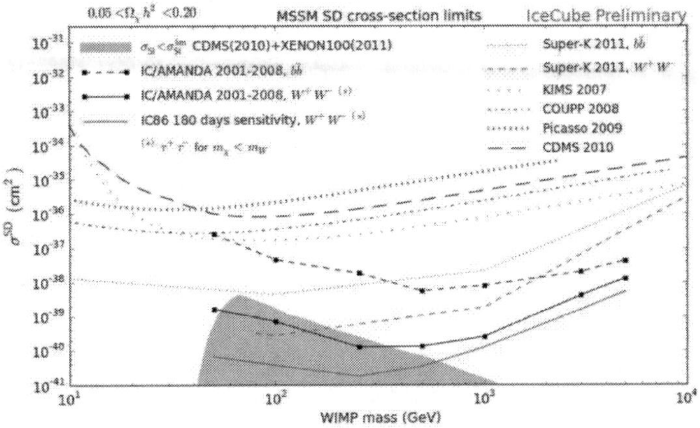

Figure 2. Limits on χp spin-dependent cross-section at 90%C.L. vs. m_χ (black markers). Also shown is a projected 86-string sensitivity(red line), and limits from other experiments (soft models in dashed lines, hard in solid lines). The model space represents a scan over the allowed MSSM parameter space, accounting for all current experimental constraints.

The first analysis is model exclusion. Here individual SUSY models are tested for consistency (at some set C.L.) with IceCube data and other existing constraints, and identified as allowed or excluded by IceCube. The second analysis constitutes a parameter estimation exercise within a given SUSY framework. This analysis involves a global fit to all existing constraints, with the inclusion of IceCube data. In this analysis, the full likelihood of each experimental result, including IceCube, is combined to give confidence intervals and Bayesian credible intervals on SUSY parameters. A detailed description of the methods used and the obtained results of both analyses will be discussed in detail in future publications.

Prospects for DM searches with the full IceCube and DeepCore

With the completion of the IceCube telescope in December of 2010, a sensitive platform for the discovery of DM candidates in the energy range of 50 GeV to 5 TeV was created. DeepCore lowers the energy threshold of the detector into the 10 GeV regime and gives the possibility to extend searches to the southern hemisphere. For solar DM searches, this effectively doubles the experimental livetime while greatly improving the sensitivity for low mass WIMP models.

Further, all analyses presented here are aiming to extend the search to include the cascade channels. This is feasible owing to the excellent atmospheric muon veto capabilities with IceCube/DeepCore, and even more rewarding as the atmospheric neutrino background in this channel is much lower. These advantages outweigh the drawbacks of a poor angular resolution of cascade events, and will allow further improvement in sensitivity.

References

[1] J. F. Beacom, N. F. Bell and G. D. Mack, 2007 *Phys. Rev. Lett.* **99**, 231301.
[2] R. Abbasi *et al.*,(IceCube Coll.,) 2009 *Phys. Rev. Lett.* **102**, 201302.
[3] Search strategies for Dark Matter in nearby Dwarf Spheroidal Galaxies with IceCube. *2011 ICRC*, arXiv:1111.2738.
[4] R. Abbasi *et al.*,(IceCube Coll.,) 2010 *Nucl. Inst. Meth.* A **618**, 139.
[5] J.F. Navarro, C.S. Frenk, S.D.M. White, 1996 *Astrophys. J.* **462**, 563.
[6] P. Meade, M. Papucci, A. Strumia and T. Volansky, 2010 *Nucl. Phys.* B **831**, 178-203.
[7] R. Abbasi *et al.*,(IceCube Coll.,) 2011 *Phys. Rev.* D **84** 022004.
[8] Search for Dark Matter in the MilkyWay with IceCube. *2011 ICRC*, arXiv:1111.2738.
[9] M. Drees, M.M. Nojiri, 1993 *Phys. Rev.* D **47** 376.
[10] D. Hooper, S. Profumo, 2007 *Phys. Rep.* **453** 29.
[11] Indirect search for Solar dark matter with AMANDA and IceCube. *2011 ICRC*, arXiv:1111.2738.
[12] R. Abbasi *et al.*,(IceCube Coll.,) 2010 *Phys. Rev.* D **81** 057101.

Journal of Physics: Conference Series **375** (2012) 012039 doi:10.1088/1742-6596/375/1/012039

Conservative upper limits on WIMP annihilation cross section from Fermi-LAT γ-rays

F Donato[1], F Calore [2] and V De Romeri [3]

[1] Dipartimento di Fisica, Torino University and INFN, Via P. Giuria 1, 10122 Torino, Italy
[2] II. Institute for Theoretical Physics, University of Hamburg, Luruper Chaussee 149, 22761 Hamburg, Germany
[3] IFIC (CSIC - Universidad de Valencia), C/ Catedratico Jose Beltran 2, E-46980 Paterna (Valencia), Spain

E-mail: donato@to.infn.it, speaker

Abstract. The spectrum of an isotropic extragalactic γ-ray background (EGB) has been measured by the Fermi-LAT telescope at high latitudes. Two new models for the EGB are derived from the subtraction of unresolved point sources and extragalactic diffuse processes, which could explain from 30% to 70% of the Fermi-LAT EGB. Within the hypothesis that the two residual EGBs are entirely due to the annihilation of dark matter (DM) particles in the Galactic halo, we obtain *conservative* upper limits on their annihilation cross section $\langle\sigma v\rangle$. Severe bounds on a possible Sommerfeld enhancement of the annihilation cross section are set as well. Finally, would $\langle\sigma v\rangle$ be inversely proportional to the WIMP velocity, very severe limits are derived for the velocity-independent term of the annihilation cross section.

1. The extragalactic γ-ray background

The high latitude ($|b| > 10^o$) γ-ray emission measured by Fermi-LAT [1], given its reduced contamination by galactic sources, is a powerful tool to set limits on the contribution of DM to the measured flux. The spectrum has been obtained subtracting from the data the sources resolved by the telescope, the (indeed model dependent) diffuse galactic emission, the cosmic ray (CR) background in the detector and the solar γ-ray emission. For each low–flux source there may be a large number of *unresolved* point sources which have not been detected because of selection effects, or too low emission.

Most of the unassociated high latitude sources are blazars, a class of Active Galactic Nuclei (AGNs), and they pile to the EGB with the largest flux [2]. Galactic resolved pulsars and Milli-Second Pulsars (MSPs) represent the second largest population in the Fermi-LAT catalog [3, 4] and they are expected to contribute significantly to the putative EGB. A non-negligible γ-ray flux seems to be guaranteed by unresolved normal star-forming galaxies [5]. Ultra-high energy CRs (UHECRs) may induce secondary electromagnetic cascades, producing γ-rays at Fermi-LAT energies [6]. Unresolved blazars and MSPs are believed to contribute at least few percent to the Fermi-LAT EGB, while predictions for star-forming galaxies and UHECRs are highly model dependent. In the following, we describe few classes of γ-ray emitters whose unresolved flux is at least few percent of the Fermi-LAT EGB. In a conservative scenario (Model I), we will subtract AGN and MSPs to the Fermi-LAT EGB as derived in Ref. [1]. A more relaxed model (Model II) will be drawn by the further subtraction of a minimal flux from star-forming galaxies and CRs at the highest energies. The derivation of each contribution is described in details in [7].

1.1. Models for the EGB

The aim of this Section is to subtract from the Fermi-LAT EGB [1] additional contributions from unresolved sources at latitudes $|b| > 10^o$. The contributions removed from the Fermi-LAT spectrum are minimal. In fact, the predictions that we will take into account for the relevant unresolved sources are the lowest ones according to the literature. For the Model I, we subtract from the Fermi-LAT EGB [1] the unresolved contributions for both BL Lacs and FSRQs, and MSPs. Model II for the EGB refers to the scenario where the additional contributions from star-forming galaxies and UHECRs add to explaining the Fermi-LAT EGB. For all the details, the interested reader is addressed to Ref. [7]. The ensuing fluxes are displayed in Fig. 1. At 100 MeV, Model II explains about 70% of the Fermi-LAT EGB, while above 1-2 GeV they count about 50% of the total.

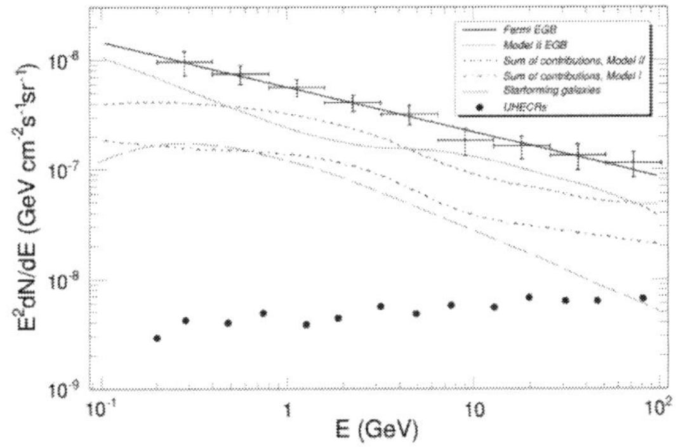

Figure 1. γ-ray spectrum for $|b| > 10^o$ latitudes. Fermi-LAT data points are displayed along with their power–law fit (solid black curve) [1]. Dots and long dashed-curve (yellow) correspond to the UHECRs and star-forming galaxies γ-ray fluxes, respectively. Dot-dashed (blue) curve: sum of BL Lacs, FSRQs and MSPs fluxes. Short-dashed (red): sum of all the unresolved components. Solid (green) curve is derived by subtracting all the contributions to the Fermi-LAT result (Model II).

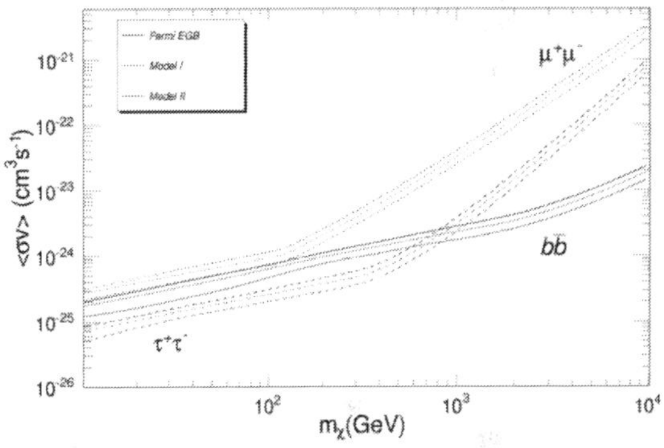

Figure 2. Upper bounds on $\langle\sigma v\rangle$ from γ-ray in the high latitude galactic halo, as a function of the DM mass. From top to bottom, solid lines refer to 90% C.L. limits from the comparison with Fermi-LAT EGB (black lines), Model I (red lines), Model II (blue lines) (see text for details). Dotted, solid and dashed lines correspond DM annihilation into $\mu^+\mu^-$, $b\bar{b}$, $\tau^+\tau^-$, respectively.

2. Upper bounds on DM annihilation cross section

We make the hypothesis that the residual fluxes in Fig. 1 are entirely provided by the γ-rays produced by thermalized WIMP DM in the halo of the Milky Way. For the prediction of the DM flux we refer to [7], and simply remind that all the considered DM density profiles provide very similar results for latitudes well above the galactic plane. We derive upper bounds at 90% C.L. on the WIMP annihilation cross section from the γ-ray Fermi-LAT EGB and the EGB residual fluxes identified as Model I and II. They are displayed in Fig. 2 and detailed in [7]. The subtraction of the minimal amount of γ-rays from unresolved sources lowers the limits on $\langle\sigma v\rangle$ by at least 50%. Our limits are *conservative*: it is very unlikely that a higher $\langle\sigma v\rangle$ be compatible with Fermi-LAT EGB. Similarly, our upper limits could be lowered only with assumptions on non-homogeneous DM distributions or, of course, by comparing to a smaller EGB residual.

2.1. Bounds on the Sommerfeld enhancement for $\langle\sigma v\rangle$

Recent claims on the excess of CR positrons [8] have stimulated the interpretation of data in terms of annihilating DM with fairly large annihilation cross sections of the order of $10^{-23} - 10^{-22}$ cm^3/s. These numbers are at least three orders of magnitude larger than the value indicated by observations of the DM abundance due to thermal production. One way to boost the annihilation cross section is through the Sommerfeld effect [9, 10, 11, 12], generically due to an attractive force acting between two particles, *i.e.* a Yukawa or a gauge interaction. In the case of DM particles, the main effect of such an attractive force would be to enhance $\langle\sigma v\rangle$ by a factor proportional to $1/\beta = c/v$, where v is the velocity of the DM particle ($1/v$ enhancement). The net result on the annihilation cross section writes as $\langle\sigma v\rangle = \mathrm{S}\ \langle\sigma v\rangle_0$, where S sizes the Sommerfeld enhancement of the annihilation amplitude. We have evaluated the Sommerfeld enhancement S using the approximation of the Yukawa potential by the Hulthen potential, for which an analytic solution is possible [13, 14] (and checked that the solution coincides with the numerical one).

In Fig. 3 we show the Sommerfeld enhanced cross sections with over-imposed the upper bounds from the residual EGB Model I and Model II (see Fig. 2). Our results show that a Sommerfeld enhancement due to a force carrier of $m_\phi < 1$ GeV (coupling $\alpha = \frac{1}{4\pi}$) is strongly excluded by Model I and II for the Fermi-LAT EGB data. For a massive force carrier (90 GeV) only the resonant peaks above the TeV mass are excluded. The result holds for $\beta = 10^{-8}$ up to $\beta = 10^{-3}$.

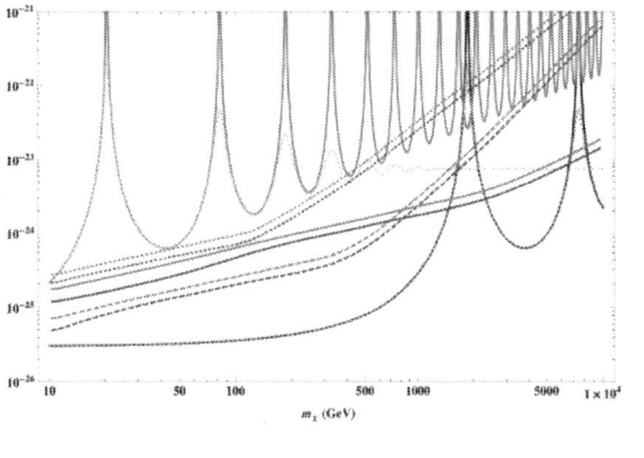

Figure 3. Sommerfeld enhanced $\langle\sigma v\rangle$ as a function of the DM mass, for $\alpha = \frac{1}{4\pi}$. Solid: $\beta = 10^{-8}$, dotted: $\beta = 10^{-3}$. The upper (lower) resonant curve is obtained for a force carrier of mass $m_\phi = 1$ GeV (90 GeV). The upper (lower) dotted, solid and dashed curves correspond to the upper bounds for EGB Model I (Model II) in $\mu^+\mu^-$, $b\bar{b}$, $\tau^+\tau^-$, respectively (see Fig. 2.)

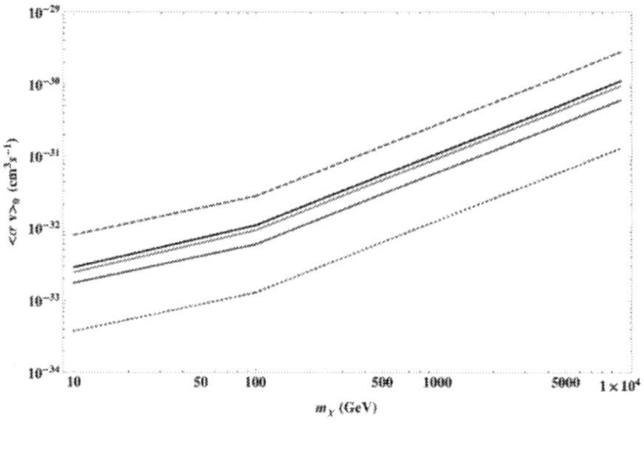

Figure 4. Bounds on $\langle\sigma v\rangle_0$ as a function of the DM mass. The central three bounds are obtained for $M_c = M_\oplus$, and from Fermi-LAT EGB (black line), Model I (red line) and Model II (blue line) respectively, from top to bottom. The upper (lower) purple lines are derived for Model II EGB and $M_c = 10^2 M_\oplus (10^{-2} M_\oplus)$.

2.2. Bounds from the high-redshift protohalos

A possible way to boost the annihilation rate is to modify the particle theory and make the ansatz that the annihilation cross section depends on the inverse of the velocity [15]. A boosted production of γ-rays in models with $\langle\sigma v\rangle \propto 1/v$ has been proposed for the first bound objects formed in the early phases of the universe [16, 17, 7]. The velocity dispersion of the first protohalos that collapse at redshift z_C is estimated to be very small ($\beta \sim 10^{-8}$) [17]. The photons arising from WIMP annihilations in very early halos can freely propagate with their energy red-shifting and reach the Earth in the range $\sim$ keV - TeV, while photons emitted out of this transparency window are absorbed by the intergalactic medium. The $1/v$ enhancement of the annihilation cross section may be simply parameterized [17]: $\langle\sigma v\rangle = \langle\sigma v\rangle_0 \frac{c}{v}$ cm^3/s. We have evaluated the energy density in photons today from WIMP annihilation in the primordial halos and compared with the experimental photon density inferred for the Fermi-LAT, Model I and Model II EGB [1], obtained by integrating the photon flux on the Fermi-LAT energy range (100 MeV - 100 GeV) [7]. The results are displayed in Fig. 4. The bounds on $\langle\sigma v\rangle_0$ are strong: for WIMP masses below 100 GeV it is forced to be $< 10^{-33}$ cm^3/s. Upper bounds grow to $< 10^{-32}$ cm^3/s for $m_\chi \simeq 1$ TeV and sets to $< 10^{-31}$ cm^3/s at 10 TeV. We make notice that they are more stringent than limits obtained from primordial light elements abundance and CMB anisotropies [18] and significantly improve the bounds of Ref. [17]. The 1/v behaviour of the $\langle\sigma v\rangle$ may be identified with the Sommerfeld effect for velocities $\beta \gg (m_\phi/m_\chi)^{1/2}$ [11]. For lower velocities, as are the ones typical for protohalos, the series of resonances appears (see Sect. 2.1) and the Sommerfeld enhancement S behaves as $1/v^2$ close to the peaks. In this case, the upper bounds on the annihilation cross section may be obtained by rescaling $\langle\sigma v\rangle$=$\langle\sigma v\rangle_0$S with a factor $1/\beta \cdot m_\phi/m_\chi$ [7, 17].

3. Conclusions

The γ-ray EGB measured by Fermi-LAT [1] likely includes contributions from galactic and extragalactic *unresolved* sources. We have discussed two residual EGB fluxes derived by the subtraction of unresolved BL Lacs, FSRQs and galactic MSPs, star-forming galaxies and UHECRs to the γ-ray EGB measured by Fermi-LAT. From our new residual EGB fluxes, we have set upper limits on the DM annihilation cross section into γ-rays. A conservative upper bound on $\langle\sigma v\rangle$ is derived by assuming that our new residual fluxes are entirely due to WIMPs pair-annihilating in the halo of our Galaxy. Furthermore, we have shown these EGB residuals bound the Sommerfeld enhancement of $\langle\sigma v\rangle$ to a factor of 3-10-50-200 for m_χ=10-100-1000-5000 GeV, respectively. In case of a Yukawa-like potential, a force carrier heavier than 1 GeV is required. Finally, within the hypothesis that $\langle\sigma v\rangle$ is inversely proportional to the WIMP velocity, very severe limits are derived for the velocity-independent part of the annihilation cross section.

References

[1] Abdo A A *et al* 2010 *Phys. Rev. Lett.* **104** 101101
[2] Abdo A A *et al* 2010 *Astrophys. J.* **720** 435
[3] Abdo A A *et al.* 2010 *Astrophys. J.* **188** 405
[4] Abdo A A *et al* 2010 *Astrophys. J. Suppl.* **187** 460–494
[5] Fields B D, Pavlidou V and Prodanovic T 2010 *Astrophys. J. Lett.* **722** L199
[6] Kalashev O, Semikoz D and Sigl G 2009 *Phys. Rev. D* **79** 063005
[7] Calore F, De Romeri V and Donato F 2011 *in press on Phys. Rev. D* **arxiv 1105.4230**
[8] Adriani O *et al* 2009 *Nature* **458** 607–609
[9] Sommerfeld A J W 1931 *Annalen der Physik* **403**
[10] Arkani-Hamed N, Finkbeiner D P, Slatyer T R and Weiner N 2009 *Phys. Rev. D* **79** 015014
[11] Lattanzi M and Silk J 2009 *Phys. Rev. D* **79** 083523
[12] Hisano J, Matsumoto S and Nojiri M M 2004 *Phys. Rev. Lett.* **92** 1
[13] Cassel S 2010 *Journal of Physics G Nuclear Physics* **37** 105009
[14] LFeng J, Kaplinghat M and Yu H B 2010 *Phys. Rev. D* **82** 083525
[15] Robertson B and Zentner A 2009 *Phys.Rev.* **D79** 083525 (*Preprint* 0902.0362)
[16] Profumo S, Sigurdson K and Kamionkowski M 2006 *Phys. Rev. Lett.* **97** 1
[17] Kamionkowski M and Profumo S 2008 *Phys. Rev. Lett.* **101** 261301
[18] Hisano J *et al.* 2011 *Phys. Rev.* **D83** 123511

12th International Conference on Topics in Astroparticle and Underground Physics (TAUP 2011) IOP Publishing
Journal of Physics: Conference Series **375** (2012) 012040 doi:10.1088/1742-6596/375/1/012040

Light neutralino dark matter

Nicolao Fornengo

Dipartimento di Fisica Teorica, Università di Torino and Istituto Nazionale di Fisica Nucleare, Sezione di Torino, via P. Giuria 1, I–10125 Torino, Italy

E-mail: nicolao.fornengo@unito.it

Abstract. We examine the status of light neutralinos in an effective MSSM at the electroweak scale. We discuss this model in the light of dark matter direct detection and show that it fits quite well the DAMA/LIBRA and CoGeNT annual modulation results, together with the recently reported CRESST excess of events when interpreted as a signal of dark matter. The analysis of light–neutralino models is implemented by the new measurements at the Tevatron, B–factories and LHC which provide significant constraints on the MSSM model. We examine in detail all these new data and show that the neutralino mass has a lower bound of 8–9 GeV, with interesting correlations with the other parameters of the model. Direct detection cross sections naturally fall in the range outlined by the direct detection positive experiments.

1. Introduction

Direct detection experiments are currently offering a possible indication of a signal due to dark matter scattering. DAMA experiments are observing an annual modulation of the detection rate along 13 years [1], with an evidence for a signal at about 9σ CL. More recently CoGeNT has reported a 2σ CL effect of annual modulation in one year cycle of data taking [2]. In this Conference, CRESST is reporting the evidence of an irreducible excess of events which does not appear to be related to known sources of background and is compatible to a signal at about 4σ CL [3, 4]. All three experimental positive indications point to a dark matter particle with a mass in the range from few GeV to few tens of GeV and scattering cross–sections on nucleons between 10^{-42} cm^2 and 10^{-39} cm^2 [1, 2, 3, 4, 5]. CDMS [6] and XENON [7] are currently reporting bounds, but the converging indication of three potential evidences for a dark matter signal is worthwhile to be investigated, also from the point of view of particle physics models able to accommodate a dark matter candidate able to explain these direct detection results.

2. The light neutralino model

A specific realization of a new–physics model with dark matter particles as light as a few GeV has been proposed and discussed in the context of supersymmetry: dark matter is here represented by neutralinos, which are light as a consequence of a small U(1)–gaugino mass parameter M_1 as compared to the SU(2)–gaugino mass parameter M_2. These models, which we will call here "Light Neutralino Models" (LNM) have been discussed and developed in a series of papers, among which Refs. [8, 9, 10, 11, 12, 13, 14, 15, 16], where the relevant phenomenology and signals have been thoroughly discussed. A small ratio M_1/M_2 allows the lightest neutralino to be very light, and basically unbounded by accelerator–physics experiments, since the chargino mass bound from LEP does not affect M_1 and the light neutralino bound from the invisible Z

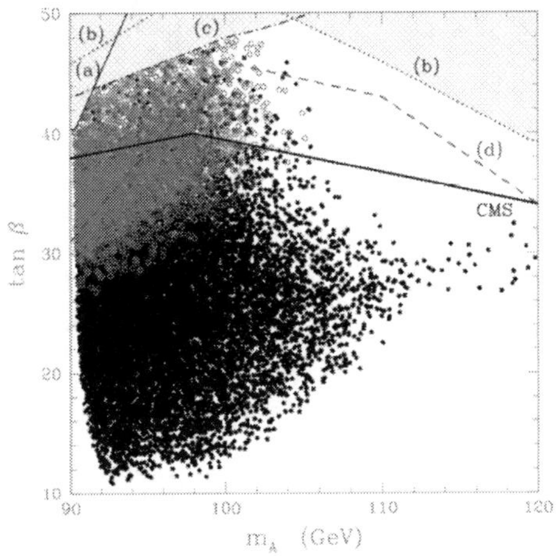

Figure 1. Scatter plot of the LNM configurations in the $m_A - \tan\beta$ plane. Circles refer to $m_\chi \leq 10$ GeV, while dots stand for heavier neutralinos. Lines (a-d) denote bounds from Tevatron (see Ref. [10] for details), while the line denoted by CMS refer to recent bounds from LHC as derived in Ref. [17]. LNM configurations shown in the plot are selected in order to satisfy the experimental bounds coming from supersymmetric and higgs searches, Z–boson invisible width, muon anomalous magnetic moment, $t \to bH^+$ decay, $b \to s + \gamma$ process and $B_s \to \mu^+\mu^-$ decay. The cosmological bound on the contribution of the light neutralinos to the DM abundance is included.

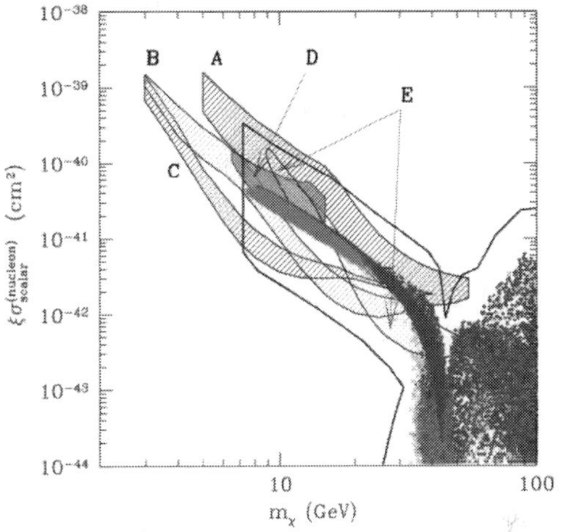

Figure 2. DM–nucleon scattering cross–section as a function of the DM mass. Regions denoted by A, B and C refer to the DAMA annual modulation effect [5] (regions deviate 7.5σ CL from no–signal; A stands for energy–independent quenching, B for an energy–dependent quenching, C for channeling). Region D refers to the CoGeNT annual modulation effect [2] as analyzed in Ref. [5] (the region deviates 1.64σ CL from no–signal). Regions E refer to the CRESST irreducible excess [3, 4] (regions at 4.5σ CL). The scatter plot shows the LNM configurations. The area outlined by the solid contour around the scatter plot shows the impact of the nuclear uncertainties on the DM–nucleon scattering cross–section.

decay is evaded, since these light neutralinos are mostly binos and their coupling to the Z boson is sufficiently suppressed [8, 10].

A lower bound on the neutralino mass actually arises from cosmology: such light neutralinos annihilate mostly into $\bar{b}b$ and $\bar{\tau}\tau$ pairs if the pseudoscalar higgs A is as light as allowed by accelerator physics searches ($m_A \gtrsim 90$ GeV) [9]: in this case the limit on the neutralino relic abundance sets a lower limit on its mass of about 7–8 GeV [9]. The sector of parameter space which is shaped out by the requirement of having neutralinos lighter than 10 GeV with the correct relic abundance has m_A in the range from 90 to 200 GeV, $\tan\beta$ larger than about 25 and the μ parameter close to its lower bound, namely in the range 100 – 200 GeV [10]. This is shown in Fig. 1, where a scatter plot of the configuration of LNM are plotted in the $m_A - \tan\beta$ plane, and where the points referring to neutralinos lighter than 10 GeV are indicated by (red) circles. The same figure also shows the impact of the first LHC results on the higgs searches.

Since LNM requires light higgses and small values of the μ parameter (which means light charginos), supersymmetric corrections to the rare decay $B \to s + \gamma$ arising from the charged-higgs/top loop need to be compensated by the chargino/stop diagram: this induces some correlation between m_A and the stop mass, which then needs to be relatively light as well, below about 800 GeV, and even below 600 GeV if the higgs is very light [10]. Bounds from LHC on stops may become very important in constraining the parameter space of the LNM: currently the bounds mostly refer to the squarks of the first two families and the LNM can be reconciled with LHC searches by allowing a split between the stops and the other squarks [15]. Light higgses are also bounded by additional rare processes, one of the most relevant being the decay $B_s \to \mu^+\mu^-$. In this case, the concurrence of light higgses, light charginos and relatively large values of $\tan\beta$ implies a specific hierarchy in the model parameters [10], which are naturally fulfilled in the LNM [10]. Once all the bounds are imposed, the lower limit on the neutralino mass slightly increases to 8–9 GeV.

The cosmological requirements of light higgses, intermediate or large values of $\tan\beta$ and small values of μ are instrumental in determining values of the direct detection scattering cross section in the range identified by the direct detection positive indication of a signal [8, 9, 10]. This can be seen in Fig. 2, where the scatter plot of the scattering cross section for the LNM configurations is shown as a function of the neutralino mass. The regions denoted by A, B and C refer to the DAMA annual modulation effect [5] (regions deviate 7.5σ CL from no–signal) under different assumptions on the nuclear quenching factors and on the presence of the channeling effect: region A refers to an energy–independent quenching, B to an energy–dependent quenching, C for channeling, as discussed in Ref. [5]. Region D refers to the CoGeNT annual modulation effect [2] as analyzed in Ref. [5] (the region deviates 1.64σ CL from no–signal) and regions E refer to the CRESST irreducible excess [3, 4] (regions at 4.5σ CL). We can notice that the predictions of the LNM are quite well compatible with the experimental results. LNM therefore offers a specific and well motivated model able to explain the direct detection positive indication of a dark matter signal.

The theoretical determination of the scattering cross section is affected by large nuclear uncertainties, which have been indicated in the figure by the area outlined by the solid contour around the scatter plot [10]. Once these uncertainties are taken into account, the agreement between theoretical predictions and experimental results is even stronger.

Light neutralinos in the LNM may also produce indirect detection signals in the form of antiprotons, gamma–rays, neutrinos or antideuterons. These signals have been studied and discussed in Refs. [11, 12, 13, 14, 18]. Antiprotons and to some extent also neutrinos may be close to either detection or exclusion of LNM models. Nevertheless, theoretical uncertainties in cosmic rays propagation may still pose limits for a clear answer in the antiproton channel [11, 12, 13]. For the neutrino signal the best chance is given by contained events in a neutrino telescope for the neutrino signal from the Sun [14]. The best discovery channel, instead, is offered by antideuterons at low energies, where a clear excess over the expected background can be seen [12, 13, 18].

Prospects for searches at LHC of the LNM are discussed in Ref. [16], where it is shown that detection of LNM benchmarks may be possible, but reconstruction of the mass value for the light neutralino is problematic. A clear indication of a small ratio M_1/M_2, specific to the LNM scheme may nevertheless be possible.

3. Conclusions

We have discussed a supersymmetric framework where light neutralinos naturally arise and are endowed with the right properties to explain the current direct detection results from DAMA [1], CoGeNT [2] and CRESST [3, 4] on the search of particle dark matter in our Galaxy. These experiments point toward a positive indication of a dark matter signal as due to a WIMP with a mass around 10 GeV and scattering cross sections on nucleons between 10^{-42} cm^2 and 10^{-39} cm^2. In our model, light neutralinos naturally arise as a consequence of a small U(1)–gaugino mass parameter M_1 as compared to the SU(2)–gaugino mass parameter M_2, which ensures neutralinos to be light, and to higgses in their allowed low–mass range, close to 90–120 GeV, which makes the annihilation and scattering cross sections to be of the right order to: (i) obtain the correct relic abundance of light neutralinos, able to explain galactic dark matter; (ii) obtain the correct scattering cross sections on nucleons, at the level required to explain the experimental positive evidences of DAMA, CoGeNT and CRESST.

Bounds from searches for supersymmetry and higgs bosons at accelerators (LEP and the first LHC results) as well as bounds from supersymmetric contributions to rare processes, like B meson decays and anomalous magnetic moment of the muon allow us to shape out the relevant sector of parameter space of the model. The interplay of all those bounds is currently setting a lower limit on the mass of light neutralinos at 8–9 GeV. Once these limits are implemented, we have shown that direct detection cross sections for light neutralinos are very predictive up to neutralino masses of 30 GeV and compatibility with direct detection positive indication of a signal is perfect. A similar analysis has been also performed in Ref. [19, 20], where results similar to ours have been obtained. We find slightly more relaxed bounds on the LNM parameter space, as a consequence of a more conservative approach. Some experimental constraints on rare processes (especially those arising from $B \to \tau\nu$ decays) do not appear to be firmly assessed yet, and therefore we did not strictly enforced them in our analysis. Also recent bounds from charginos searches at the LHC may start to pose relevant bounds on the LNM parameter space [21].

Finally, we have also commented that indirect detection signals may be potentially around the corner (or, alternatively, close to set significant limits), although they are currently limited by astrophysical uncertainties [11, 12, 13, 14]. The best chance for discovery may be offered by the indirect signal in antideuterons, for which good prospects will be offered by the upcoming AMS and GAPS detectors [18, 12, 12].

References

[1] Bernabei R *et al* 2010 *Eur. Phys. J.* C **67** 39 [arXiv:1002.1028 [astro-ph.GA]]
[2] Aalseth C E *et al* 2011 *Phys. Rev. Lett.* **107** 141301 [arXiv:1106.0650 [astro-ph.CO]]
[3] Petricca F 2011 these Proceedings
[4] Angloher G *et al* 2011 arXiv:1109.0702 [astro-ph.CO]
[5] Belli P *et al* 2011 *Phys. Rev.* D **84** 055014 [arXiv:1106.4667 [hep-ph]]
[6] Ahmed Z *et al* 2010 *Science* **327** 1619 [arXiv:0912.3592 [astro-ph.CO]]
[7] Aprile E *et al* 2011 *Phys. Rev. Lett.* **107** 131302 [arXiv:1104.2549 [astro-ph.CO]]
[8] Bottino A, Fornengo N and Scopel S 2003 *Phys. Rev.* D **67** 063519 [hep-ph/0307303]
[9] Bottino A, Donato F, Fornengo N and Scopel S 2003 *Phys. Rev.* D **68** 043506 [hep-ph/0304080]
[10] Fornengo N, Scopel S and Bottino A 2011 *Phys. Rev.* D **83** 015001 [arXiv:1011.4743 [hep-ph]]
[11] Bottino A, Donato F, Fornengo N and Scopel S 2004 *Phys. Rev.* D **70** 015005 [hep-ph/0401186]
[12] Bottino A, Donato F, Fornengo N and Scopel S 2008 *Phys. Rev.* D **77** 015002 [arXiv:0710.0553 [hep-ph]]
[13] Bottino A, Donato F, Fornengo N and Scopel S 2008 *Phys. Rev.* D **77** 083520 [arXiv:0806.4099 [hep-ph]]
[14] Niro V, Bottino A, Fornengo N and Scopel 2009 *Phys. Rev.* D **80** 095019 [arXiv:0909.2348 [hep-ph]]
[15] Scopel S, Choi S, Fornengo N and Bottino A 2011 *Phys. Rev.* D **83** 095016 [arXiv:1102.4033 [hep-ph]]
[16] Choi S, Scopel S, Fornengo N and Bottino A 2011 arXiv:1108.2190 [hep-ph] (to appear in *Phys. Rev.* D)
[17] Baglio J and Djouadi A 2011 arVix:1103.6247 [hep-ph]
[18] Donato F, Fornengo N and Maurin D 2008 *Phys. Rev.* D **78** 043506 [arXiv:0803.2640 [hep-ph]]
[19] Calibbi L, Ota T and Takanishi Y 2011 *JHEP* **1107** 013 [arXiv:1104.1134 [hep-ph]]
[20] Ota T 2011 these Proceedings
[21] Aad G *et al* 2011 arXiv:1110.6189 [hep-ex]

Journal of Physics: Conference Series **375** (2012) 012041 doi:10.1088/1742-6596/375/1/012041

Light neutralino in the MSSM: An update with the latest LHC results

Lorenzo Calibbi[1], Toshihiko Ota[1] and Yasutaka Takanishi[2,3]
[1]Max-Planck-Institut für Physik (Werner-Heisenberg-Institut)
Föhringer Ring 6, 80805 München, Germany
[2]Physik-Department, Technische Universität München, 85748, Garching, Germany

E-mail: calibbi@mppmu.mpg.de, toshi@mppmu.mpg.de, yasutaka@mpi-hd.mpg.de

Abstract. We discuss the scenario of light neutralino dark matter in the minimal supersymmetric standard model, which is motivated by the results of some of the direct detection experiments — DAMA/LIBRA, CoGeNT, and CRESST. We update our previous analysis with the latest results of the LHC. We show that new LHC constraints disfavour the parameter region that can reproduce the results of DAMA/LIBRA and CoGeNT.

1. Introduction

One of the most dramatic progresses in our understanding of nature in the last decade is the establishment of the dark sector of the universe, which was brought by precision cosmological observations. However, even basic properties — mass and interactions — of Dark Matter (DM) have not been fully understood yet. Numerous experimental efforts to detect DM have been made, and some of the direct detection experiments claim a positive signal that might be originated by DM. The DAMA/LIBRA and CoGeNT collaborations show an annual modulation of signal events, which can be interpreted as the change of the relative velocity of the detectors against the DM halo. In this conference, the long-awaited result of the CRESST collaboration was presented, which indicates a significant excess of events [1]. The results of these experiments suggest a light DM field ($M_\chi \sim 10$ GeV) with a relatively large scattering cross-section with nucleons ($\sigma \sim 10^{-41}$ cm^2); see e.g., [2]. In this talk, we discuss the scenario of light neutralino DM in the minimal supersymmetric standard model (MSSM). Since the lightest supersymmetric particle is protected by R-parity and thus is stable, the lightest neutralino is a good candidate for DM. This possibility has been extensively discussed in the prior studies (cf. [3, 4] and references therein). In our recent paper [5], we inspected the compatibility in the MSSM parameter space between the light neutralino DM scenario and particle physics constraints such as B and K meson decays, neutralino production rate, and Higgs boson searches. Here, we update our analysis with the latest results of the LHC. We will show that the light neutralino DM scenario that is consistent with the results of DAMA/LIBRA and CoGeNT comes into disfavour with the latest LHC bounds, especially constraints from the neutral Higgs boson search.

[3] Present address: Max-Planck-Institut für Kernphysik, Saupfercheckweg 1, 69117 Heidelberg, Germany

2. Light neutralino in the MSSM

What does a light neutralino with mass of around 10 GeV require of the MSSM parameters? The mass of the neutralino is mainly controlled by three parameters, the Bino mass M_1, Wino mass M_2, and higgsino mass μ. Since two of them, M_2 and μ, are related to chargino masses, they must be larger than $\sim$ 100 GeV to satisfy their LEP mass bounds. Therefore, the only possibility is to take $M_1 \sim 10$ GeV to obtain the light neutralino. Consequently, the lightest neutralino in this scenario becomes mostly Bino. The model that we deal with is described by the following nine parameters:

$$\tan\beta, \quad M_1, \quad M_2, \quad M_3, \quad a_0, \quad \mu, \quad m_A, \quad m_{\tilde{q}}, \quad m_{\tilde{\ell}}, \tag{1}$$

where $\tan\beta$ is defined as the ratio of the two Higgs VEV's, M_3 is the gluino mass, a_0 is the universal coefficient for tri-linear scalar couplings, m_A is the CP-odd Higgs mass, and $m_{\tilde{q}}$ and $m_{\tilde{\ell}}$ are the common soft SUSY breaking masses for squarks and sleptons, respectively. We scan them to search for the parameter region in which the light neutralino DM scenario is consistent with the particle physics observations[1].

We assume that the neutralino was thermally produced in the early universe. Since such a light particle is overproduced in the thermal process, an efficient annihilation process is necessary to reproduce the correct relic density. There are two main annihilation processes for light neutralino DM: Sfermion mediation and CP-odd Higgs mediation. It is known that the parameter space in which sfermion mediation becomes the main annihilation process cannot be consistent with the results suggested by DAMA/LIBRA, CoGeNT and CRESST [9] (see also Ref. [10, 11]). Therefore, following Ref. [12], we adopt the Higgs mediation process as the main annihilation process. In order to enhance the Higgs-mediated annihilation, the following three conditions are required of the model parameters: (i) light m_A, (ii) small μ to have a significant amount of higgsino component in the lightest neutralino, and (iii) large $\tan\beta$ to enhance the main annihilation process to a $b\bar{b}$ pair. This parameter choice enhances not only the annihilation process but also many flavour physics processes. In addition, it magnifies the signal of neutral Higgs bosons decaying to lepton pairs at hadron colliders. Therefore, one must carefully inspect the compatibility in the parameter choice between the cosmological requirements and the particle physics constraints.

We employ the following flavour physics processes and categorize them into two groups by their parameter dependence:

- **Group I:** $B \to \tau\nu$, $B \to D\tau\nu$, $D_s \to \tau\nu$, and $K \to \mu\nu$. They essentially depend only on Higgs parameters, $m_{H^\pm}$ and $\tan\beta$.
- **Group II:** $b \to s\gamma$ and $B_s \to \mu^+\mu^-$. These processes depend not only on the Higgs parameters but also on the SUSY parameters such as $m_{\tilde{q}}$ and a_0.

Group I observables complemetarily constrain the parameter space of $m_{H^\pm}$ and $\tan\beta$, which are regardless of the detail of the SUSY parameters [13]. The key observables are $B \to \tau\nu$ and $K \to \mu\nu$. We adopt the following values

$$0.52 < R_{B\tau\nu} < 2.61 \quad \text{and} \quad 0.985 < R_{\ell 23}(K \to \mu\nu) < 1.013, \tag{2}$$

as the allowed range of these observables. $B \to \tau\nu$ excludes intermediate values of $\tan\beta$. On the other hand, $K \to \mu\nu$ disfavours larger values of $\tan\beta$. Therefore, the allowed parameter region with middle-to-high values of $\tan\beta$ is sharply narrowed by combination of these two constraints. After a numerical scan of the parameters Eq. (1), we found that the allowed parameter regions form two separated clusters: (i) A narrow strip at high $\tan\beta$ ($32 \lesssim \tan\beta \lesssim 38$) and (ii) A low

[1] We use the public codes SuSpect [6], micrOMEGAs [7], and SuperIso [8] in our numerical studies.

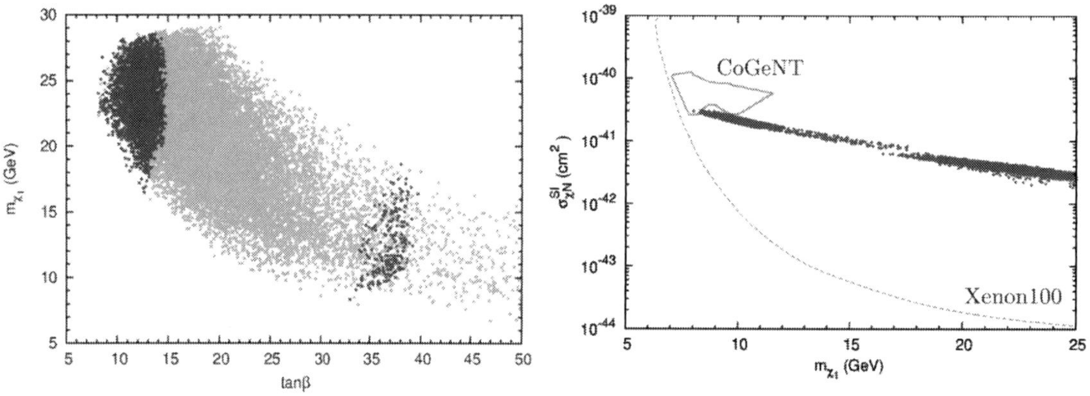

Figure 1. Left: Allowed parameter region on the plane of the lightest neutralino mass and $\tan\beta$. After taking into account all the constraints, two separated regions are left (blue points): (i) A high $\tan\beta$ strip and (ii) A low $\tan\beta$ region. To obtain a light neutralino with $M_{\tilde{\chi}_1^0} < 15$ GeV, a large value of $\tan\beta$ is necessary. Right: Neutralino-nucleon scattering cross-section as a function of mass of the lightest neutralino. The prediction of the high $\tan\beta$ strip roughly coincides with the results of DAMA/LIBRA and CoGeNT. The best-fit region of CRESST is located around $(M_{\tilde{\chi}_1^0}, \sigma_{\rm SI}) \simeq$(20 GeV, 10^{-42} cm^2) [1].

$\tan\beta$ region ($7 \lesssim \tan\beta \lesssim 15$). The mass of the lightest neutralino is inversely proportional to $\tan\beta$, because a lighter neutralino requires a more efficient annihilation process that is enhanced by a large value of $\tan\beta$. As shown in the left plot of Fig. 1, a neutralino with mass lighter than 15 GeV is viable only on the high $\tan\beta$ strip.

Next, we discuss the neutralino-nucleon scattering cross-section. DAMA/LIBRA and CoGeNT suggest relatively large values for the cross-section. In both parameter regions (i) and (ii), the neutralino-nucleon scattering process is mediated by Higgs bosons, and its parameter dependence is similar to that of the neutralino annihilation process. Consequently, the direct detection cross-section becomes large at the light neutralino region on which a large annihilation cross-section is required. As shown in the right plot in Fig. 1, the parameter choice of the high $\tan\beta$ strip approximately reproduces the results of DAMA/LIBRA and CoGeNT.

Since the high $\tan\beta$ strip is standing on the edge of the experimental constraints with delicate tune of the model parameters Eq. (1), it might be easily excluded by an improvement of some of the constraints. Here, we update our previous analysis [5] with the latest LHC results. The bound to the branching ratio of $B_s \to \mu^+\mu^-$ is improved by LHCb with an integrated luminosity of 0.3 fb^{-1} [14], which is constrained to be smaller than 1.5×10^{-8} at 95 % CL. As shown in the left plot in Fig. 2, this new bound excludes all the points in the high $\tan\beta$ region, which satisfy the Higgs mass bounds at LEP. Another crucial constraint is brought from the search for the neutral Higgs bosons decaying to lepton pairs. The latest results of CMS [15] and ATLAS [16] with an integrated luminosity of around 1 fb^{-1} shave a large area of the high $\tan\beta$ region, as shown in the right plot in Fig. 2. The successful run of the LHC even allows us to access the low $\tan\beta$ region.

3. Conclusions

We have discussed the scenario of light neutralino dark matter in the MSSM, which is motivated by the direct detection experiments DAMA/LIBRA, CoGeNT, and CRESST. In our previous study, we showed that a narrow parameter region with high values of $\tan\beta$ was consistent with

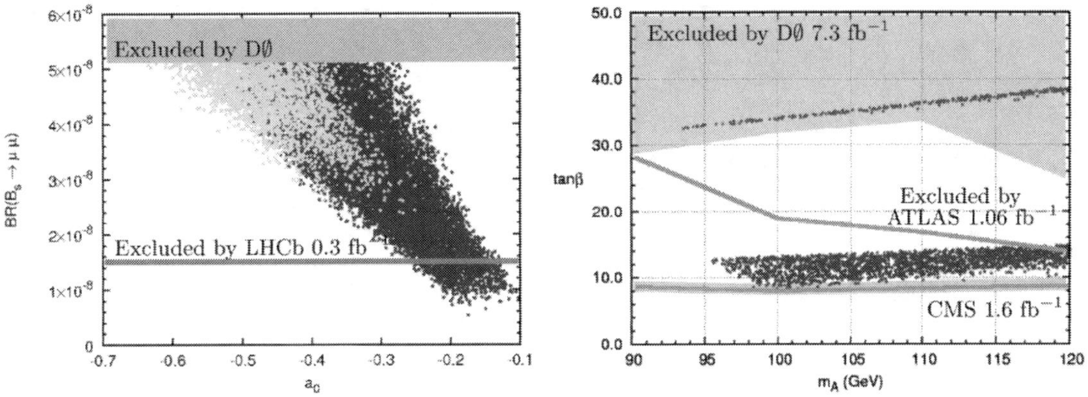

Figure 2. Left: Branching ratio of $B_s \to \mu^+\mu^-$ as a function of a_0 with the points on the high $\tan\beta$ strip. Light blue points satisfy the LEP bounds on Higgs boson masses, which are excluded by the latest LHCb result (orange line). Right: Allowed points overlayed with the constraints to the Higgs parameters m_A and $\tan\beta$ from the neutral Higgs boson search at hadron colliders. The latest results of both CMS and ATLAS (also DØ) exclude the high $\tan\beta$ strip at 95% CL.

the results of DAMA/LIBRA and CoGeNT, and it could satisfy the cosmological requirements and the particle physics constraints. In this talk, we have updated our numerical analysis with the latest LHC results. It turns out that the high $\tan\beta$ region comes into disfavour with the new bound to $B_s \to \mu^+\mu^-$ reported by LHCb. In addition, the neutral Higgs boson search at CMS and ATLAS brings a critical constraint to this scenario, which excludes a large part of the parameter region with small m_A and large $\tan\beta$. There is still discussion on the background estimation for the relevant process [17], and thus we should wait for the consensus. It is expected that the next official release from LHC will finally examine the viability of this scenario.

References

[1] Petricca F 2011 talk at TAUP 2011
[2] Arina C 2011 talk at TAUP 2011
[3] Munoz C 2011 talk at TAUP 2011
[4] Fornengo N 2011 talk at TAUP 2011
[5] Calibbi L *et al.* 2011 *JHEP* **1107** 013 (*Preprint* 1104.1134)
[6] Djouadi A *et al.* 2007 *Comput. Phys. Commun.* **176** 426–455 (*Preprint* hep-ph/0211331)
[7] Belanger G *et al.* 2007 *Comput. Phys. Commun.* **176** 367–382 (*Preprint* hep-ph/0607059)
[8] Mahmoudi F 2009 *Comput. Phys. Comun.* **180** 1579–1613 (*Preprint* 0808.3144)
[9] Dreiner H K *et al.* 2009 *Eur. Phys. J.* **C62** 547–572 (*Preprint* 0901.3485)
[10] Vasquez D A *et al.* 2011 *Phys.Rev.* **D84** 095015 (*Preprint* 1108.1338)
[11] Cumberbatch D T *et al.* 2011 (*Preprint* 1107.1604)
[12] Fornengo N *et al.* 2011 *Phys. Rev.* **D83** 015001 (*Preprint* 1011.4743)
[13] Antonelli M *et al.* (FlaviaNet Working Group on Kaon Decays) 2008 (*Preprint* 0801.1817)
[14] Bettler M O 2011 (*Preprint* 1110.2411)
[15] CMS Collaboration 2011 CMS-PAS-HIG-11-020 http://cms.web.cern.ch/org/cms-papers-and-results
[16] ATLAS Collaboration 2011 ATLAS-CONF-2011-132 https://atlas.web.cern.ch/Atlas/GROUPS/PHYSICS/CONFNOTES
[17] Baglio J *et al.* 2011 (*Preprint* 1103.6247)

12th International Conference on Topics in Astroparticle and Underground Physics (TAUP 2011) IOP Publishing
Journal of Physics: Conference Series **375** (2012) 012042 doi:10.1088/1742-6596/375/1/012042

Dark matter in a SUSY left-right model

A. Vicente

Laboratoire de Physique Théorique, CNRS – UMR 8627, Université de Paris-Sud 11, F-91405 Orsay Cedex, France

E-mail: `Avelino.Vicente@th.u-psud.fr`

Abstract. Supersymmetric left-right models are well motivated extensions of the Minimal Supersymmetric Standard Model since they automatically contain the ingredients to explain the observed neutrino masses and mixings. Here we study a SUSY model in which the left-right symmetry is broken by triplets at a high scale leading to automatic R-parity conservation at low energies. The sparticle spectra in this model differ from the usual constrained MSSM expectations and these changes affect the relic abundance of the lightest neutralino. We discuss two examples: (1) the standard stau co-annihilation region, and (2) flavoured co-annihilation.

1. Introduction

There are good theoretical and phenomenological reasons to consider left-right (LR) extensions [1, 2, 3] of the MSSM. Besides the restoration of parity at high energies, two motivations can be especially highlighted. First, LR models contain all the ingredients to explain the observed neutrino masses and mixings thanks to the well-known seesaw mechanism [4, 5, 6, 7]. And second, the gauging of $B - L$ provides a natural framework to explain the origin of R-parity, which can be understood as the low energy remnant of the high energy gauge symmetry.

Here we will concentrate on the LR model proposed in [8, 9]. The main difference with respect to previous proposals [5, 10, 11] was the presence of additional triplets with zero $B - L$ charge. The main advantage of this setup, in the following called the ΩLR model, is the existence of R-parity conserving minima in the tree level scalar potential.

In the ΩLR model [8, 9] new superfields appear with masses below the GUT scale. Furthermore, the gauge group is extended to $SU(3)_C \times SU(2)_L \times SU(2)_R \times U(1)_{B-L}$ at energies above the LR breaking scale. Both facts significantly affect the RGE running of all parameters. Therefore, even though the LR breaking scale is very far from the reach of current accelerators, there are interesting effects in the phenomenology caused by the running of the soft supersymmetry breaking parameters and the subsequent deformation of the SUSY spectra with respect to the CMSSM expectation [12, 13].

Astrophysical observations and the data from WMAP [14] put on solid grounds the existence of non-baryonic dark matter in the universe. The most popular candidate is the lightest neutralino in R-parity conserving supersymmetry. However, all solutions in the CMSSM parameter space [15] require a fine-tuning [16] among some masses in order to enhance the annihilation/co-annihilation cross-sections, decrease the lightest neutralino relic abundance, $\Omega_{\tilde{\chi}_1^0} h^2$, and reproduce the observed dark matter relic density [14]. Therefore, it is not surprising that even small changes in the SUSY spectra can dramatically change the resulting $\Omega_{\tilde{\chi}_1^0} h^2$. The description of these changes in the ΩLR model is the main subject of this work.

2. The model

In this section we present the model originally defined in [8, 9]. For further details see [12, 13].

Below the GUT scale the gauge group of the model is $SU(3)_c \times SU(2)_L \times SU(2)_R \times U(1)_{B-L}$. Besides the particle content of the MSSM with the addition of (three) right-handed neutrino(s) ν^c, some additional superfields are introduced. First, two generations of Φ superfields, bidoublets under $SU(2)_L \times SU(2)_R$, are introduced. Furthermore, triplets under (one of) the $SU(2)$ gauge groups are added with gauge quantum numbers $\Delta(1,3,1,2)$, $\Delta^c(1,1,3,-2)$, $\bar{\Delta}(1,3,1,-2)$, $\bar{\Delta}^c(1,1,3,2)$, $\Omega(1,3,1,0)$ and $\Omega^c(1,1,3,0)$.

With these representations and assuming parity conservation, the most general superpotential compatible with the symmetries is

$$\begin{aligned} \mathcal{W} &= Y_Q Q\Phi Q^c + Y_L L\Phi L^c - \frac{\mu}{2}\Phi\Phi + fL\Delta L + f^* L^c\Delta^c L^c + a\Delta\Omega\bar{\Delta} + a^*\Delta^c\Omega^c\bar{\Delta}^c \\ &+ \alpha\Omega\Phi\Phi + \alpha^*\Omega^c\Phi\Phi + M_\Delta\Delta\bar{\Delta} + M_\Delta^*\Delta^c\bar{\Delta}^c + M_\Omega\Omega\Omega + M_\Omega^*\Omega^c\Omega^c \,. \end{aligned} \quad (1)$$

The breaking of the LR gauge group to the MSSM gauge group takes place in two steps: $SU(2)_R \times U(1)_{B-L} \to U(1)_R \times U(1)_{B-L} \to U(1)_Y$. First, the neutral component of the Ω^c triplet takes a VEV (vacuum expectation value), $\langle\Omega^{c\,0}\rangle = \frac{v_R}{\sqrt{2}}$, breaking $SU(2)_R$. Next, the group $U(1)_R \times U(1)_{B-L}$ is broken by $\langle\Delta^{c\,0}\rangle = \langle\bar{\Delta}^{c\,0}\rangle = \frac{v_{BL}}{\sqrt{2}}$. Neutrino masses are generated at this stage via the superpotential term $f^* L^c\Delta^c L^c$ which, after $U(1)_{B-L}$ gets broken, leads to Majorana masses for the right-handed neutrinos and a type-I seesaw mechanism. Therefore, we define the seesaw scale M_{Seesaw} as the lightest eigenvalue of the matrix $M_S \equiv f^* v_{BL}$.

3. Low energy spectrum and dark matter

In general, with $v_{BL} \leq v_R < m_{\text{GUT}}$, the ΩLR has a lighter spectrum than the CMSSM for the same parameters. Since in the ΩLR model the running of the gauge couplings is changed with respect to the MSSM case, also gaugino masses are changed. Consider for example M_1. At 1-loop leading-log order we find

$$M_1(m_{\text{SUSY}}) = M_{1/2}\left[X_1 + X_2\left(-3l_1 + l_2\right)\right] \quad (2)$$

where $l_1 = \ln\left(\frac{m_{\text{GUT}}}{v_R}\right)$, $l_2 = \ln\left(\frac{v_R}{v_{BL}}\right)$ and $X_{1,2} > 0$ are two numerical factors. Eq. (2) shows that $M_1(m_{\text{SUSY}})$ can be decreased with respect to the CMSSM value by using a large l_1 and $l_2 = 0$ and increased only in the case $l_2 > 3l_1$. This change has a strong impact on dark matter phenomenology, since the lightest neutralino is mostly bino in most parts of parameter space.

We present now our results concerning dark matter phenomenology in the ΩLR model. We will discuss the main differences with respect to the CMSSM for (1) the standard stau co-annihilation region [17], and (2) flavoured co-annihilation [18]. The numerical evaluation of the parameters at the SUSY scale was done with `SPheno` [19, 20, 21], using the complete 2-loop RGEs at all scales. These were derived with the Mathematica package `SARAH` [22, 23, 24]. Finally, we used `micrOmegas` [26] to obtain the value of $\Omega_{\tilde{\chi}_1^0} h^2$

3.1. Stau coannihilation

As discussed, lowering the values of v_{BL} and v_R leads in general to a lighter spectrum compared to the pure CMSSM case. Since the mass of the bino decreases faster than the mass of the stau, the quantity $\Delta_m = m_{\tilde{\tau}_1} - m_{\chi_1^0}$, and thus the neutralino relic abundance, gets increased for low v_{BL} and v_R. This can only be compensated by reducing m_0 [13].

Two examples of stau co-annihilation regions are shown in Figure 1. As expected, a shift of the allowed region towards smaller values of m_0 is found. Morever, the region can disappear for sufficiently low v_{BL} and v_R. The observation of a particle spectrum consistent with the CMSSM stau co-annihilation region could thus be turned into lower limits on v_{BL} and v_R.

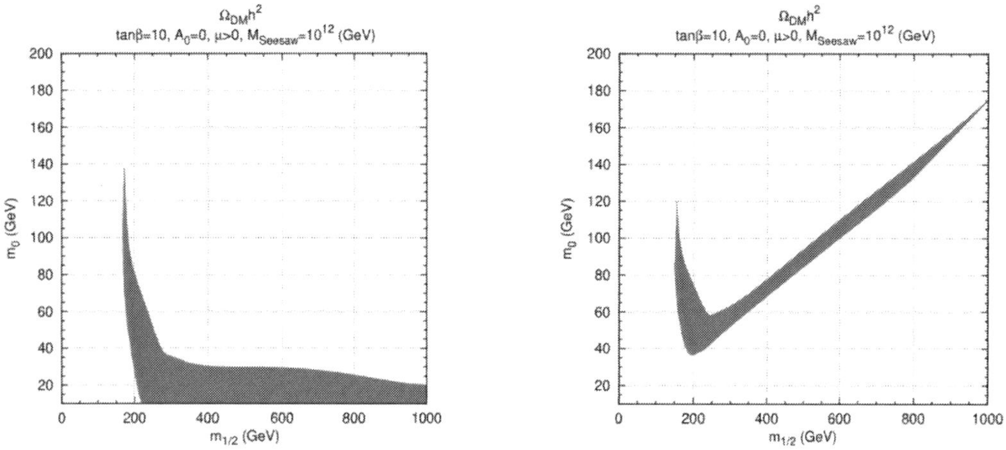

Figure 1. Dark Matter 3 σ allowed regions for $\tan\beta = 10$, and $v_{BL} = v_R = 1.5 \times 10^{15}$ GeV (left) and $v_{BL} = v_R = 10^{16}$ GeV (right).

3.2. *Flavoured coannihilation*

As shown in [12], the ΩLR model allows for large flavour violating effects in the right slepton sector. This implies that the $\tilde{\tau}_R \simeq \tilde{\tau}_1$ can be made ligther by flavour contributions, leading to co-annihilation with the lightest neutralino in points of parameter space where it would be impossible without flavour effects. Moreover, flavour violating processes become relevant in the determination of the relic density and they must be taken into account. This is the so-called flavoured co-annihilation mechanism [18].

When searching for such points in parameter space one must take into account the strong limits for $\text{Br}(\mu \to e\gamma)$ given by the MEG experiment [25]. A fine-tuning of the parameters is required in order to cancel this observable and, at the same time, reduce the $\tilde{\tau}_R$ mass sufficiently. Figure 2 shows two examples. On the left panel the cancellation of $\text{Br}(\mu \to e\gamma)$ is obtained with $\delta = \pi$ and $\theta_{13} = 8°$, while on the right panel with $\theta_{13} = 9.5°$.

4. Conclusions

We have studied the dark matter phenomenology of the ΩLR model, a supersymmetric left-right model that breaks the LR symmetry at high energies. The change in the running of the parameters at energies above the breaking scale has a strong impact on the low energy spectra, modifying the usual dark matter predictions of the CMSSM.

The shift of the stau co-annihilation region and its potential disappearance has been studied. The fact that the lightest neutralino is usually lighter in the ΩLR model than in the CMSSM reduces the allowed stau co-annihilation region, which can even vanish for low v_R and/or v_{BL}.

The model also allows for flavoured co-annihilation. Examples have been given of points in parameter space where the $\tilde{\tau}_R$ mass is reduced by means of flavour effects, leading to the observed dark matter relic abundance.

Acknowledgements

This work has been done in collaboration with J. N. Esteves, M. Hirsch, W. Porod, J. C. Romão and F. Staub. A.V. acknowledges support from the ANR project CPV-LFV-LHC NT09-508531. A.V. was also partially suported by Marie Curie Early Initial Training Network Fellowships of

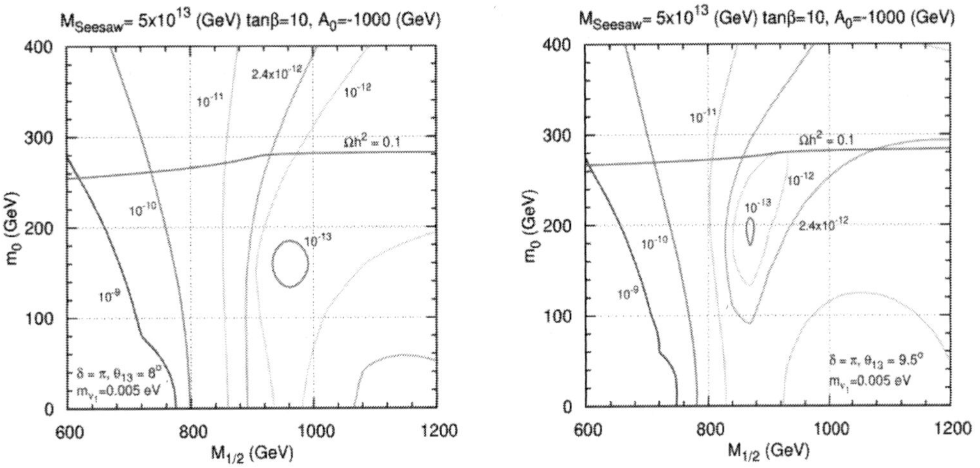

Figure 2. $\Omega_{\tilde{\chi}_1^0} h^2$ and $\mathrm{Br}(\mu \to e\gamma)$ contour plots in the m_0-$M_{1/2}$ plane for $\delta = \pi$, $\theta_{13} = 8^0$ (left) and $\theta_{13} = 9.5^0$ (right). Both examples were obtained with $v_R = 10^{15}$ GeV and $v_{BL} = 10^{14}$ GeV.

the European Community's Seventh Framework Programme under contract number (PITN-GA-2008-237920-UNILHC) and by the DFG project number PO-1337/1-1.

References

[1] J. C. Pati, A. Salam, Phys. Rev. D **10** (1974) 275 [Erratum-ibid. D **11** (1975) 703].
[2] R. N. Mohapatra, J. C. Pati, Phys. Rev. D **11** (1975) 2558.
[3] G. Senjanović, R. N. Mohapatra, Phys. Rev. D **12** (1975) 1502.
[4] P. Minkowski, Phys. Lett. B **67** (1977) 421. T. Yanagida, in *KEK lectures*, ed. O. Sawada and A. Sugamoto, KEK, 1979; M Gell-Mann, P Ramond, R. Slansky, in *Supergravity*, ed. P. van Niewenhuizen and D. Freedman (North Holland, 1979).
[5] R.N. Mohapatra, G. Senjanović, *Phys. Rev. Lett.* **44** (1980) 912.
[6] J. Schechter, J. W. F. Valle, Phys. Rev. D **22** (1980) 2227.
[7] T. P. Cheng, L. F. Li, Phys. Rev. D **22** (1980) 2860.
[8] C. S. Aulakh, K. Benakli, G. Senjanović, Phys. Rev. Lett. **79** (1997) 2188 [arXiv:hep-ph/9703434].
[9] C. S. Aulakh, A. Melfo, A. Rasin, G. Senjanović, Phys. Rev. D **58** (1998) 115007 [arXiv:hep-ph/9712551].
[10] R. N. Mohapatra, G. Senjanović, Phys. Rev. **D23** (1981) 165.
[11] M. Cvetič, J. C. Pati, Phys. Lett. B **135** (1984) 57.
[12] J. N. Esteves, J. C. Romao, M. Hirsch, A. Vicente, W. Porod, F. Staub, JHEP **1012** (2010) 077 [arXiv:1011.0348 [hep-ph]].
[13] J. N. Esteves, J. C. Romao, M. Hirsch, W. Porod, F. Staub, A. Vicente, arXiv:1109.6478 [hep-ph].
[14] E. Komatsu *et al.* [WMAP Collaboration], Astrophys. J. Suppl. **192** (2011) 18 [arXiv:1001.4538 [astro-ph.CO]].
[15] M. Drees, M. M. Nojiri, Phys. Rev. D **47** (1993) 376 [arXiv:hep-ph/9207234].
[16] K. Griest, D. Seckel, Phys. Rev. D **43** (1991) 3191.
[17] J. R. Ellis, T. Falk, K. A. Olive, Phys. Lett. B **444** (1998) 367 [arXiv:hep-ph/9810360].
[18] D. Choudhury, R. Garani, S. K. Vempati, arXiv:1104.4467 [hep-ph].
[19] W. Porod, Comput. Phys. Commun. **153** (2003) 275 [arXiv:hep-ph/0301101].
[20] W. Porod, F. Staub, arXiv:1104.1573 [hep-ph].
[21] F. Staub, T. Ohl, W. Porod, C. Speckner, arXiv:1109.5147 [hep-ph].
[22] F. Staub, arXiv:0806.0538 [hep-ph].
[23] F. Staub, Comput. Phys. Commun. **181** (2010) 1077 [arXiv:0909.2863 [hep-ph]].
[24] F. Staub, Comput. Phys. Commun. **182** (2011) 808 [arXiv:1002.0840 [hep-ph]].
[25] J. Adam *et al.* [MEG Collaboration], Phys. Rev. Lett. **107** (2011) 171801 [arXiv:1107.5547 [hep-ex]].
[26] G. Belanger, F. Boudjema, A. Pukhov, A. Semenov, Comput. Phys. Commun. **176** (2007) 367 [arXiv:hep-ph/0607059].

Journal of Physics: Conference Series **375** (2012) 012043 doi:10.1088/1742-6596/375/1/012043

Dark matter stability from non-Abelian discrete flavor symmetries

E. Peinado

AHEP Group, Institut de Física Corpuscular – C.S.I.C./Universitat de València
Edificio Institutos de Paterna, Apt 22085, E–46071 Valencia, Spain

E-mail: epeinado@ific.uv.es

Abstract. We present a mechanism for the dark matter (DM) stability in the framework of a non-Abelian flavour symmetry renormalizable model. The same non-Abelian discrete flavor symmetry which accounts for the observed pattern of neutrino oscillations, spontaneously breaks to a Z_2 subgroup which renders DM stable. The simplest scheme leads to a scalar doublet DM potentially detectable in nuclear recoil experiments, inverse neutrino mass hierarchy, hence a neutrinoless double beta decay rate accessible to upcoming searches.

1. Introduction

Non-baryonic dark matter (DM) is one of the most compelling problems of modern cosmology. Despite the fact that its existence is well established by cosmological and astrophysical probes, its nature remains elusive. Still, observations can constrain the properties of dark matter and give some hints about its identity. For instance, a fundamental requirement for a viable dark matter candidate is the stability over cosmological times. This suggests the existence of an exact or slightly-broken symmetry protecting or suppressing its decay.

It has been shown recently that such symmetry can also be related to the flavor structure of the Standard Model [1]. The model proposed in [1] is based on an A_4 symmetry with four $SU(2)$ Higgs doublets. After the electroweak symmetry breaking, the A_4 (even permutation of four objects) group is spontaneously broken into a Z_2 subgroup which is responsible for the DM stability. The leptonic sector is also extended. It consist in four right-handed neutrinos and the light neutrino masses are generated through the type-I seesaw mechanism and obey an inverted hierarchy mass spectrum with $m_{\nu_3} = 0$ and vanishing reactor angle $\theta_{13} = 0$.

We have also considered models with a different matter content for the right-handed neutrinos with the same DM stability mechanism but with different neutrino phenomenology [2] or a model based on the dihedral group D_4 where the some flavour changing neutral currents are present and constrain the DM sector [3]. Other models with flavor symmetries but with decaying DM have also been considered, see for instance [4, 5]. For a model with stable DM but with the same behavior of the DM, in the sense that instead of being totally inert, the DM couples to some right-handed neutrinos in a similar way in our model see [6].

2. The Model and DM stability

The model consist of the SM matter fields plus three Higgs doublets which transform as a triplet under A_4, and four right-handed neutrinos transforming as a singlet and a triplet of the flavor

Table 1. Summary of relevant model quantum numbers

	L_e	L_μ	L_τ	l_e^c	l_μ^c	l_τ^c	N_T	N_4	H	η
$SU(2)$	2	2	2	1	1	1	1	1	2	2
A_4	1	$1'$	$1''$	1	$1''$	$1'$	3	1	1	3

group. In table 1 we present the relevant quantum numbers for the matter fields. The group of even permutations of four objects, A_4, has two generators: S, and T, which obeys the properties $S^2 = T^3 = (ST)^3 = I$. S is a Z_2 generator while T is a Z_3 generator. A_4 has four irreducible representations, three singlets 1, $1'$, and $1''$ and one triplet. The generators of A_4 for each irreducible representation are

$$\begin{array}{lll} 1 & S=1 & T=1 \\ 1' & S=1 & T=\omega^2 \\ 1'' & S=1 & T=\omega \end{array} \quad 3: \; S = \begin{pmatrix} 1 & 0 & 0 \\ 0 & -1 & 0 \\ 0 & 0 & -1 \end{pmatrix} \; T = \begin{pmatrix} 0 & 1 & 0 \\ 0 & 0 & 1 \\ 1 & 0 & 0 \end{pmatrix} \tag{1}$$

where $\omega^3 = 1$. The Yukawa Lagrangian of the model is

$$\begin{aligned} \mathcal{L} &= y_e L_e l_e^c H + y_\mu L_\mu l_\mu^c H + y_\tau L_\tau l_\tau^c H + + y_1^\nu L_e (N_T \eta)_1 + y_2^\nu L_\mu (N_T \eta)_{1''} + \\ &\quad + y_3^\nu L_\tau (N_T \eta)_{1'} + y_4^\nu L_e N_4 H + M_1 N_T N_T + M_2 N_4 N_4 + \text{h.c.} \end{aligned} \tag{2}$$

This way H is responsible for quark and charged lepton masses, the latter automatically diagonal[1]. Neutrino masses arise from H and η. One solution for the minimum of the Higgs potential [1] is

$$\langle H^0 \rangle = v_h \neq 0, \qquad \langle \eta_1^0 \rangle = v_\eta \neq 0, \qquad \langle \eta_{2,3}^0 \rangle = 0\,, \tag{3}$$

which means the vev alignment for the A_4 triplet of the form $\langle \eta \rangle \sim (1, 0, 0)$. This alignment is invariant under the S generator[2], see eq. (1), which means that the minimum of the potential breaks spontaneously A_4 into a Z_2 subgroup generated by S. All the fields in the model singlets under A_4 are even under the residual Z_2, the triplets transform as:

$$\begin{array}{lllllll} N_1 & \to & +N_1\,, & \quad & \eta_1 & \to & +\eta_1 \\ N_2 & \to & -N_2\,, & & \eta_2 & \to & -\eta_2 \\ N_3 & \to & -N_3\,, & & \eta_3 & \to & -\eta_3. \end{array} \tag{4}$$

The DM candidate is the lightest particle charged under Z_2 i.e., the lightest combination of the scalars η_2 and η_3, which we will denote generically by $\eta_{\rm DM}$. We list below all interactions of $\eta_{\rm DM}$:

(i) Yukawa interactions

$$\eta_{\rm DM}\, \overline{\nu}_i N_{2,3}\,, \tag{5}$$

where $i = e, \mu, \tau$.

(ii) Higgs-Vector boson couplings

$$\eta_{\rm DM}\eta_{\rm DM} ZZ\,, \quad \eta_{\rm DM}\eta_{\rm DM} WW\,, \quad \eta_{\rm DM}\eta_{2,3}^{\pm} W^{\pm} Z\,, \quad \eta_{\rm DM}\eta_{2,3}^{\pm} W^{\pm}\,, \quad \eta_{\rm DM} A_{2,3} Z\,. \tag{6}$$

[1] For quark mixing angles generated through higher dimension operators see reference [7].
[2] H is in the 1 representation of A_4 and its vev also respect the generator S.

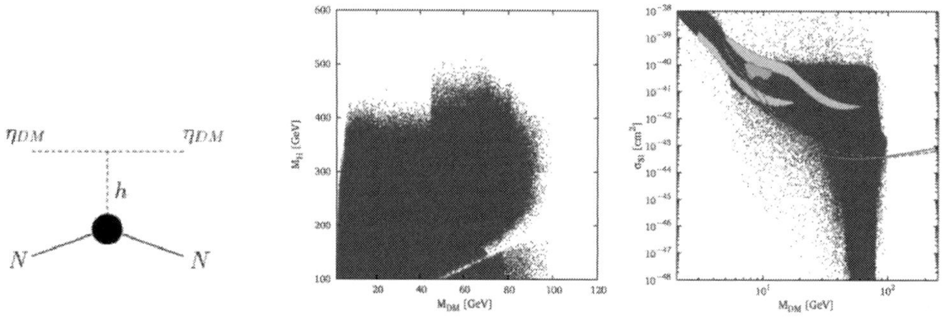

Figure 1. Feynman diagram for elastic scattering (left). Regions in the plane DM mass (MDM) – lightest Higgs boson MH allowed by collider constraints and leading to a DM relic abundance compatible with WMAP measurements (center). Spin-independent DM scattering cross section off-protons as a function of the dark matter mass (right). The orange regions delimited by the dashed (solid) line show the DAMA/LIBRA annual modulation regions including (neglecting) the channeling effect. The green region corresponds to the CoGeNT data. Dashed and dotted red lines correspond to the upper bound from CDMS. XENON100 bounds are shown as a solid red line.

(iii) Scalar interactions from the Higgs potential:

$$\begin{array}{ll} \eta_{\rm DM}\, A_1 A_2 h\,, & \eta_{\rm DM}\, A_1 A_3 h_1\,, \\ \eta_{\rm DM}\, A_1 A_2 h_1\,, & \eta_{\rm DM}\, A_1 A_3 h\,, \\ \eta_{\rm DM}\, A_2 A_3 h_3\,, & \eta_{\rm DM}\, h_1 h_3 h \\ \eta_{\rm DM}\eta_{\rm DM} h h\,, & \eta_{\rm DM}\eta_{\rm DM} h_1 h_1\,. \end{array} \tag{7}$$

After electroweak symmetry breaking, the Higgs fields acquire vacuum expectation values, v_h and v_η for the singlet and the first component of the triplet respectively. Additional terms are obtained from those in Eq. (7) by replacing $h \to v_h$ and $h_1 \to v_\eta$. These vertices are relevant for direct detection [8] see Figure 1. The phenomenology of dark matter of this model has been studied in detail in [8]. The model accommodates WMAP and collider constraints and is consistent with XENON100 and CDMS on one hand and CoGeNT or DAMA/LIBRA on the other hand [8], see Fig. 1.

3. Neutrino masses phenomenology

The model contains four heavy right-handed neutrinos. It is a special case, called (3,4), of the general type-I seesaw mechanism[9]. After electroweak symmetry breaking, it is characterized by Dirac and Majorana mass-matrix:

$$m_D = \begin{pmatrix} x_1 & 0 & 0 & x_4 \\ x_2 & 0 & 0 & 0 \\ x_3 & 0 & 0 & 0 \end{pmatrix}, \quad M_R = \begin{pmatrix} M_1 & 0 & 0 & 0 \\ 0 & M_1 & 0 & 0 \\ 0 & 0 & M_1 & 0 \\ 0 & 0 & 0 & M_2 \end{pmatrix}. \tag{8}$$

where x_1, x_2, x_3 and x_4 are respectively proportional to y_1^ν, y_2^ν, y_3^ν and y_4^ν of Eq. (2) and are of the order of the electroweak scale, while $M_{1,2}$ are assumed to be close to the unification scale. Light neutrinos get Majorana masses by means of the type-I seesaw relation and the light-neutrinos mass matrix has the form:

$$m_\nu = -m_{D_{3\times4}} M^{-1}_{R_{4\times4}} m^T_{D_{3\times4}} \equiv \begin{pmatrix} y^2 & ab & ac \\ ab & b^2 & bc \\ ac & bc & c^2 \end{pmatrix}. \tag{9}$$

This texture of the light neutrino mass matrix has a null eigenvalue $m_3 = 0$ corresponding to the eigenvector $(0, -b/c, 1)^T$, [3] implying a vanishing reactor mixing angle $\theta_{13} = 0$ and inverse hierarchy. The model implies a neutrinoless double beta decay effective mass parameter in the range 0.03 to 0.05 eV at 3 σ, within reach of upcoming experiments.

4. Conclusions

We have studied a model where the stability of the dark matter particle arises from a flavor symmetry. The A_4 non-Abelian discrete group accounts for both the observed pattern of neutrino mixing and for DM stability. We have analyzed the constraints that follow from electroweak precision tests, collider searches and perturbativity. We have also analyzed the prospects for direct and indirect dark matter detection and found that, although the former already excludes a large region in parameter space, we cannot constrain the mass of the DM candidate. In contrast, indirect DM detection is not yet sensitive enough to probe our predictions. However, forecasted sensitivities indicate that Fermi-LAT should start probing them in the near future.

All of the above relies mainly on the properties of the scalar sector responsible for the breaking of the gauge and flavor symmetries. The motivation of our approach is to link the origin of dark matter to the origin of neutrino mass and the understanding of the pattern of neutrino mixing, two of the most outstanding challenges in particle physics today. At this level one may ask what are the possible tests of this idea in the neutrino sector. We found an inverted neutrino mass hierarchy, hence a neutrinoless double beta decay rate accessible to upcoming searches, while $\theta_{13} = 0$ giving no CP violation in neutrino oscillations. Note however that the connection of dark matter to neutrino properties depends strongly on how the symmetry breaking sector couples to the leptons.

Acknowledgments

I thank my collaborators Sofiane Boucenna, Martin Hirsch, Stefano Morisi, Marco Taoso and Jose W. F. Valle for the work presented here. This work was supported by the Spanish MICINN under grants FPA2008-00319/FPA, FPA2011-22975 and MULTIDARK CSD2009-00064 (Consolider-Ingenio 2010 Programme), by Prometeo/2009/091 (Generalitat Valenciana), by the EU ITN UNILHC PITN-GA-2009-237920 and by CONACyT (Mexico).

References

[1] M. Hirsch, S. Morisi, E. Peinado and J. W. F. Valle, Phys. Rev. D **82** (2010) 116003 [arXiv:1007.0871 [hep-ph]].
[2] D. Meloni, S. Morisi and E. Peinado, Phys. Lett. B **697** (2011) 339 [arXiv:1011.1371 [hep-ph]].
[3] D. Meloni, S. Morisi, E. Peinado, S. Morisi and E. Peinado, Phys. Lett. B **703** (2011) 281 [arXiv:1104.0178 [hep-ph]].
[4] Y. Daikoku, H. Okada and T. Toma, arXiv:1010.4963 [hep-ph].
[5] Y. Kajiyama and H. Okada, Nucl. Phys. B **848** (2011) 303 [arXiv:1011.5753 [hep-ph]].
[6] D. A. Eby and P. H. Frampton, arXiv:1111.4938 [hep-ph].
[7] R. d. A. Toorop, F. Bazzocchi and S. Morisi, arXiv:1104.5676 [hep-ph].
[8] M. S. Boucenna, M. Hirsch, S. Morisi, E. Peinado, M. Taoso and J. W. F. Valle, JHEP **1105** (2011) 037 [arXiv:1101.2874 [hep-ph]].
[9] J. Schechter and J. W. F. Valle, Phys. Rev. D **22** (1980) 2227.

[3] Note that if we were to stick to the minimal (3,3)-type-I seesaw scheme, with just 3 SU(2) singlet states, one would find a projective nature of the effective tree-level light neutrino mass matrix with two zero eigenvalues, hence phenomenologically inconsistent. That is why we adopted the (3,4) scheme.

Aspects of production and kinetic decoupling of non-thermal dark matter

G. Arcadi and P. Ullio

SISSA, Via Bonomea 265, I-34136 Trieste, Italy and
Istituto Nazionale di Fisica Nucleare, Sezione di Trieste, I-34136 Trieste, Italy

E-mail: arcadi@sissa.it, ullio@sissa.it

Abstract. We reconsider non-thermal production of WIMP dark matter in a systematic way and using a numerical code for accurate computations of dark matter relic densities. Candidates with large pair annihilation rates are favored, suggesting a connection with the anomalies in the lepton cosmic-ray flux detected by Pamela and Fermi. Focussing on supersymmetric models we will consider the impact of non-thermal production on the preferred mass scale for dark matter neutralinos. We have also developed a new formalism to solve the Boltzmann equation for a system of coannihilating species without assuming kinetic equilibrium and applied it to the case of pure Winos.

1. General framework

In this work we present a numerical code, now added to the package DARKSUSY [1], aimed to determine the dark matter relic density when this is produced by the decay of heavy long lived states with weak enough interactions with ordinary matter to be out-of-equilibrium also in the early stages of the history of the Universe. These states, referred to for simplicity as cosmological moduli X_i, dominate the early stages of the history of the Universe and lately decay into DM as well as SM particles with a sharp increase of entropy density. When this decay occurs at lower temperatures than the standard freeze-out temperature $T_{\text{t.f.o.}}$ this non-thermal component dominates the dark matter relic density being the component already present in the thermal bath diluted by the entropy injection. Starting from the simplest case of only one decaying field without referring to a definite particle physics scenario we have numerically recovered an interesting approximate trend occurring when, during the production process the dark matter pair annihilation rate $\Gamma = n_\chi \langle \sigma v \rangle$ is larger than the expansion rate H. In such a case pair annihilations are very efficient and instantaneously decrease the number density of χ to the critical density level corresponding to $\Gamma \simeq H$ when the annihilations stop. The non-thermal relic density is given by a simple scaling law as a function of the one obtained by assuming the DM candidate a thermal relic freezing-out at $T_{\text{t.f.o.}}$.

$$\Omega_\chi^{\text{NT}} h^2 \simeq \frac{T_{\text{t.f.o.}}}{T_{\text{RH}}} \Omega_\chi^T h^2 \tag{1}$$

with T_{RH} is a characteristic temperature, numerically determined by the code, identifying the end of the decay process of the cosmological moduli. It can be analytically estimated by the condition $\Gamma = H(T_{\text{RH}})$ where Γ is the decay rate of the heavy field. Notice that a DM candidate

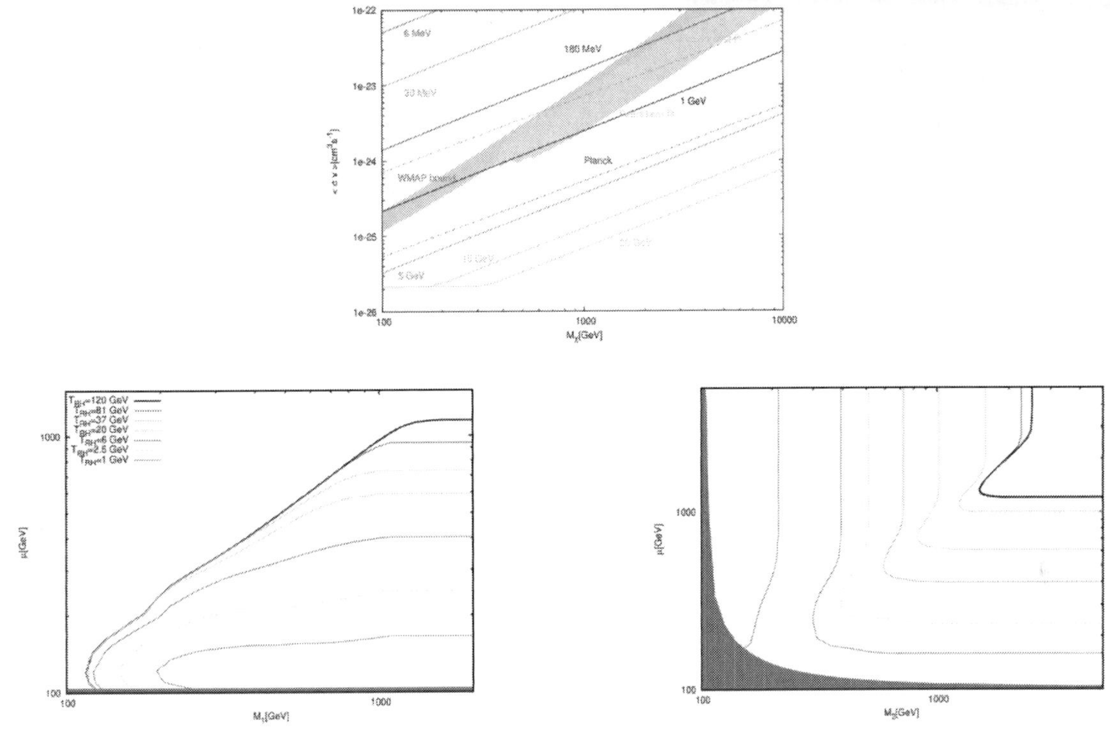

Figure 1. Upper panel: Models fitting the cosmological value of the DM relic density in the plane $(m_\chi, \langle\sigma v\rangle)$. Colored regions are the Pamela and Fermi best fit. The lines WMAP and Planck represent the current bound on the annihilation cross section from CMB and the Planck projected sensitivity. Lower panels: Curves of the cosmological relic density in two-dimensional slices of the M_1, M_2, μ parameter space for a few values of $T_{\rm RH}$. The filled areas correspond to regions violating the LEP lower bound on the chargino mass.

χ whose thermal relic density is small compared to the DM density because the annihilation rate is too large, may become a viable dark matter candidate for an appropriate value of $T_{\rm RH}$. This suggest an interesting connection with anomalies detected by Pamela and Fermi which could be explained by an annihilating cross section 2-3 orders of magnitude larger than the thermal one.

2. Application to the MSSM

Let's now consider a definite particle physics scenario, i.e. the MSSM with neutralino DM. In the aim of determining the DM mass scale favored by non-thermal production we restrict ourselves to a class of models referred as split supersymmetry [2], characterized by very high mass scale for the scalar superpartners. In this way we could compute the non-thermal relic density for several values of the reheating temperature as a function of the Bino and the Wino mass parameters M_1 and M_2, and of the Higgs superfield parameter μ.

The scenario depicted in the previous section is fitted by Wino-like or Higgsino-like DM. As shown in figure 1 the cosmological value of the DM relic density can be achieved for masses of the order of one hundred GeV, contrary to the thermal case where Wino and Higgsino masses in the TeV range are required.

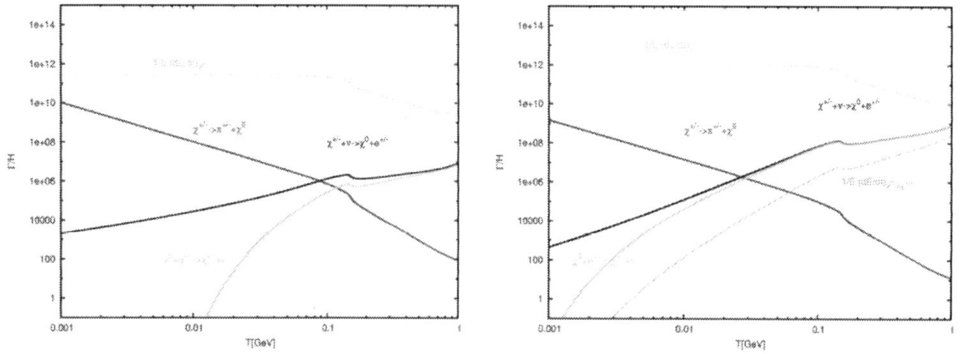

Figure 2. Ratios between the effective energy loss rate $-1/E \cdot dE/dt$ (dashed lines), or of the scattering/decay rate Γ (solid lines), to the Universe expansion rate H, for a few processes involving charginos and neutralinos. The panels refer to a G2-MSSM DM model with $m_\chi = 103.5$ GeV and $T_{RH} = 100$ MeV; the plot on the left-hand-side refers to non-relativistic particles, $E/m_\chi = 1.005$, while that on the right-hand side corresponds to a sample relativistic case, $E/m_\chi = 10$.

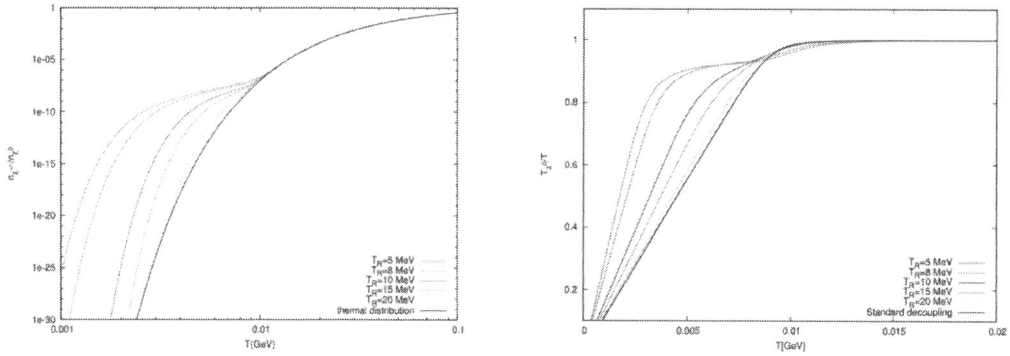

Figure 3. Left panel: ratio of the chargino number density over the neutralino number density for several values of $T_{\rm RH}$. Right panel: Ratio $T_{\chi 0}/T$ as function of the temperature of the thermal bath for same values of T_{RH}. Plots are obtained for a Wino with mass equal to 200 GeV, however results depend only on the chargino-neutralino mass splitting which is about 160 MeV in the scenario under consideration.

3. Kinetic equilibrium and decoupling

The results presented until now are obtained under the assumption that DM is in kinetic equilibrium with the SM states of the thermal bath along all its production process. In case that DM is part of a system of coannihilating particles, this allows to describe it by a unique Boltzmann equation for sum of their number densities with an effective annihilation cross section defined as a weighted sum of the thermally averaged cross sections of the coannihilating particles. The validity of the kinetic equilibrium assumption is typically guaranteed by invoking a crossing symmetry among annihilations and scattering with SM particles. This assumption may not hold in all the models. Furthermore the picture is more complicated in presence of non-thermal production. Non thermal dark matter is produced at relativistic energies and thermalizes with the heat only if it has an efficient rate of energy loss though interactions with the heat bath.

How to address this issue varies from one model to another. We have considered the case of a class of models referred as G2-MSSM [3]. Here DM is represented by a system of two coannihilating species: the lightest neutralino and chargino (both Wino-like) which are quasi degenerate in mass. The remaining spectrum is at a very high mass scale, $O(10-100)$ TeV with the exception of the other gauginos, below the TeV scale, but not relevant for coannihilations. The issue of relaxing the condition of kinetic equilibrium have been approached in two steps. First of all we have verified the thermalization of the two species soon after production. This depends on their energy loss rates due to interactions with the SM particles of the thermal bath. Thermalization occurs essentially instantaneously due to the following facts: at temperatures higher than order of 1 GeV charginos are promptly converted into neutralinos with the latter losing energy very efficiently. At lower temperatures charginos are able to lose energy before being converted and guarantee the thermalization of the two species system as long as inelastic neutralino scatterings are efficient. This picture is summarized in figure 2.

Once thermalization is accomplished, the evolution of the system until kinetic decoupling have been followed by writing two Boltzmann equations for the distribution functions of the two species. These equation can be integrated by expanding the collisional operator in an analogous manner to [4]:

$$\begin{aligned} \frac{dn_{\chi^0}}{dt} + 3\,H\,n_{\chi^0} &= \left(\widetilde{\Gamma}_{\chi^0\leftrightarrow\chi^\pm} + \Gamma_{\chi^\pm}\right)\left[g_{\chi^0} n_{\chi^\pm} - g_{\chi^\pm} n_{\chi^0} \exp\left(-\frac{\Delta m_\chi}{T}\right)\right] \\ &\quad -\langle\sigma v\rangle_{\chi^0\chi^0}\left[n^2_{\chi^0} - (n^{eq}_{\chi^0})^2\right] - \langle\sigma v\rangle_{\chi^0\chi^\pm}\left[n_{\chi^0} n_{\chi^\pm} - n^{eq}_{\chi^0} n^{eq}_{\chi^\pm}\right] \\ \frac{dn_{\chi^\pm}}{dt} + 3\,H\,n_{\chi^\pm} &= \left(\widetilde{\Gamma}_{\chi^0\leftrightarrow\chi^\pm} + \Gamma_{\chi^\pm}\right)\left[g_{\chi^\pm} n_{\chi^0} \exp\left(-\frac{\Delta m_\chi}{T}\right) - g_{\chi^0} n_{\chi^\pm}\right] \\ -\langle\sigma v\rangle_{\chi^+\chi^+}\left[n^2_{\chi^\pm} - (n^{eq}_{\chi^\pm})^2\right] &- \langle\sigma v\rangle_{\chi^+\chi^0}\left[n_{\chi^+} n_{\chi^0} - n^{eq}_{\chi^\pm} n^{eq}_{\chi^0}\right] + \sum_i \frac{B_{X_i}}{m_{X_i}} \Gamma_{X_i} \rho_{X_i}\,. \end{aligned} \tag{2}$$

with $\widetilde{\Gamma}_{\chi^0\leftrightarrow\chi^\pm}$ and $\Gamma_{\chi^\pm}$, being, respectively the scattering and the decay rate. The last term in the second equation is the non-thermal production term and is defined in [5]. The same formalism describes the dark matter distribution close to kinetic decoupling. Indeed, integrating the second moment of the distribution function it is possible to define the effective temperature T_{χ_0} (we define only one temperature being charginos always kept in equilibrium by electromagnetic interactions) which coincides with the thermal bath temperature until there is kinetic equilibrium while scales as $T_{\chi_0} \propto T^2$ after kinetic decoupling. In the second panel of figure 3 we show the behaviour of T_{χ_0} for different values of the reheating temperature. The impact of the decay of the cosmological moduli is twofold (figure 3): the chargino decays tend to populate the system with neutralinos that are on average more energetic than for a thermal distribution, delaying the onset of the regime $T_{\chi^0} \propto T^2$ and making the transition into this regime to be less sharp; at the same time, if $T_{\rm RH}$ is so low that reheating increases significantly the expansion rate of the Universe H at the time of kinetic decoupling the departure from $T_{\chi_0} = T$ tends to be anticipated.

In conclusion we point out that the developed formalism for the study of dark matter close to kinetic decoupling can be applied to other particle physics frameworks, also within standard cosmology.

[1] P. Gondolo, J. Edsjö, P. Ullio, L. Bergström, M. Schelke and E. A. Baltz, JCAP **0407** (2004) 008 [arXiv:astro-ph/0406204].

[2] N. Arkani-Hamed, S. Dimopoulos, G. F. Giudice and A. Romanino, Nucl. Phys. B **709** (2005) 3 [arXiv:hep-ph/0409232]. G. F. Giudice and A. Romanino, Nucl. Phys. B **699** (2004) 65 [arXiv:hep-ph/0406088].

[3] B. S. Acharya, K. Bobkov, G. L. Kane, J. Shao and P. Kumar, Phys. Rev. D **78** (2008) 065038 [arXiv:0801.0478 [hep-ph]]. B. S. Acharya, P. Kumar, K. Bobkov, G. Kane, J. Shao and S. Watson, JHEP **0806** (2008) 064 [arXiv:0804.0863 [hep-ph]].

[4] T. Bringmann and S. Hofmann, JCAP **0407** (2007) 016 [arXiv:hep-ph/0612238]. T. Bringmann, New J. Phys. **11** (2009) 105027 [arXiv:0903.0189 [astro-ph.CO]].

[5] G. Arcadi and P. Ullio, Phys. Rev. D **84** (2011) 043520 [arXiv:1104.3591 [hep-ph]].

Journal of Physics: Conference Series **375** (2012) 012045 doi:10.1088/1742-6596/375/1/012045

Invisible Higgs and scalar dark matter

Yann Mambrini

Laboratoire de Physique Théorique Université Paris-Sud, F-91405 Orsay, France

E-mail: yann.mambrini@th.u-psud.fr

Abstract. In this proceeding, we show that when we combined WMAP and the most recent results of XENON100, the invisible width of the Higgs to scalar dark matter is negligible ($\lesssim 10\%$), except in a small region with very light dark matter ($\lesssim 10$ GeV) not yet excluded by XENON100 or around 60 GeV where the ratio can reach 50% to 60%. The new results released by the Higgs searches of ATLAS and CMS set very strong limits on the elastic scattering cross section.

1. Introduction

Two of the most important issues in particle physics phenomenology are the nature of the dark matter and the mechanism to realize spontaneously the electroweak symmetry breaking of the Standard Model (SM). The observations made by the WMAP collaboration show that the matter content of the universe is dark, making up about 85% of the total amount of matter whereas the XENON collaboration recently released its constraints on direct detection of dark matter. These constraints are the most stringent in the field nowadays, and begin to exclude a significant part of the parameter space of the Weakly Interacting Massive Particle (WIMP) paradigm. On the other front, the accelerator collaborations ATLAS, CMS and D0/CDF [1] have obtained important results concerning the Higgs searches. It is obvious that the Higgs hunting at LHC is intimately linked with measurement of elastic scattering on nucleon, especially in Higgs-portal like models where the Higgs boson is the key particle exchanged through annihilation/scattering processes. It has already been shown recently that a combined LEP/TEVATRON/XENON/WMAP analysis can restrict severely the parameter space allowed in generic constructions [2]. In this work, we apply such analysis in the specific context of a scalar singlet dark matter extension of the SM and show that most of the region allowed by WMAP will be excluded/probed by LHC and XENON100 by the end of next year.

2. The model

The simplest extension of the SM is the addition of a real singlet scalar field. Although it is possible to generalize to scenarios with more than one singlet, the simplest case of a single additional singlet scalar provides a useful framework to analyze the generic implications of an augmented scalar sector to the SM. The most general renormalizable potential involving the SM Higgs doublet H and the singlet S is

$$\mathcal{L} = \mathcal{L}_{\rm SM} + (D_\mu H)^\dagger (D^\mu H) + \frac{1}{2}\mu_H^2 H^\dagger H - \frac{1}{4}\lambda_H H^4$$

$$+ \frac{1}{2}\partial_\mu S \partial^\mu S - \frac{\lambda_S}{4} S^4 - \frac{\mu_S^2}{2} S^2 - \frac{\lambda_{HS}}{4} S^2 H^\dagger H$$
$$- \frac{\kappa_1}{2} H^\dagger H S - \frac{\kappa_3}{3} S^3 - V_0 \quad (1)$$

where D_μ represents the covariant derivative.

Different aspects of scalar singlet extension of the SM has already been studied in [3] whereas a nice preliminary analysis of its dark matter consequences can be found in [4]. Some authors also tried to explain the DAMA/LIBRA and/or CoGeNT excess [5] whereas other authors probed the model by indirect searches [6], or looked at the consequences of earlier XENON data [7] or at the LHC [8] or all combined [9].

Recently the XENON100 collaboration released new data, the most stringent in the field of dark matter detection[1]. Moreover, recently CRESST experiment released their analysis in the low mass region [12] and seems to converge with DAMA/LIBRA and CoGeNT toward a possible light dark matter signal for a mass around 10 GeV [13]. In the meantime, if $m_S \lesssim M_H/2$ the invisible width decay[2] of the Higgs $H \to SS$ could perturbe the Higgs searches at LHC based on SM Higgs branching ratio (see Eq. 2). However, one can easily understand that there exists a tension between the direct detection measurement and the invisible branching ratio. Indeed, for decreasing mass of DM ($m_S \lesssim$ 100 GeV), the spin independent cross section increases. Quantitatively speaking, one needs to compare the invisible Higgs width ($H \to SS$)

$$\Gamma_H^{\rm inv} = \frac{\lambda_{HS}^2 M_W^2}{32\pi g^2 M_H^2}\sqrt{M_H^2 - 4m_S^2} \quad (2)$$

with the spin independent scattering cross section on the proton

$$\sigma_{S-p}^{\rm SI} = \frac{m_p^4 \lambda_{HS}^2 (\sum_q f_q)^2}{16\pi (m_p + m_S)^2 M_H^4} \quad (3)$$

Combining Eq. (2) and (3) one obtains

$$\frac{\Gamma_H^{\rm inv}}{\sigma_{S-p}^{\rm SI}} = \frac{(m_S + m_p)^2 M_H^2 M_W^2 \sqrt{M_H^2 - 4m_S^2}}{2g^2 f^2 m_p^4} \quad (4)$$

Some points with high invisible width still survive. They correspond to two distinct regions:

- A region with very light scalar ($m_S \lesssim$ 10 GeV) still not yet excluded by the precision of XENON100 experiments due to its high threshold. This corresponds to very large invisible branching ratio.
- A region with 50 GeV $\lesssim m_S \lesssim$ 70 GeV with branching ratio which can reach 60% – 70% which is the region taken in consideration in [15].

We show the effects of combining WMAP and XENON100 data in Fig. 1 and 2. As one can see, in Fig. 2 except for these two particular regions, the majority of points respecting WMAP and XENON100 constraints give very low invisible width. As a conclusion, we can affirm that the Higgs searches at LHC with a scalar dark matter is not affected for $m_H \gtrsim$ 150 GeV It means that we can use the standard Higgs limit searches of ATLAS and CMS and apply them in the

[1] Keeping an eye on the results of CoGeNT collaboration, recent works showed there exists a tension between XENON100 and CoGeNT [10], or not [11]. We thus safely decided not to discuss in detail the CoGeNT issue in our analysis.

[2] See the works in [14] for an earlier study of invisible width of the Higgs. Moreover, during the revision of this study, several independant works confirming our results were published in [15].

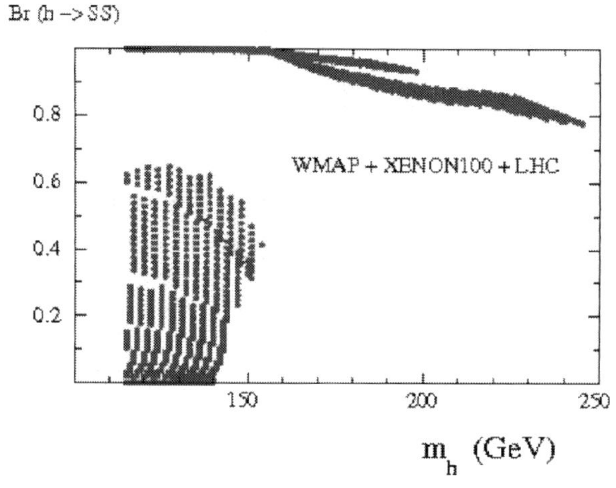

Figure 1. Branching ratio for the invisible Higgs decay into scalars as a function of the Higgs mass. The scan is over (m_H, m_S, λ_{hs}) and subject to the WMAP, XENON100 and ATLAS/CMS constraints.

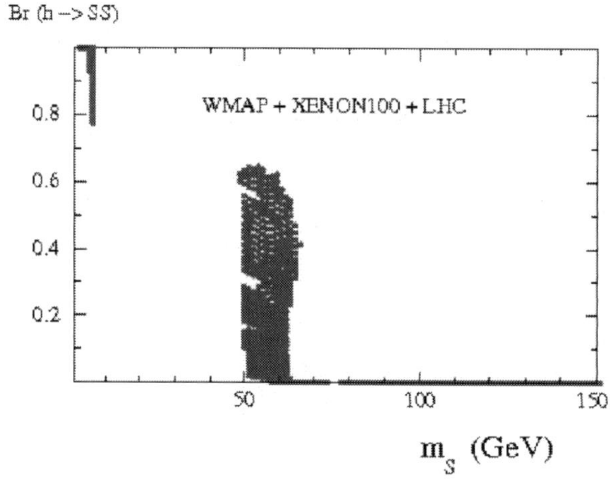

Figure 2. Same as Fig. 1 now as function of the dark matter mass

model in this region of the parameter space, they are only slightly affected by the presence of the scalar dark matter[3].

Due to the last data released recently by CRESST collaboration [12] it is interesting to notice that some points in the parameter space around $m_S \simeq 10$ GeV are not yet excluded by the latest XENON100 constraints as can be seen in the upper left corner of Fig. 2. These points generates a Higgs completely invisible at the LHC. This corresponds to the region near Br$(H \to SS) \simeq$ 100% in Fig. 1.

It is also interesting to point out that the Higgs-portal construction is similar by several aspects to the Z'-portal model of dark matter [17]: as any Higgs searches restrict severely the parameter space of the model, any Z' searches at LHC should be use in complementarity with

[3] In our numerical study, we obviously took into consideration the invisible Higgs width to apply the CMS and ATLAS constraints

direct detection searches to probe the entire parameter space allowed by WMAP. A vectorial dark matter will also give similar results [16] At the same time, the analysis should be done is SUSY scenario where light Higgses are the main annihilation channel, leading to sever direct detection constraints [18]. Writing the conclusion of this work, we noticed that authors just looked at some consequences of recent Higgs searches at LHC in the NMSSM case [19] and extended scalar sectors [20].

References

[1] REPORT ATLAS-CONF-2011-112; REPORT ATLAS-CONF-2011-157; REPORTCMS PAS HIG-11-011; D0 Note 6229-CONF; REPORT FERMILAB-CONF-10-257-E
[2] Y. Mambrini and B. Zaldivar, JCAP **1110** (2011) 023.
[3] V. Silveira, A. Zee, Phys. Lett. **B161**, 136 (1985); In the context of mirror matter, the Higgs portal interaction also appeared in R. Foot, H. Lew, R. R. Volkas, Phys. Lett. **B272**, 67 (1991); J. McDonald, Phys. Rev. **D50** (1994) 3637-3649; J. McDonald, Phys. Rev. Lett. **88** (2002) 091304; C. P. Burgess, M. Pospelov, T. ter Veldhuis, Nucl. Phys. **B619** (2001) 709-728; B. Patt, F. Wilczek, [hep-ph/0605188]; K. A. Meissner, H. Nicolai, Phys. Lett. **B648** (2007) 312-317; H. Davoudiasl, R. Kitano, T. Li, H. Murayama, Phys. Lett. **B609** (2005) 117-123; S.-h. Zhu, X.-G. He *et al.* Phys. Rev. **D79** (2009) 023521; X.-G. He *et al.* Phys. Lett. **B688** (2010) 332-336; X.-G. He, S.-Y. Ho, J. Tandean, H.-C. Tsai, Phys. Rev. **D82** (2010) 035016; S. Kanemura, S. Matsumoto, T. Nabeshima, N. Okada, Phys. Rev. **D82** (2010) 055026; M. Aoki, S. Kanemura, O. Seto, Phys. Lett. **B685** (2010) 313-317; W.-L. Guo, Y.-L. Wu, JHEP **1010** (2010) 083; V. Barger, Y. Gao, M. McCaskey, G. Shaughnessy, Phys. Rev. **D82** (2010) 095011; J. R. Espinosa, M. Quiros, Phys. Rev. **D76** (2007) 076004; H. Sung Cheon, S. K. Kang, C. S. Kim, JCAP **0805** (2008) 004; A. Melfo *et al.* Phys. Rev. D **84** (2011) 034009.
[4] V. Barger *et al.* Phys. Rev. **D77** (2008) 035005; T. E. Clark, B. Liu, S. T. Love, T. ter Veldhuis, Phys. Rev. **D80**, 075019 (2009); R. N. Lerner, J. McDonald, Phys. Rev. **D80**, 123507 (2009); O. Lebedev, H. M. Lee, [arXiv:1105.2284].
[5] S. Andreas, T. Hambye, M. H. G. Tytgat, JCAP **0810** (2008) 034; S. Andreas, C. Arina, T. Hambye, F.-S. Ling, M. H. G. Tytgat, Phys. Rev. **D82** (2010) 043522; M. H. G. Tytgat, [arXiv:1012.0576].
[6] C. E. Yaguna, JCAP **0903** (2009) 003; A. Goudelis, Y. Mambrini, C. Yaguna, JCAP **0912** (2009) 008; C. Arina, M. H. G. Tytgat, JCAP **1101** (2011) 011.
[7] Y. Cai, X.-G. He, B. Ren, Phys. Rev. **D83** (2011) 083524; A. Biswas, D. Majumdar, [arXiv:1102.3024]; M. Farina *et al.* [arXiv:1104.3572].
[8] C. Englert, T. Plehn, D. Zerwas, P. M. Zerwas, Phys. Lett. **B703**, 298 (2011); K. Ghosh, B. Mukhopadhyaya, U. Sarkar, Phys. Rev. D **84** (2011) 015017; C. Englert *et al.* [arXiv:1112.3007].
[9] Y. Mambrini, [arXiv:1108.0671]; A. Djouadi, O. Lebedev, Y. Mambrini, J. Quevillon, [arXiv:1112.3299]; M. Kadastik, K. Kannike, A. Racioppi, M. Raidal, [arXiv:1112.3647];
[10] T. Schwetz, J. Zupan, [arXiv:1106.6241]; M. T. Frandsen *et al.* Phys. Rev. D **84** (2011) 041301.
[11] D. Hooper, C. Kelso, [arXiv:1106.1066].
[12] G. Angloher *et al.*, [arXiv:1109.0702].
[13] C. Kelso, D. Hooper, M. R. Buckley, [arXiv:1110.5338]; J. Kopp, T. Schwetz, J. Zupan, [arXiv:1110.2721].
[14] M. C. Bento, O. Bertolami, R. Rosenfeld, Phys. Lett. **B518** (2001) 276-281; M. C. Bento, O. Bertolami, R. Rosenfeld, L. Teodoro, Phys. Rev. **D62** (2000) 041302; R. E. Shrock and M. Suzuki, 110B, 250 (1982); D. A. Demir, [arXiv:1110.3815].
[15] M. Raidal, A. Strumia, [arXiv:1108.4903]; X.-G. He, J. Tandean, [arXiv:1109.1277].
[16] T. Hambye, JHEP **0901**, 028 (2009); T. Hambye, M. H. G. Tytgat, Phys. Lett. **B683**, 39 (2010); J. Hisano, K. Ishiwata, N. Nagata, M. Yamanaka, Prog. Theor. Phys. **126**, 435 (2011); O. Lebedev, H. M. Lee, Y. Mambrini, [arXiv:1111.4482].
[17] Y. Mambrini, JCAP **1107** (2011) 009; M. T. Frandsen, F. Kahlhoefer, S. Sarkar, K. Schmidt-Hoberg, [arXiv:1107.2118 [hep-ph]]; D. Feldman, B. Kors, P. Nath, Phys. Rev. **D75** (2007) 023503; Y. Mambrini, JCAP **1009**, 022 (2010); Y. Mambrini, JCAP **0912** (2009) 005; K. Cheung, J. Song, Phys. Rev. Lett. **106** (2011) 211803; E. Dudas, Y. Mambrini, S. Pokorski, A. Romagnoni, JHEP **0908** (2009) 014.
[18] D. Das, A. Goudelis, Y. Mambrini, JCAP **1012** (2010) 018; S. Bhattacharya *et al.* Phys. Rev. **D81** (2010) 075009; A. Djouadi, Y. Mambrini, JHEP **0612** (2006) 001.
[19] U. Ellwanger, [arXiv:1108.0157].
[20] X. G. He, G. Valencia, [arXiv:1108.0222].

Journal of Physics: Conference Series **375** (2012) 012046 doi:10.1088/1742-6596/375/1/012046

Electroweak lights from Dark Matter annihilations

Andrea De Simone

Institut de Théorie des Phénomènes Physiques, École Polytechnique Fédérale de Lausanne, CH-1015 Lausanne, Switzerland

E-mail: andrea.desimone@epfl.ch

Abstract. The energy spectra of Standard Model particles originated from Dark Matter annihilations can be significantly altered by the inclusion of electroweak gauge boson radiation from the final state. A situation where this effect is particularly important is when a Majorana Dark Matter particle annihilates into two light fermions. This process is in p-wave and hence suppressed by the small value of the relative velocity of the annihilating particles. The inclusion of electroweak radiation eludes this suppression and opens up a potentially sizeable s-wave contribution to the annihilation cross section. I will discuss the impact of this effect on the fluxes of stable particles resulting from the Dark Matter annihilations, which are relevant for Dark Matter indirect searches.

1. Introduction

The fluxes of stable Standard Model particles that originate from the annihilation (or decay) of Dark Matter (DM) in the galactic halo are the primary observable for DM indirect searches. The radiation of ElectroWeak (EW) gauge bosons from the final state of the annihilation process turns out to have a great influence on the energy spectra of stable particles and hence on the predictions for fluxes to be measured at Earth [1, 2] (see also Ref. [3]). In particular, there are three situations where the effect of including the EW corrections is especially important:

(i) when the low-energy regions of the spectra, which are largely populated by the decay products of the emitted gauge bosons, are the ones contributing the most to the observed fluxes of stable particles;

(ii) when some particle species are absent if EW corrections are not included, e.g. antiprotons from W/Z decays in an otherwise purely leptonic channel;

(iii) when the $2 \to 3$ annihilation cross section, with soft gauge boson emission, is comparable or even dominant with respect to the $2 \to 2$ cross section.

In this talk I will focus on case (iii). A possible situation realizing it occurs when one considers the DM as a gauge-singlet Majorana particle χ of mass M_χ annihilating into light fermions f of mass $m_f \ll M_\chi$; it is well known that in this case the $\chi\chi \to f\bar{f}$ cross section is suppressed. In fact, one can perform the usual expansion of the cross section $v\sigma = a + b\,v^2 + \mathcal{O}(v^4)$, where $v \sim 10^{-3}$ is the relative velocity (in units of c) of the DM particles in our galaxy today. The v-independent term, corresponding to the annihilation of particles in s-wave, is constrained by helicity arguments to be proportional to $(m_f/M_\chi)^2$, and hence very small for light final state fermions. The second term, corresponding to p-wave annihilation, suffers from the v^2 suppression. For a DM particle singlet under the SM gauge group, the radiation of EW gauge

Published under licence by IOP Publishing Ltd

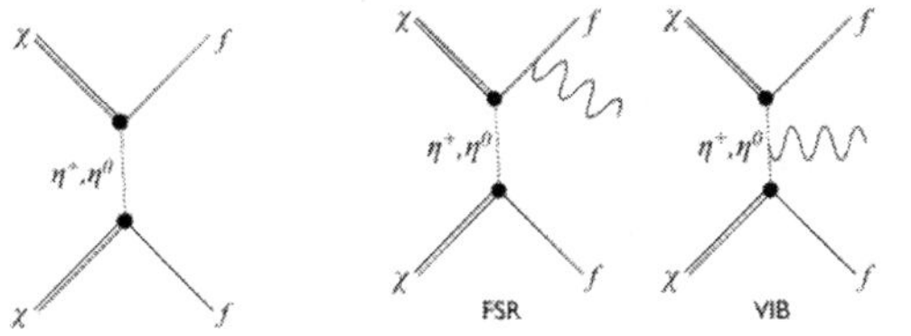

Figure 1. Diagrams for 2-body (left panel) and 3-body (right panel) DM annihilations.

bosons from the final state and from the internal propagator of the annihilation process eludes the suppressions and opens up a potentially sizeable s-wave contribution to the cross section (see Ref. [4] for the case of photon radiation and Ref. [5] for gluon radiation).

There is actually another situation realizing the condition 3 above, where we expect therefore the EW corrections to have a great impact [6]. Relaxing the hypothesis that χ is a gauge singlet, and considering the possibility that the DM is part of a multiplet charged under the EW interactions, then the DM annihilates predominantly in s-wave into W^+W^-, if kinematically allowed. However, now even the initial state of the channel $\chi\chi \to f\bar{f}$ can radiate a gauge boson; this process also lifts the helicity suppression and contributes to the s-wave cross section, becoming competitive with the di-boson channel. We refer the reader to Ref. [6] for more details.

2. The model

In this section, I present the (toy) model I will use to describe more concretely the relevance of the EW corrections in DM annihilations. Let us add to the particle content of the SM a Majorana spinor χ with mass M_χ, singlet under the SM gauge group and playing the role of DM, and a scalar $SU(2)$-doublet $S = (\eta^+, \eta^0)^T$, with mass $M_S > M_\chi$. The field S provides the interactions of the DM with the generic fermion of the SM, described by the left-handed doublet $L = (f_1, f_2)$. In fact, the total Lagrangian of the model is

$$\mathcal{L} = \mathcal{L}_{\rm SM} + \frac{1}{2}\bar{\chi}(i\partial\!\!\!/ - M_\chi)\chi + (D_\mu S)^\dagger(D^\mu S) - M_S^2 S^\dagger S + [y_L \bar{\chi}(Li\sigma_2 S) + {\rm h.c.}]\,. \tag{1}$$

The parameter M_S will be traded for the ratio $r \equiv M_S^2/M_\chi^2$. The stability of the DM can be achieved e.g. by endowing χ and S with odd parity under a Z_2 symmetry, while the rest of the SM spectrum is even. In a supersymmetric framework, one recognizes the same interactions of a Bino with fermions and their supersymmetric scalar partners. We will restrict our attention to the massless limit $m_{f_1} = m_{f_2} = 0$. The generalization of our calculations to non-zero fermion masses will not be addressed here.

3. Helicity suppression and its removal

As it is well-known, the first non-zero contribution to the tree-level cross section of the annihilation of a Majorana fermion into a pair of massless left-handed fermions (left panel of Fig. 1) is velocity dependent, and hence suppressed, and is approximately given by (see Ref. [2] for the exact expression)

$$v\sigma(\chi\chi \to f\bar{f}) \sim \frac{1}{M_\chi^2}\frac{v^2}{r^2}\,. \tag{2}$$

The inclusion of higher-order processes with the emission of soft weak gauge bosons evades the helicity suppression and turns on an unsuppressed s-wave contribution to the DM annihilation cross section. Let us consider, for definiteness, the 3-body process with the emission of a Z boson $\chi\chi \to \bar{f}_L f_L Z$ (right panel of Fig. 1). Following Ref. [4], we will call "FSR" (final state radiation) the processes where a gauge boson is radiated from an external leg, while we refer to

Journal of Physics: Conference Series **375** (2012) 012046 doi:10.1088/1742-6596/375/1/012046

the emission from internal virtual particles as "VIB" (virtual internal bremsstrahlung). I am not going to write down any detailed expression for the amplitude, and I refer the interested reader to Ref. [2] where the exact analytical results can be found; here, I would just like to discuss the structure of the amplitude, where the various contributions can be organized as follows

$$\mathcal{M} \sim \frac{1}{M_\chi}\mathcal{O}(v)\left[\mathcal{O}\left(\frac{1}{r}\right)\Big|_{\rm FSR} + \mathcal{O}\left(\frac{1}{r^2}\right)\Big|_{\rm FSR}\right] + \frac{1}{M_\chi}\left[\mathcal{O}\left(\frac{1}{r^2}\right)\Big|_{\rm VIB} + \mathcal{O}\left(\frac{1}{r^2}\right)\Big|_{\rm FSR}\right] . \quad (3)$$

At this point we can learn an important lesson: the opening of the s-wave originates from diagrams of both FSR and VIB type, at $\mathcal{O}(1/r^2)$ in the amplitude; limiting the expansion up to $\mathcal{O}(1/r)$ in the amplitude would cause the process to stay in the p-wave. An order-of-magnitude estimate for the 3-body cross section, showing the leading dependence on the expansion parameters, can be obtained straightforwardly

$$v\sigma(\chi\chi \to f\bar{f}Z) \sim \frac{\alpha_W}{M_\chi^2}\left[\mathcal{O}\left(\frac{v^2}{r^2}\right) + \mathcal{O}\left(\frac{v^2}{r^3}\right) + \mathcal{O}\left(\frac{1}{r^4}\right)\right] , \quad (4)$$

where the weak coupling $\alpha_W = g^2/(4\pi)$ for the gauge boson emission has been restored. The estimates in Eqs. (2) and (4) allow to gather an understanding in simple terms of the situation we are studying. While the 2-body annihilation cross section behaves like v^2/r^2, the 3-body FSR and VIB diagrams give rise to both s-wave and p-wave terms; the latter cannot compete with the 2-body cross section because of the extra α_W factor; however, the former, free from the v^2 suppression, can overcome the $2 \to 2$ cross section if r is not too large.

Because of the importance of this point, and being the distinction between FSR and VIB not able to disentangle clearly the s-wave contribution from the p-wave one, let us introduce now a specific notation. Having in mind an expansion in powers of $1/r$ in the amplitude – as sketched in Eq. (3) – we will call "leading order" (LO) the lowest order term $\mathcal{O}(1/r)$ in this expansion, which originates from lowest order FSR-type diagrams. As shown above, in the LO approximation the annihilation cross section proceeds through p-wave. Only higher order terms are able to remove the helicity suppression.

4. Energy spectra of final stable particles

Let us now look at the energy spectra of stable particles produced by DM annihilations, with the inclusion of EW bremsstrahlung, focusing in particular on positrons, antiprotons, photons and neutrinos. For simplicity, I consider only the case where $L = (\nu_{e\,L}, e_L)$ and the primary annihilation channels for $\chi\chi \to I$, including EW bremsstrahlung, are

$$I = \{e_L^+ e_L^-,\ \nu_{e\,L}\bar{\nu}_{e\,L},\ e_L^+ e_L^- \gamma,\ e_L^+ \nu_{e\,L} W^-,\ e_L^- \bar{\nu}_{e\,L} W^+,\ e_L^+ e_L^- Z,\ \nu_{e\,L}\bar{\nu}_{e\,L} Z\} . \quad (5)$$

Using the techniques described in Ref. [2] one is able to combine the analytical description of the primary annihilation channels with the numerical simulations for subsequent hadronization and decay and it is possible to extract the energy distributions of each stable particle f

$$\frac{d\mathcal{N}_f}{d\ln x} \equiv \frac{1}{\sigma_0}\frac{d\sigma(\chi\chi \to f + \text{anything})}{d\ln x} , \qquad f = \{e^+, e^-, \gamma, \nu, \bar{\nu}, p, \bar{p}\} , \quad (6)$$

where $x \equiv E^{(f)}_{\rm kinetic}/M_\chi$, $E^{(f)}_{\rm kinetic}$ is the kinetic energy of the particle f (the difference between total and kinetic energies is obviously relevant only for the (anti)protons). The normalization is chosen to be the tree-level cross section of the 2-body processes $\sigma_0 = \sigma_{\rm tree}(\chi\chi \to e_L^+ e_L^-) + \sigma_{\rm tree}(\chi\chi \to \nu_{e\,L}\bar{\nu}_{e\,L})$. The plot in Figure 2 shows the resulting $d\mathcal{N}_f/d\ln x$ for $e^+, \gamma, \nu = (\nu_e + \nu_\mu + \nu_\tau)/3, \bar{p}$ for a specific, but representative, choice of parameters: $M_\chi = 1\,\text{TeV}$, $M_S = 4\,\text{TeV}$ and $v = 10^{-3}$; for

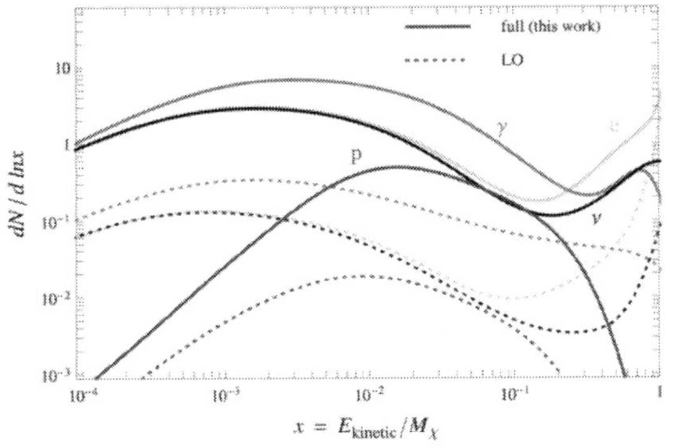

Figure 2. The spectra $dN_f/d\ln x$, as defined in Eq. (6), for e^+ (green), γ (red), $\nu = (\nu_e + \nu_\mu + \nu_\tau)/3$ (black) and $\bar{p}$ (blue), from the annihilation $\chi\chi \to e_L^+ e_L^-, \nu_{e\,L}\bar{\nu}_{e\,L}$ with the corresponding weak boson emission corrections. for the case $M_\chi = 1\,\text{TeV}, M_S = 4\,\text{TeV}, v = 10^{-3}$ (solid lines). For comparison, we show the spectra (dashed lines) in the LO approximation.

comparison, the situation where only the LO term is taken into account is also shown (dashed lines). The energy spectra result to be much larger than those obtained in the LO approximation, by factors $\mathcal{O}(10-100)$. This is a consequence of having a sizeable s-wave annihilation channel opened at the next-to-leading order in the $1/r$ expansion. We have also studied the propagation of these fluxes of stable particles through the galactic halo and verified that the enhancement effect under consideration does not get spoiled by galactic propagation.

5. Conclusions

In this talk, I have discussed the relevance of the EW corrections in theories where the cross section for DM annihilation into 2-body final states is suppressed. A gauge-singlet Majorana DM annihilating into two light SM fermions is one such case.

It is found that an efficient lifting of the helicity suppression, opening up a potentially large s-wave, is achieved by including EW bremsstrahlung and that the resulting energy spectra of stable particles, in the annihilation region, get substantially enhanced by this effect by factors $\mathcal{O}(10-100)$. This effect crucially affects the predictions for fluxes to be measured at Earth.

These results have a wider generality than the specific model I have considered here. Reliable computations of energy spectra of stable particles and predictions for their fluxes at Earth – the key observable for DM indirect searches – cannot prescind from including the effects of EW radiation.

Acknowledgments

This contribution is based on work done in collaboration with P. Ciafaloni, M. Cirelli, D. Comelli, A. Riotto and A. Urbano. I acknowledge support by the Swiss National Science Foundation under contract 200021-125237.

References

[1] P. Ciafaloni, D. Comelli, A. Riotto, F. Sala, A. Strumia, A. Urbano, JCAP **1103**, 019 (2011) [arXiv:1009.0224].
[2] P. Ciafaloni, M. Cirelli, D. Comelli, A. De Simone, A. Riotto, A. Urbano, JCAP **1106**, 018 (2011) [arXiv:1104.2996].
[3] M. Garny, A. Ibarra, S. Vogl, JCAP **1107**, 028 (2011) [arXiv:1105.5367].
[4] L. Bergström, Phys. Lett. B **225**, 372 (1989); T. Bringmann, L. Bergström, J. Edsjö, JHEP **0801**, 049 (2008) [arXiv:0710.3169]; L. Bergström, T. Bringmann, J. Edsjö, Phys. Rev. D **78**, 103520 (2008) [arXiv:0808.3725].
[5] M. Drees, G. Jungman, M. Kamionkowski, M. M. Nojiri, Phys. Rev. **D49**, 636 (1994) [hep-ph/9306325].
[6] P. Ciafaloni, M. Cirelli, D. Comelli, A. De Simone, A. Riotto, A. Urbano, JCAP **1110**, 034 (2011) [arXiv:1107.4453].

12th International Conference on Topics in Astroparticle and Underground Physics (TAUP 2011) IOP Publishing
Journal of Physics: Conference Series **375** (2012) 012047 doi:10.1088/1742-6596/375/1/012047

keV sterile neutrino dark matter and neutrino model building

Alexander Merle

Department of Theoretical Physics, School of Engineering Sciences, KTH Royal Institute of Technology – AlbaNova University Center, Roslagstullsbacken 21, 106 91 Stockholm, Sweden

E-mail: amerle@kth.se

Abstract. A sterile neutrino with a mass around the keV scale could be an interesting candidate for warm dark matter. Although there are several scenarios and production mechanisms known in which such a particle could yield the correct abundance, there are astonishingly few models around that can actually yield an explanation for the appearance of a keV-like scale. We here review three main classes of such mass models for keV sterile neutrino dark matter, based on split seesaw, on $L_e - L_\mu - L_\tau$ symmetry, and on the Froggatt-Nielsen mechanism, respectively.

1. Introduction

One hot topic in contemporary astroparticle physics is the question about the nature of the dark matter (DM) in the Universe. One can imagine different scenarios [1]: DM could be highly relativistic today (*hot dark matter*, HDM), it could be non-relativistic (*cold dark matter*, CDM), or in between (*warm dark matter*, WDM). Out of these, CDM is considered to be the standard scenario [2], while HDM is ruled out as the dominant DM component by structure formation [3]. However, WDM is still compatible with computer simulations on structure formation [4], and also recent analyses of the current data seem to point towards WDM [5, 6].

A well-motivated candidate particle for WDM is a sterile neutrino with a mass at the keV scale: First, sterile (right-handed) neutrinos are required in many models for a non-zero neutrino mass, and second, the mass of the sterile neutrinos is generated by a Majorana mass term which is *not* bound to the electroweak scale. Hence, a sterile neutrino could very well have a mass of a few keV. This immediately suggests interesting connections to neutrino model building.

Before discussing some model building aspects, i.e., how to find an explanation for the appearance of a keV-scale mass, let us shortly comment on a few *production mechanisms* of keV sterile neutrinos. It is clear that, in order to constitute the DM today, the WDM candidates have to be produced in the right amount in the early Universe. Indeed, there are several production mechanisms available as, e.g., non-thermal production as used in the νMSM [7, 8], by having a primordial abundance (e.g. due to inflaton decay) [9, 10], or by thermal overproduction and subsequent dilution by entropy production [11]. On the other hand, it is required to find a dedicated explanation for the keV-scale itself, since it is different from other "natural" scales in particle physics. While we focus on such *mass mechanisms* here, it is the ultimate goal to find models containing explanation for both, the keV scale as well as the WDM production mechanism.

2. Mass models for keV sterile neutrino DM

Note that, in this section, when we talk about *mass mechanisms*, then this refers to possible models that indeed give (to some extent) an explanation for a keV-like mass, in the form of, e.g., a suppression mechanism or a suitable form of the mass spectrum. Relating certain quantities and thereby explaining patterns is the maximum one can do, but at the moment there is no method known to predict absolute scales. Contrary to mass mechanisms, there are also *production mechanisms* (sometimes also called "models"), a term we use to refer to settings that include keV sterile neutrinos without providing an explanation for the keV scale. Scenarios containing production mechanisms are extremely useful for concrete calculations, while models containing mass mechanisms try to explain the scales involved.

2.1. Split seesaw

The first mass mechanism discussed here is based on the *split seesaw mechanism* [12]. This mechanism is very powerful in suppressing and separating scales, due to an exponential factor involved. The trick is to consider a 5-dimensional theory that is compactified on a S^1/Z_2 manifold, with a 5D coordinate $y \in [0, l]$ which separates the Standard Model (SM) brane at $y = 0$ from the ultraviolet brane at $y = l$. The right-handed (RH) neutrinos can, as SM singlets, penetrate the bulk and hence their zero modes acquire a wave function the depends exponentially on y:

$$\Psi_{Ri}^{(0)} = \sqrt{\frac{2m_i}{e^{2m_i l} - 1}} \frac{1}{\sqrt{M}} e^{m_i y} \Psi_R^{4\mathrm{D}}(x) \;\Rightarrow\; M_{Ri} = \kappa_i v_{B-L} \frac{2m_i/M}{e^{2m_i l} - 1}. \tag{1}$$

It is exactly this exponential profile that leads to an exponential suppression of the 4D-masses of the zero models! Furthermore, the 5D masses $\{m_i\}$ can be relatively close together, e.g., a moderate splitting of $(m_1, m_2) \simeq (24, 2.3)l^{-1}$ is enough to achieve an extreme splitting of the physical masses, $(M_1, M_2) \sim (\mathrm{keV}, 10^{12}\mathrm{GeV})$, which exemplifies the strength of this mechanism. Note that this mechanism can be refined using a flavour symmetry, such as A_4 [13]: The symmetry can be used to generate the small splitting of the 5D masses in the first place, which is then used as "input" spectrum for the split seesaw.

One further bonus of this model is that the seesaw mechanism is guaranteed to work: Due to the Yukawa couplings receiving similar exponential suppression factors, the exponentials actually drop out of the seesaw formula. Hence, in the seesaw, the neutrinos only "feel" the natural (high) scale v_{B-L} of the right-handed neutrino masses, and there arises no problem due to the appearance of a low keV scale in the seesaw denominator.

2.2. $L_e - L_\mu - L_\tau$ symmetry

A model that achieves a mass splitting by an $L_e - L_\mu - L_\tau$ flavour symmetry was first suggested in general by Ref. [14] and then applied to the keV sterile neutrino case in [15]. Later on, this independently worked out in detail in [16]: As known from earlier studies [17], exact $L_e - L_\mu - L_\tau$ symmetry leads to light neutrino mass patterns of the form $(0, m, m)$, which means that one neutrino is massless while the other two are degenerate. It was spotted in Ref. [16] that this idea could easily be extended to the RH neutrino sector, which yields a similar pattern $(0, M, M)$ for the heavy neutrino masses.

A pattern like $(0, m, m)$ is not compatible with neutrino oscillation data, and hence the flavour symmetry $L_e - L_\mu - L_\tau$ must be broken. This breaking is often parametrized by so-called *soft breaking terms*. These terms do not respect the initial flavour symmetry and they can be chosen more or less arbitrarily. The point is that any choice will have two generic effects, namely making the massless neutrino massive, by bringing in a new scale, while simultaneously lifting the degeneracy of the non-zero masses,

$$(0, M, M) \to (\mathcal{O}(S), M - \mathcal{O}(S), M + \mathcal{O}(S)). \tag{2}$$

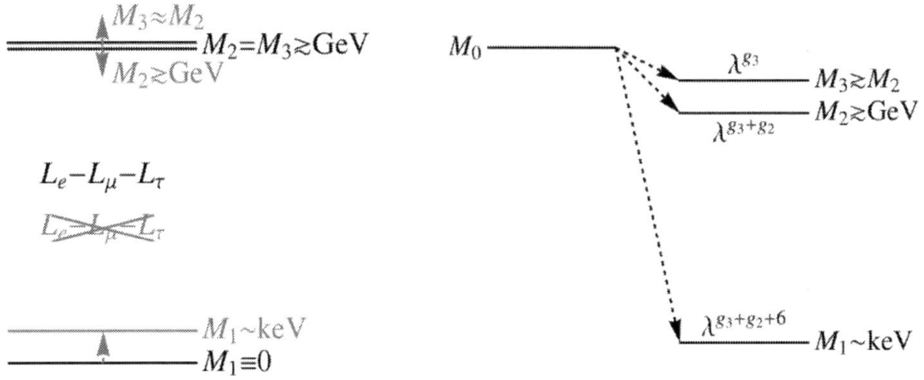

Figure 1. The mass shifting schemes for the $L_e - L_\mu - L_\tau$ and the Froggatt-Nielsen models. (Figures taken from Ref. [18].)

This is exemplified in the left panel of Fig. 1. Since the breaking scale must be small compared to the symmetry preserving scale, $S \ll M$, as otherwise we would not speak of a symmetry in the first place, this mechanism yields a perfect motivation for $S \sim$ keV, while $M \gtrsim \mathcal{O}(\text{GeV})$ or heavier.

2.3. The Froggatt-Nielsen mechanism

The third model containing an explanation for the keV scale [18] was based on the Froggatt-Nielsen (FN) mechanism [19]. By introducing an unknown high energy sector, this mechanism leads to generation-dependent suppression factors in the mass matrices, see right panel of Fig. 1. It leads to an extremely strong (exponential) suppression of certain masses, and hence to the appearance of strong hierarchies. Thus, FN has become very popular for the explanation of quark mass patterns, but it is equally suited for keV sterile neutrinos, since a strong mass splitting is required by many production mechanisms.

Similar to what happens in the model presented in Sec. 2.1, the FN mechanism guarantees seesaw to work. This is because any global $U(1)$ charges of the RH neutrinos, and in particular the one of $U(1)_{\text{FN}}$, are forced to cancel in the seesaw mass matrix, thereby avoiding any small scales in the denominator. In addition it has been shown in Ref. [18] that the FN mechanism, although usually involving a certain degree of arbitrariness, is actually quite constrained when applied to keV sterile neutrino scenarios. This allows for predictivity, as well as compatibility with only some frameworks. For example, an embedding in an $SO(10)$ Grand Unified Theory is strongly disfavoured as compared to $SU(5)$, and FN scenarios are incompatible with Left-Right symmetry, which was the main example used in the production mechanism of Ref. [11]. In turn, models based on the FN mechanism are falsifiable not only by more information on low energy neutrino data, but also by verifying one of the settings and/or production mechanisms which are incompatible with FN.

3. Conclusions

A sterile neutrino with a mass at the keV scale is a very interesting candidate for WDM, and its existence is well motivated from particle physics, too. Although many scenarios and production mechanisms have been discussed in the literature, there is a lack of models than can explain the appearance of the keV scale in the first place. This gap is currently being filled, and three such examples have been presented here. We can hopefully look forward to many more interesting ideas for new mass and production mechanisms of keV sterile neutrino dark matter.

Acknowledgements
AM thanks the organizers of the TAUP conference 2011 for doing an excellent job, and he also thanks all the participants he could share interesting discussions with. Furthermore, AM is particularly grateful to his collaborators M. Lindner and V. Niro. The work of AM is supported by the Göran Gustafsson foundation.

References
[1] F. D. Steffen, Eur. Phys. J. C **59** (2009) 557 [arXiv:0811.3347].
[2] E. Komatsu *et al.* [WMAP Collaboration], Astrophys. J. Suppl. **192** (2011) 18 [arXiv:1001.4538].
[3] A. Del Popolo, Astron. Rep. **51** (2007) 169 [arXiv:0801.1091].
[4] P. Bode, J. P. Ostriker and N. Turok, Astrophys. J. **556** (2001) 93 [arXiv:astro-ph/0010389].
[5] H. J. de Vega and N. G. Sanchez, Mon. Not. Roy. Astron. Soc. **404** (2010) 885 [arXiv:0901.0922].
[6] E. Papastergis, A. M. Martin, R. Giovanelli and M. P. Haynes, Astrophys. J. **739** (2011) 38 [arXiv:1106.0710].
[7] T. Asaka, S. Blanchet and M. Shaposhnikov, Phys. Lett. B **631** (2005) 151 [arXiv:hep-ph/0503065].
[8] T. Asaka, M. Laine and M. Shaposhnikov, JHEP **0701** (2007) 091 [arXiv:hep-ph/0612182].
[9] A. Anisimov, Y. Bartocci and F. L. Bezrukov, Phys. Lett. B **671** (2009) 211 [arXiv:0809.1097].
[10] F. Bezrukov and D. Gorbunov, JHEP **1005** (2010) 010 [arXiv:0912.0390].
[11] F. Bezrukov, H. Hettmansperger and M. Lindner, Phys. Rev. D **81** (2010) 085032 [arXiv:0912.4415].
[12] A. Kusenko, F. Takahashi and T. T. Yanagida, Phys. Lett. B **693** (2010) 144 [arXiv:1006.1731].
[13] A. Adulpravitchai and R. Takahashi, JHEP **1109** (2011) 127 [arXiv:1107.3829].
[14] R. N. Mohapatra, Phys. Rev. D **64**, 091301 (2001) [hep-ph/0107264].
[15] M. Shaposhnikov, Nucl. Phys. B **763** (2007) 49 [arXiv:hep-ph/0605047].
[16] M. Lindner, A. Merle and V. Niro, JCAP **1101** (2011) 034 [arXiv:1011.4950].
[17] L. Lavoura and W. Grimus, JHEP **0009** (2000) 007 [arXiv:hep-ph/0008020].
[18] A. Merle and V. Niro, JCAP **1107** (2011) 023 [arXiv:1105.5136].
[19] C. D. Froggatt and H. B. Nielsen, Nucl. Phys. B **147** (1979) 277.

12th International Conference on Topics in Astroparticle and Underground Physics (TAUP 2011) IOP Publishing
Journal of Physics: Conference Series **375** (2012) 012048 doi:10.1088/1742-6596/375/1/012048

Principal properties of the velocity distribution of dark matter particles near the Solar System

Anton N. Baushev
DESY, 15738 Zeuthen, Germany
Institut für Physik und Astronomie, Universität Potsdam, 14476 Potsdam-Golm, Germany

Abstract. The velocity distribution of the dark matter particles on the outskirts of the Solar System remains unclear. We show that under very common assumptions it should be highly anisotropic and have a sharp maximum near $v \sim 500$ km/s. The distribution is totally different from the Maxwell one. We analyze the influence of the distribution function on the results of dark matter detection experiments. It is found that the direct detection signal should differ noticeably from the one calculated from the Maxwell distribution with $\langle v \rangle \simeq 220$ km/s, which is conventional for direct detection experiments. Moreover, the sharp distinction from the Maxwell distribution can be very essential to the observations of dark matter annihilation.

1. Introduction

One of the most evident manifestations of the dark matter existence is the detection of huge invisible halos (with density profiles $\rho \sim r^{-2}$) surrounding galaxies. Our Galaxy also has such a halo. We symbolize the orbital radius of the Solar System, the average velocity of the Galaxy rotation at this radius, and the escape velocity at it by $r_\odot$, $v_\odot$, and $v_{\rm esc}$, respectively. We also denote the radial and tangential components of a dark matter particle velocity by v_r and $v_\rho \equiv \sqrt{v_\phi^2 + v_\theta^2}$. Velocity distribution of the dark matter particles (hereafter DMPs) inside the halo is poorly known; it is usually supposed to be Maxwell with a cut-off when $v > v_{\rm esc}$ [1].

$$f(v) = \frac{4N}{\sqrt{\pi} v_\odot} \left(\frac{v}{v_\odot}\right)^2 \exp\left(-\frac{v^2}{v_\odot^2}\right), \quad v < v_{\rm esc} \tag{1}$$

We accept $r_\odot = 8$ kpc, $v_\odot = 220$ km/s, $v_{\rm esc} = 575$ km/s. N is a normalizing constant, and for the chosen parameters $N \simeq 1.003$. It is worthy of noting that $v_{\rm orb}$ remains almost constant throughout the halo.

Distribution (1) faces with difficulties. In fact, in the framework of collisionless dynamics (1) can be naturally obtained from profile $\rho \sim r^{-2}$ subject to the condition, however, that function f is isotropic, i.e. f depends only on $|v|$ (so called isothermal model). This assumption seems highly improbable. Indeed, if (1) is true, the majority of the dark matter particles has large specific angular momentum $\mu \equiv \mathfrak{M}/m_\chi = [v \times r]$. The average angular momentum of the particles (and, consequently, of the halo) is zero. However, the root-mean-square momentum is $\sqrt{\langle \mu^2 \rangle} \simeq 1800$ kpc $\cdot$ km/s at $r_\odot$. Moreover, since in this model $\mu \sim r v_{\rm orb}$ and $v_{\rm orb}$ is constant in the halo if $\rho \sim r^{-2}$, the root-mean-square angular momentum reaches an incredibly huge value $\sqrt{\langle \mu^2 \rangle} \sim 2 \cdot 10^4$ kpc $\cdot$ km/s at the edge of the halo ($r \sim 100$ kpc). Meanwhile, according to

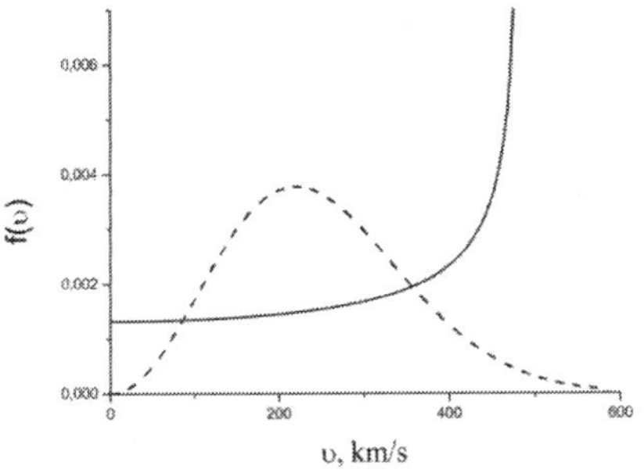

Figure 1. Anisotropic (6) (solid line) and Maxwell ((1), dashed line) distributions of dark matter particles over the velocity absolute value.

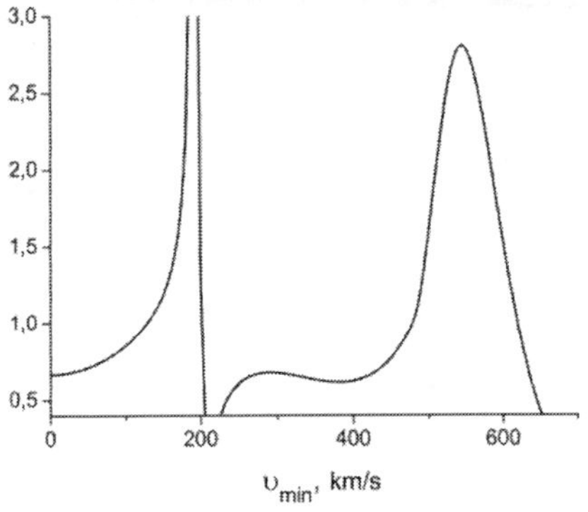

Figure 2. The ratio between the double amplitudes $2A = I_{\max} - I_{\min}$ of direct detection signals (9) calculated for anisotropic distribution (8) and Maxwell distribution (1).

modern cosmological conceptions, not only the total angular momentum of the halo but also the momentum of each particle should have been negligibly small on the linear stage of the structure formation. The halo could gain some angular momentum later, as a result of tidal perturbations or merging of smaller halos; numerical simulations show, however, that it cannot be large [2].

Some results of stellar dynamics are frequently used in order to show that the dark matter particles could gain a large angular momentum during the Galaxy evolution. The parallels between DMP and stellar dynamics, however, are not universally true. The point is that stars are compact objects, and their gravitational field can be strong, at least, locally. On the contrary, the small-scale gravitational field of the dark matter is always small. Therefore important relaxation mechanisms of stellar systems like close pair approaches or an interaction with the interstellar medium are completely ineffective for DMPs. So called violent relaxation [3] is perhaps the only way to impart significant angular momentum to the dark matter particles. However, it acts also on the halo stars; moreover, its efficiency decreases with radius, and it should stronger affect the star distribution since the stellar halo is more compact.

To summarize: all halo objects initially had $v_\rho \simeq 0$ and later gained some angular momentum because of various processes like relaxation or tidal effects. All the mechanisms increased $\langle v_\rho^2 \rangle$ of the halo stars, at least, as much as of the DMPs, while some mechanisms affected only the stars, and not the DMPs. Consequently, velocity distribution of the oldest stellar halo population should be closer to the Maxwellian one, than the distribution of the dark matter particles. In particular, the tangential velocity dispersion $\sigma(v_\rho)$ of the DMPs cannot be larger than that of the halo stars at the same radius.

Modern observations of the oldest halo stars – subdwarfs – confirm the above reasoning [4]. Their tangential dispersion $\sigma(v_\rho) \equiv \sigma_0 \simeq 80$ km/s, which corresponds to $\sqrt{\langle \mu^2 \rangle} \simeq 900$ kpc·km/s, is two times lower than in (1). Moreover, the distribution widely differs from the Maxwellian: $\sigma(v_r)$ is much larger than $\sigma(v_\rho)$. Consequently, $\sigma(v_\rho)$ of the dark matter particles on the outskirts of the Solar System does not exceed $\sigma_0 = 80$ km/s and can be even smaller. Second, the observations show that the halo stars have not yet relaxed and their orbits are rather prolate. On the other hand, if distribution (1) is correct, the ellipticity of the majority of DMP orbits is small, and then the dark matter is the only class of halo objects that move almost circularly. It seems much more natural to assume the opposite.

Journal of Physics: Conference Series **375** (2012) 012048 doi:10.1088/1742-6596/375/1/012048

2. Calculations

For our Galaxy we accept $R = 90$ kpc, which corresponds to the total mass of the Galaxy $M = 10^{12} M_\odot$. The module of gravitational potential on the edge of the halo is equal $\Phi = GM/R$. It is easy to see that $\Phi = v_\odot^2$. Since $\rho \sim r^{-2}$, the mass inside some radius can be found as $\frac{r}{R}M$, gravitational field is equal $\vec{g} = G\frac{M}{rR}$, and we obtain the gravitational potential inside the halo:

$$\phi = -\Phi\left(1 + \ln\frac{R}{r}\right) \tag{2}$$

Let us start our consideration from the case when the DMPs have no angular momentum at all. Then their trajectories are radial ($v = |v_r|$), and the task becomes one-dimensional. Therefore the particle distribution in the halo can be entirely described by a single function $\psi(r, v)$, so that $\psi(r, v)drdv$ gives the total mass of dark matter in the element of phase space $drdv$. In our case $r \propto r^{-2}$, and the halo density is bound with the function ψ by a trivial relation

$$4\pi r^2 \rho(r) = \int_0^\infty \psi(r, v)dv = \frac{M}{R} \tag{3}$$

Each dark matter particle executes a radial oscillation around the galactic centre. We denote by r_0 the maximum distance it moves off the centre. We introduce a distribution function ξ of the particles throughout parameter r_0, so that $\xi(r_0)dr_0$ is the total mass of DMPs which apoapsis lies in the interval $[r_0; r_0 + dr_0]$. The r-coordinate of these particles varies between 0 and r_0, and they give a yield into the halo density over all this interval. We can easily find function ξ if we suppose that the halo boundary is sharp, i.e. the density obeys the law $\rho \sim r^{-2}$ up to radius R and is equal to zero just after it. Then the maximum velocity the halo particles may have at radius r is

$$v_{\max} = \sqrt{2\Phi \ln\frac{R}{r}} = v_\odot\sqrt{2\ln\frac{R}{r}} \tag{4}$$

It is a matter of direct verification to prove that function

$$\xi(r_0) = \frac{M}{\sqrt{\pi}R\sqrt{\ln\frac{R}{r_0}}} \tag{5}$$

satisfies condition (3). This function has a peculiarity at $r_0 = R$. So the main part of DMPs comes to us from the very edge of the halo. Normalized velocity distribution on the outskirts of the Solar System is equal to

$$f(v) = \frac{2}{\pi\sqrt{v_{\max}^2 - v^2}} \simeq \frac{2}{\pi\sqrt{(2.2v_\odot)^2 - v^2}} \tag{6}$$

The angular momentum of the dark matter particles, however, hardly can be exactly equal to zero [5]. If a particle possesses some specific momentum μ, its velocity in gravitational field (2) is equal to:

$$v_\rho = \frac{\mu}{r}; \quad |v_r| = \sqrt{2\Phi\ln\frac{r_0}{r} - \mu^2\left(\frac{1}{r^2} - \frac{1}{r_0^2}\right)} \tag{7}$$

As we have already discussed in the Introduction, specific angular momentum of the majority of the dark matter particles hardly can be larger than 900 kpc · km/s. Substituting $r_0 = R$, $r = r_\odot$, we can see that the second term under the root is equal to 110^2 km^2/s^2, while the first one is $\sim 500^2$ km^2/s^2. Hence near the Solar System the influence of the angular momentum on the radial dynamics is still negligible for the majority of DMPs, and we can use (5) up to $r_\odot$

as before. Thus the DMP distribution throughout v_r in this approximation coincides with (6). However, the particles have also some distribution throughout v_ρ. For simplicity we will suppose that $f \propto \exp(-v_\rho^2/2\sigma_0^2)$ where $\sigma_0 = 80$ km/s, though the distribution can be much narrower. Then the normalized DMP distribution near the Solar System can be closely approximated by

$$f(v) = \frac{\exp\left(-\dfrac{v_\rho^2}{2\sigma_0^2}\right)}{2\pi^2\sigma_0^2\sqrt{v_{\max}^2 - v_r^2}} \tag{8}$$

where $v_r \in [-v_{\max}; v_{\max}]$, $v_{\max} = 484$ km/s. Distribution (8) is strongly anisotropic and actually describes two colliding beams of particles.

3. Discussion

Fig. 1 represents distributions (6) and (1) (solid and dashed lines, respectively). One can see that (6) is much narrower and has much higher average velocity. The physical reason of it is obvious: in the case of Maxwell distribution (1) the particles move almost circularly, which is why only a few of DMPs from the edge of the halo reach the Solar orbit. On the contrary, in the case considered in this article the majority of DMPs come from the halo edge and thus are much more accelerated by the gravitational field. Consequently, the question of what of the distributions, (1) or (6), is correct, can be reduced to whether the particles from the halo edge can reach the Solar orbit or not. In addition to the arguments presented in the Introduction we note that, according to (7), a particle falling from $r = R$ should have a specific angular momentum $\mu \sim 4000$ kpc · km/s, lest the particle can reach $r = 8$ kpc. This value is huge, it far exceed not only the characteristic momentum of halo objects, but even the momentum of the disk, and thus looks very unlikely. So particles from the edge of the halo freely reach the Earth, and their spectrum should be closer to (6).

The difference between (1) and (6) is important for various aspects of the dark matter physics. In the case of direct dark matter search the signal, roughly speaking, can be represented as a product of a part depending almost not at all on the DMP distribution and an integral [6]

$$I(v) = \int_{v_{\min}}^{\infty} \frac{\tilde{f}(v)}{v}\, d\vec{v} \tag{9}$$

Here $v_{\min}$ is the minimal DMP speed, to which the detector is sensitive, $\tilde{f}(v)$ is the distribution in the Earth's frame of reference, obtained from (1) or (6) by a Galilean transformation (see details in [1], section 3.3). Because of the Earth's orbital motion I varies with a year period, and it is this variation that is observed in the direct detection experiments. Fig. 2 shows the ratio between the double amplitudes $2A = I_{\max} - I_{\min}$ of direct detection signals calculated for anisotropic distribution (8) and Maxwell distribution (1), as a function of $v_{\min}$. One can see a very significant difference. The difference between distributions (1) and (6) can also be important for the indirect dark matter search.

References

[1] Cerdeño D. G., Green A. M., 2010, *Particle Dark Matter : Observations, Models and Searches*, Published by Cambridge University Press, Cambridge, UK ; New York : 2010. ISBN: 9780521763684, p. 347, arXiv:1002.1912
[2] Macciò A. V., Dutton A. A., van den Bosch F. C., Moore B., Potter D., Stadel J., 2007, MNRAS, 378, 55
[3] Lynden-Bell D., 1967, MNRAS, 136, 101
[4] Smith M. C., et al., 2009, MNRAS, 399, 1223
[5] Baushev A. N., 2012, arXiv:0911.3109, accepted to MNRAS
[6] Bélanger G., Nezri E., Pukhov A., 2009, Phys. Rev. D, 79, 015008

AUTHOR INDEX

AUTHOR INDEX

AUTHOR INDEX

Institute of Physics
Dirac House, Temple Back
Bristol BS1 6BE UK

CURRAN ASSOCIATES INC
proceedings.com